BIOGENESIS

BIOGENESIS

THEORIES OF LIFE'S ORIGIN

Noam Lahav

New York Oxford

Oxford University Press

1999

Oxford University Press

Oxford New York
Athens Auckland Bangkok Bogotá Buenos Aires Calcutta
Cape Town Chennai Dar es Salaam Delhi Florence Hong Kong Istanbul
Karachi Kuala Lumpur Madrid Melbourne Mexico City Mumbai
Nairobi Paris São Paulo Singapore Taipei Tokyo Toronto Warsaw

and associated companies in
Berlin Ibadan

Published by Oxford University Press, Inc.
198 Madison Avenue, New York, New York 10016

Oxford is a registered trademark of Oxford University Press

Library of Congress Cataloging-in-Publication Data
Lahav, Noam, 1929–
Biogenesis : theories of life's origin / by Noam Lahav.
p. cm.
Includes bibliographical references and index.
ISBN 0-19-511754-9; 0-19-511755-7 (pbk.)
1. Life—Origin. 2. Life (Biology). I. Title.
QH325.L24 1999
576.8'3—dc21 97-19694

3 5 7 9 8 6 4 2

Printed in the United States of America
on acid-free paper

To Hana,
Ran, Yael, and Ruth
and their families
with love

Preface

Pattee (1995) noted that "when a problem persists, unresolved, for centuries in spite of enormous increase in our knowledge, it is a good bet that the problem entails the nature of knowledge itself. The nature of life is one of these problems." However, complex problems such as the origin of life may take long periods of time to decipher. And thus the discussion of the problem of the "nature of knowledge itself" in this context should be postponed until more research is performed.

This book deals largely with the most important aspects of the search into the origin of life, as unfolded to me during more than two decades of my involvement in this fascinating quest. The book begins with speculations made by the Greek philosophers and ends with speculations made by contemporary scientists. But during the time elapsed between those ancient and modern scholars, progress has taken these speculations and turned them into sound scientific foundations for the understanding of the origins of life. The historical background of part I is necessary not only to familiarize the readers with some important aspects of scientific thinking but also to expose them to its thinkers, their biases, mistakes, fashions, and successes. Like biology, which cannot be understood without its evolutionary history, the study of the origin of life cannot be fully understood without its history.

Part II serves as an introduction to the chemical, biological, and molecular-biological aspects of part IV. It focuses mainly on those topics that are discussed in part IV as the rationale of the search into the origin of life. Therefore, it is not a substitute for textbooks such as Lehninger et al., 1993, and Stryer, 1995, in biochemistry, and Lodish et al., 1995, and Alberts et al., 1994, in molecular biology. Furthermore, it serves as a point of departure for the bold back-extrapolation from extant biology (including biochemistry and molecular biology) into the realm of experiments, theories, and speculations, as well as scenarios and computer models surrounding the origin of life, which are the essence of part IV.

Part III deals mainly with the cosmogeological topics directly connected to the study of the origin of life. These include the formation and properties of the earth, as well as its atmosphere and the primordial ocean.

Part IV is the climax of our effort to comprehend the origin of life. The interdisciplinary enquiry described and discussed here has been conducted in conjunction with the hypothetical terrestrial scenarios the rational behind which has been developed by scientists since the beginning of the modern search into the origin of life; the enquiry is based on general principles and thus can also serve as a starting point for the study of the evolution of different forms of life elsewhere in the universe. The present book focuses on the transition from inanimate to animate where the conventional top-down and bottom-up approaches are further divided according to the involvement of the prebiotic environment in the reactions and processes undergone by organic molecules. The main environmental parameters under consideration are temperature, hydration, irradiation, and minerals, including their rhythms and relevant chemical features.

My presentation of the material was inspired by two methodological approaches. The first is the talmudic methodology, in which legends are interwoven into serious legalistic discussions (Halacha). Thus, many citations are dispersed throughout the scientific text of the book, as are relevant ideas and notes related to the origin-of-life research. The second methodology stems from the *statu nascendi,* which characterizes the study of the origin of life today. Given the controversies over practically every aspect of this research, the various points of view published in the scientific literature need to be presented and discussed. Moreover, an inherent feature of the search for the origin of life is the lack of a paradigm, as well as the disagreements and controversies. Therefore, in order to emphasize this preliminary, preparadigmatic state of the field, I tried to incorporate quotations and ideas of the scientists themselves, thus enabling the reader to be present (with some imaginative effort) in various group discussions. This would help familiarizing the reader with many of the members of the small community of "origin-of-lifers."

I am grateful to NASA and to Sherwood Chang, who in 1975, 1980, and 1992 let me join the Exobiology group at Ames Research Center, thus introducing me to the fascinating field of the origin of life. I am grateful to Gilda Loew for enabling me to write one part of the manuscript and much editorial work of this book at the Molecular Research Institute, and for discussions. And I thank Leslie Orgel for enabling me to learn about template-directed reactions in his laboratory at the Salk Institute.

I thank Ran Lahav for numerous discussions, philosophical insights, and encouragement, and for his help in the writing and editing process. I thank Ruth Eilon for discussions and help in the writing and editing. I thank Eli Eilon for discussions.

I thank Günter Wächtershäuser for discussions, for kindly sharing some of his manuscripts with me before publication, and for commenting on various parts of the manuscript. I thank Avshalom Elitzur, Iris Fry, Antonio Lazcano, Marie-Christine Maurel, Tzvi Mazeh, and Eli Zamski for reading various parts of an early version of the manuscript and providing me with their criticism, and for discussions. I thank the members of the Rehovot Origin of Life (ROOL) Club—namely, Shneor Lifson, Doron Lancet, Avshalom Elitzur, Tsahi Pilpel, Daniel Segré, Gad Yagil, Etty Kohavi, and Ora Kedem—for the fruitful group discussions.

A special note of appreciation is due to Shlomo Nir, with whom I have collaborated in the development and computerization of the coevolution model. I thank Eva Jablonka and Anastasia Kanavarioti for discussions, and Shmuel Yariv, Isac Lapides, Renata Reisfeld, Lisa Heller-Kallai, and David Avnir for discussions. I appreciate the graphic work of Shoshana Zioni, Amit Mishali, and Tal Wiesel, and the first editorial work of Camille Wainstein.

Other colleagues with whom I have exchanged ideas and know-how over the years are too many to mention individually. I thank them all.

My students over the years in the course entitled Workshop on the Origin of Life helped me gain a better insight into some of the central topics in the study of the origin of life. I thank them for their questions, comments, and enthusiasm.

A special note of appreciation is due to Kirk Jensen for directing me in the first stages of the preparation of the text and to Mary Beth Branigan for taking care of all the editorial procedures at Oxford University Press.

I appreciate the invaluable help of Dvora Trop during the long years of writing this book. I thank Hana Ben-Or and Adriana Szekely for library research, Shlomo Tal and Sue Solomon for technical help, and the Hebrew University for financial support for the graphic work.

I appreciate the opportunity given to me over the years by the Hebrew University of Jerusalem, NASA, the Salk Institute for Biological Studies, and the Molecular Research Institute, to enjoy the intellectual atmosphere and academic freedom that I needed to write this book.

And last but not least, I thank Hana and our children for their support, encouragement, patience, and love.

Mevaseret Zion, Israel N.L.
January 1998

CONTENTS

BIOGENESIS

Prologue

> Beginnings hold an endless fascination for the inquiring human mind.
>
> David Deamer and Gail R. Fleischaker, *Origins of life*

The expanding universe

"Why is the sky dark at night?" The question is not as trivial as it may seem, because if the universe is infinite and eternal, according to Newton's model, then starlight should reach our planet from all directions, and the sky should also be illuminated at night. This problem is sometimes called "Olbers' Paradox," even though it is not a paradox, nor was it Willhelm Olbers (1758–1840) who addressed it first, in 1823. This problem had first been contemplated by Thomas Digges in 1576 and was known to later astronomers, including Johannes Kepler, Edmund Halley, and Sir William Herschel. It was Edgar Allan Poe, however, the founder of popular scientific journalism, who suggested a partial solution to this problem as early as 1848. According to his proposal, the universe has a finite age, though it was generally assumed it is inherently static. Thus, the light from remote stars did not have enough time to reach the earth and is beyond our horizon now. Forty years later this suggestion was supported by Lord Kelvin (1824–1907) but was not accepted, and it was later forgotten.

When Albert Einstein (1879–1955) attempted to apply his newly developed general relativity theory to cosmology in 1917, he was disappointed to find out that the resulting picture was either an expanding or shrinking universe. This world picture did not fit into the Newtonian model of a static, infinite, and eternal universe. In order to adjust his results to that model, he invented and added another term to his equations—the "cosmological constant." So deep was the Newtonian paradigm in Einstein's mind that he missed the discovery of the modern picture of the universe, a decade before it was first observed. He abandoned the idea, however, after Edwin Hubble's discovery (1889–1953) that the universe is expanding. According to Einstein, this was the "greatest blunder" of his life. Quite interestingly, several cosmologists have recently suggested that this constant did not die out, and they are trying to incorporate it into their modern cosmological picture.

The main features of the modern world-picture were suggested in 1922 by the theorist Alexander Friedmann, who developed the concept of an expanding universe. But the turning point of the paradigm shift from the Newtonian world to modern cosmology was the observation made by Hubble in 1929 with the 100-inch telescope on Mount Wilson in California. Hubble, an attorney-turned-astronomer, measured the spectral lines of various galaxies and based his interpretation of these measurements on the Doppler effect, which was discovered 87 years earlier by the physicist Christian Johann Doppler (1803–1853). According to the Doppler effect, the spectral lines of a moving galaxy shift in proportion to its velocity. Spectral measurements of the energy of the light emitted from a galaxy thus enable an observer to calculate its velocity in relation to his or her own position. Hubble was the first to demonstrate that the galaxies are *moving away from each other,* where their running-away velocities increase with their distance from the observer. This feature means that all the galaxies began their excursion in space from one point, sometime in the past. By measuring their velocity using the Doppler effect, it was possible to extrapolate this movement backward in time and to calculate the time elapsed since the galaxies started their excursion, from a common point, at time zero.

The Newtonian world was thus replaced by a dynamic, expanding universe, with a final size and age, and the so-called Olbers' Paradox could be explained now in the framework of this model. The new model, together with the relativity theory and quantum mechanics, made up the basis for the recent theories on the formation and structure of the universe. According to this theory, the universe was started some 15 to 20 billion years ago (according to more recent measurements it may be 10 billion) by an enormous burst of energy called "The Big Bang." This burst is the source of all the energy and matter in the universe. Running the expansion of the universe backward to the very moment of its creation revealed that it would have to have been compressed into an infinite small volume, with infinite density and temperature. In the mathematical jargon, the universe at that very moment was a *singularity.* To some extent it may be dealt with by the quantum theory of gravity.

Ironically, the name of this theory, which soon became not only popular among cosmologists but also deeply incorporated into modern thought of many other disciplines, including art and poetry, was given by one of its opponents, the astrophysicist Fred Hoyle, in 1950. And, similar to the fate of the name "impressionism" given in 1874 by an art critic to the founders of impressionism in art as a term of derision, so was the fate of the phrase "the big bang": In both cases the new world-picture, which was put down by the critics, became very popular soon afterward.

It should be noted that the big bang theory has recently been challenged again by several physicists. Is it conceivable that this event that is seemingly without cause, the big bang, could be brought within a wider and more general framework?

The first minutes

In the beginning, there was the big bang—the beginning of the space-time, matter, and energy; the beginning of the cosmic era. Scientists are still struggling with the most fundamental question: How did all these come into being from nothingness? The current physical theories can characterize this process only from an age of 10^{-35} seconds; they are not applicable to a universe younger than this small figure. From this age of less-than-no-time, the dramatic events that took place in the primordial universe as a function of time are partially understood.

Our intuition will never be able to comprehend an age of 10^{-35} seconds. For comparison—the time needed for a photon to cross a distance equal to the proton's diameter is 10^{-23} seconds! We shall never be able to visualize this primordial entity, in which all the matter and energy of the present immense universe was concentrated in a volume smaller than that of an atom. For instance, at the age of 10^{-23} seconds, the temperatures of this concentrated universe were higher than 10^{30} degrees; the density at that time was on the order of 10^{50} grams per cubic centimeter!

From its very beginning the universe has been in a process of expansion, reminiscent of an inflating balloon. This expansion, known as inflation, which is hypothesized to have been very rapid at the age of close to 10^{-30} seconds, was followed by a drop in temperature and density, and a series of processes through which the universe as we know it was gradually shaped. At the moment of its formation, the universe was dominated by electromagnetic radiation. As the temperature dropped, elementary particles—the building blocks of matter—could form.

According to Einstein's famous equation, $E = mc^2$. Thus, an elementary particle of mass m can be formed from photons if their energy is at least mc^2. The first particles to be synthesized were the hadrons—the heavy elementary particles such as protons, neutrons, and mesons. But, because of the high temperature, they collided with each other as soon as they formed and were eliminated, being back-transformed into photons. During its first minutes, the young universe was dominated by electromagnetic radiation; this is the *era of radiation.*

As the expansion and temperature drop continued, some of the particles formed in this gigantic tumult were not eliminated by collisions and were thus able to accumulate in the young universe. At the same time, the radiation density decreased. This is the era of the hadrons, which started to accumulate in the young universe. As the expansion continued, the formation of leptons, the light particles such as electrons, neutrons, and muons, became possible. The temperature at this stage was 10^{10} degrees, and the density 10^{10} grams per cubic centimeter; the volume of the universe was about that of a bean.

At an age of 35 minutes, the average temperature of the universe was 300 million degrees. The radiation fluxes continued to decrease gradually, and the amount of matter continued to increase. During the time interval between this age and the age of approximately 1 million years, the charged particles continued to coalesce and to form atoms, while the rate of decomposition of the atoms decreased. The first atom to be formed was hydrogen (H), and from it, helium (He). The new era was characterized by domination of matter over radiation; it is the *era of matter.*

The synthesis of elements

The formation of heavier elements, such as carbon (C) or iron (Fe), needed more time and additional processes. The time is measured in billions of years: The processes were destined to take place inside stars.

The expanding universe became less and less homogeneous with time. Its matter, at this stage predominantly hydrogen, coalesced into large bodies that later became stars and galaxies, due to gravitational force. The young stars went through a process of contraction, in which the potential gravitational energy was converted into heat, increasing the temperature and pressure inside the star. As matter inside the stars continued to contract, the temperature and pressure gradually increased until a thermonuclear fusion reaction became possible. In this reaction four hydrogen atoms are fused to form one helium atom plus energy:

$$4H \rightarrow He + \text{energy}$$

This reaction occurred in the inner part of each star, the "nucleus," and the energy released counterbalanced the gravitational effect, preventing a collapse toward the center of the star. The hydrogen thus fueled the thermonuclear reaction, in which the star became richer and richer in the product of the reaction—helium—at the expense of hydrogen.

After a time period measured in millions of years, the amount of nuclear fuel—hydrogen—decreased and was no longer capable of counterbalancing the gravitational movement of matter toward the nucleus. The pressure and temperature of the contracting matter inside the star increased, and fusion of helium atoms to form heavier elements began. The temperature needed to fuse helium nuclei is several hundreds of millions of degrees. In this reaction two helium nuclei (4) form one beryllium (Be) nucleus (8) and an additional amount of energy:

$$He + He \rightarrow Be + \text{energy}$$

The beryllium nucleus can then fuse with another helium nucleus to form a carbon (C) nucleus (12), plus energy:

$$Be + He \rightarrow C + \text{energy}$$

Thus, the element that would later become the central *biogenic element* was synthesized.

The next evolutionary pathway of the star under consideration depends on its mass: If the mass is relatively small, the process of carbon synthesis would go on, and the star would eventually turn into a *white dwarf* made of a very dense core of carbon nuclei and electrons. The star can exist for billions of years in this state, fading slowly while emanating small amounts of energy.

If the mass of the star is at least eight times larger than that of our sun, its fate is entirely different from that of a white dwarf. Because of its large mass, the fusion temperature is obtained at a lower density, upon exhaustion of the nuclear fuel. The shrinking process continues due to the gravitational acceleration, and both the temperature and pressure inside the star increase beyond the corresponding temperature and pressure of small stars; under such conditions, carbon nuclei can fuse to form nuclei of heavier elements, while emitting energy. This energy, however, decreases with the weight of the nucleus that is formed. For instance, the fusion of two carbon nuclei (12 + 12) to form one magnesium (Mg; 24) releases about one-tenth of the energy released during the fusion of two protons. Therefore, the fusion energy is not high enough to counterbalance the gravitational collapse; matter continues to accelerate toward the star's center, and the concomitant increase in temperature brings about the formation of heavier elements. In the ensuing thermal processes, elements up to and including iron (Fe) are formed. The formation of heavier elements can be obtained only at higher temperatures. This reaction takes place at the next stage.

The presence of electrons, with their great repulsive forces between each other, does not allow the core of the star to collapse. But as the contraction of the star continues, the high temperature enables the fusion of protons with electrons to form neutrons. When electrons are removed because of their fusion with protons, the inner part of the star collapses. This process is extremely rapid and very energetic; nuclei collide with one another and fuse to form elements heavier than iron, such as mercury (Hg), silver (Ag), and uranium (U). The neutrons immediately form a very dense nucleus made almost exclusively of neutrons—the *neutron star.* The outer portion of the star now accelerates very rapidly toward the center, to collide violently with the neutron star just formed and to be ejected forcefully out to the *interstellar space.* This enormous explosion occurs very rapidly; matter continues to be ejected from the exploding star for several weeks, and the star releases huge amounts of energy equal to the total amount of energy released by our sun over billions of years.

The exploding star is called a supernova; the ejected elements now become interstellar dust, which, together with the hydrogen and helium already in interstellar space, create the basis for a new cycle of stars (and planets, as we shall see in a moment), during the next billions of years. The process in which elements have formed is called "nucleosynthesis." During these processes, which continue to this day in our galaxy and in others, all the chemical elements beyond H and He, including the multitude of radioisotopes present today, were formed.

The prospects ahead

The universe will continue its expansion, but nobody knows for how long. Its fate depends on whether its density is less or greater than a particular critical value. If the cosmic density is less than that critical value, then the universe will go on expanding forever, slowly cooling off and dying out. We may find comfort in the estimated time period during which this very slow death will come about, on the order of several tens of billions of years.

If the cosmic density is greater than the critical value, however, then the expansion of our universe is destined to gradually slow down, billions of years from now, and then reverse its direction. By the end of this contraction, the temperature and density will attain the values they had initially during the big bang. Our universe will thus come to a violent end with a spectacular show of the reverse of the big bang scenario. And maybe another universe would form instead, in a cosmic transfiguration on a gigantic scale, thus perpetuating the rise and fall of universes in an endless series.

Our only consolation in view of this cosmic eternity is that because of the law of conservation of mass and energy, "we" shall be present there, forever.

And in a remote region of our universe, in the Milky Way galaxy, which is just one of about 400 billion galaxies, interstellar dust hovering around our proto-sun started to accrete, more than 4.6 Ga (billion years) ago. The planets of our solar system gradually emerged from the many planetesimals that accreted slowly while revolving around the sun. Our planet, which initially had a molten surface, gradually cooled down.

In the sun, however, thermonuclear fusion of hydrogen to form helium, positrons, and electromagnetic radiation have continued since the beginning. As with other stars, the heat output of our sun increased with age: At first the radiation intensity was smaller than at present, by about 25%. Gradually this intensity increased to its present rate. Photons of visible light have reached the surfaces of all planets ever since their accretion, bringing about photochemical reactions in various molecules; these are destined to be the ultimate energy source of biology on our planet for billions of years to come.

Many heavenly bodies, such as meteorites and comets, bombed our planet, lifting rocks and dust into the primitive atmosphere and enriching its young surface with water and organic molecules from space. Gradually, the rate of these bombardments decreased. The atmosphere became less opaque between bombardments, the temperature dropped, and water vapors condensed to form the liquid water of the first puddles, lakes, and seas. On the surface of this planet, a series of reactions and processes started, named *prebiotic evolution,* in which the first living entities were formed. These entities continued slowly but incessantly evolving through the eons, diversifying into myriads of species, and forming the entire biosphere, including humankind.

How can we explain these latter processes, which are thought to have started on the young earth some 4 billion years ago? Do they continue today? Are we alone in the universe?

Part I

HISTORY OF THE SEARCH INTO THE ORIGIN OF LIFE

On The Shoulders of So Many

If we admit a priori that science is just acquisition of knowledge, that is, building an inventory of all observable phenomena in a given disciplinary domain—then, obviously, any science is empirical.

René Thom, "Causality and finality in theoretical biology"

1

From Myths to Logos to Stagnation

From Thales (Seventh Century B.C.) to Aristotle (Fourth Century B.C.) to the Pre-Redi Era (the Seventeenth Century A.D.)

The Greek philosophers and their views of life

The scientific methodology was originated by the Greek philosophers, some 2,600 years ago. Curiosity was not forbidden by Greek traditions and beliefs. Moreover, their philosophers explained the world in rational terms of mechanisms and natural processes and entities, not by divine actions. Apparently, two of the trivial questions they asked were what differentiates living from nonliving entities, and how plants and animals were formed. The Greek philosophers developed different ideas and proposals with regard to the problems they studied, but the discussions and intellectual struggles among the various schools of thought exemplify a most fundamental feature of the scientific method, namely, "The philosopher proposes a solution in the form of an hypothesis, not as a revelation of a truth that cannot be challenged without committing a sin" (Haezrahi, 1970; p. 28; my translation). In this ideal spirit, the scientific process is supposed to progress. The actual pathways of scientific research and progress are more complex and intricate than this ideal pattern, as we shall see later.

The most important Greek philosophers involved in the establishment of the schools of thought regarding life and its origin are reviewed briefly in the next section.

Main philosophers and their schools of thought

Thales (ca. 625 to ca. 547 B.C.)

This scholar from Miletus is traditionally believed to have been the first Greek philosopher. By emphasizing reason as the key to the understanding of nature, and by carefully observing the latter, he tried to discover the laws of nature. Thus, he is considered the founder of the scientific methodology.

Since his original writings were lost, our knowledge of his teachings comes from secondary sources. According to these sources, the magnet is a living entity, since it causes iron bodies to move. Amber is also a living entity, since it can move other substances. Moreover, according to Thales water is the very substance of which the world is made; it is the ultimate substance out of which all things are generated and into which everything eventually perishes (Girvetz et al., 1966).

It is interesting to compare Thales to Abraham, who preceded him by about 1,300 years. Both Abraham and Thales believed in the unity of the world in which they lived: For Thales it was the unity of substance, the primary material—water. For Abraham it was the unity of the universe, the belief in one God.

Anaximander (ca. 611 to ca. 547 B.C.)

The most important pupil of Thales was Anaximander from Miletus. His writings were also lost, and what we know about him also comes from secondary sources. This philosopher, who is said to have been the first to prepare maritime navigation maps, presented a rather explicit scheme of the origin of life, which seemingly has in it elements of the Darwin evolution theory. According to his school the following order of events explains the beginning of life:

1. The first living creatures were formed from water (itself an elementary substance, according to Thales) and had thorny shells. Upon maturation, they lost their shells and changed their way of life.
2. Animals were created from water as a result of its evaporation by the sun.
3. Man was born by another living creature and developed inside a fish, like the shark. He moved to land only when he was able to take care of himself.

Does Anaximander's school have certain Darwinian evolution features? The scholars deny this interpretation. Rather than evolution of one kind of creature from another one, Anaximander taught that adaptation to different conditions was needed in order to survive. Thus, the thorny shell had to protect the young creature from water; the adaptation was behavioral, not organic.

Anaximander's teaching is an example of an attempt to explain natural phenomena without resorting to gods and supernatural agents. Life, according to him, was formed *spontaneously*, and *abiogenetically*, in the sea, without even an implication of evolution. The road to Darwin is still very far away; and the seeming constancy of kinds of animals one generation after the other is apparently supported by everyday observations.

Xenophanes (ca. 576 to ca. 490 B.C.)

A contemporary of Anaximander, Xenophanes of Colophon, taught that everything, including humankind, originated from water and earth. Moreover, earth and sea had changed places in the past and would do so again in the future. Xenophanes was probably the first to address the presence of fossils in rocks. His explanation was that in the past the land was covered by sea, where plants and animals were formed from mud and rocks and later came out of the water. Thus, the beginning and end of all living creatures is in the earth. Life is formed from nonliving substance; this is *abiogenesis*.

Disputes among the Greek philosophers were probably as common as they are among scientists today. Thus, it is said that the philosopher Heraclitus of

> Ephesus, a contemporary of Xenophanes, dismissed Xenophanes' work, ridiculing his wisdom. (Magner, 1994, p. 14)

Anaxagoras (ca. 500 to 428 B.C.)

According to this scholar of Clazomenae, the universe is made of an infinite number of elementary substances, which he called *spermata* (seeds). Each such unit is characterized by an infinite number of qualities. These seeds give rise to living forms upon reaching the soil or the mud on the earth. Thus, rather than spontaneous bursting forth from the earth, Anaxagoras coined the word *Panspermia* (seeds everywhere) for the omnipresent seeds of his proposal. This word was destined to be used more than two millennia later with a different meaning, as we shall see later.

> In contrast to its modern image as a liberal democracy, Athens was a conservative city. It is told that when . . . Anaxagoras was overheard teaching that the sun was not really a god but only a red-hot stone, many of the citizens became incensed and he was imprisoned. Had it not been for his friend Pericles, who arranged for his escape . . . Anaxagoras would surely have been put to death by the Athenians. (Abel, 1973)

Empedocles (ca. 492 to ca. 432 B.C.)

Empedocles of Akragas extended the list of elements of which the world is made, from just one (water, according to Thales) to four: Earth, Water, Air, and Fire. These elements, which he called *rhizomata* (roots), were eternal and immutable. All the manifestations of nature and their occurrence were the result of the association and dissociation of these elements. The processes were thus two: *love* for association and *strife* for dissociation.

The origin of life is explained by the association of soil and humidity under the influence of heat. Plants were the first to have been formed out of these combinations. These were followed by the formation of limbs of animals, which wandered alone at first, separated from each other, but striving to combine with each other. The first combinations were very strange; for instance, a horse head combined with a cow's body. It took a prolonged struggle of the limbs to find the correct association among themselves, and the appearance of males and females, according to Empedocles, was a later development. Empedocles' proposition seems to contain elements of selection and a struggle for existence but not of evolution.

Leucippus (exact time is not clear) and Democritus (ca. 460 to ca. 370 B.C.)

Leucippus is believed to have been the first philosopher to develop the concept of atoms. His pupil, Democritus of Abdera, is generally acknowledged as the first person to use the term *atom* (indivisible) to explain all the observed phenomena of the world. To him, "nothing exists except atoms and space; everything else is opinion" (Lederman, 1993, p. 59). According to this school, there are four kinds of atoms: *stone atoms,* dry and heavy; *water atoms,* wet and heavy; *air atoms,* light and cold; *fire atoms,* light and hot. Combinations of these kinds of atoms make up all known and as yet unknown materials. Democritus also argued that there is an infinite number of worlds like ours. His rationale is most interesting: Since there is an infinite number of atoms, there should also be an infinite number of worlds made up of these atoms.

"The first animals arose on the surface of the earth that was covered with warm mud. . . . The first living creatures came into being without articulate forms, the moisture being productive of life" (Robinson, 1968, p. 216). The spontaneous and initial development of living creatures, according to Democritus, results when soil atoms are combined with fire atoms. According to the interpretation of Lederman (1993, p. 59) this takes place accidentally, following the postulate "Everything existing in the universe is the fruit of chance and of necessity." The characteristics of these combinations are determined by the mechanical motion of the particles.

Referring to the famous quotation regarding "chance and necessity," Lederman (1993, p. 60) observes that the same paradox is found in quantum mechanics.

Epicurus (342 to 270 B.C.)

Epicurus was the most prominent philosopher to follow Democritus's atomic teaching. About a hundred years after Democritus, he continued his teaching about the formation of living creatures, many of which were formed in wet soils. He also negated any spiritual influence on these processes. Moreover, according to him, the notion of the motion of the atoms was a built-in feature, independent of any divine forces.

Aristotle (384–322 B.C.)

The most prominent opponent of Democritus's atomic world was Aristotle. This world, with its implied randomness and lack of teleology (*telos*—end, goal; *logos*—reason, doctrine; thus, end-directed or goal-directed processes; see also Eschenmoser, 1994) did not fall into line with Aristotle's philosophical teachings and his intuitive teleological-metaphysical view. Indeed, the Aristotelian school, which dominated Western civilization for almost 2,000 years, overshadowed Democritus's atomic world with its implications. The revival of the atomic world in its new form was started only during the eighteenth century, by Dalton.

Aristotle was apparently the first philosopher to describe the *graduality* of living creatures and to address the gradual transition between nonliving matter and plants and between plants and animals. Some people think that this implies evolution. But most scholars think that the basic concepts of Aristotle have nothing to do with evolution. According to these scholars, he considered the different species of living forms as immutable and eternal, like nature itself; thus, his gradualism is static. Indeed, Aristotle rejected the "apparent evolution" of Empedocles.

As noted by Allan (1970, pp. 62–63), Aristotle's view of nature has three central features, namely (1) all living things can be arranged on a scale according to the attributes they display, where humans are at the top of this scale; (2) natural processes are interpreted in terms of teleology; (3) nature proceeds gradually from lifeless things to animal life without exact demarcation of the borderline between animate and inanimate.

In view of the tendency of various scientists to portray Aristotle as a dogmatic scholar, it should be noted that Küppers (1990) describes Aristotle as an open-minded scholar, far from the dogmatic person caricatured by later generations.

The belief in the eternity of nature may partially explain what we now conceive as the little attention that was paid by those philosophers to the age of natural substances or events. Thus, the Greek historian Herodotus (fifth century B.C.) examined layers of the sediments in the Nile delta and estimated this delta to be many thousands years old. This estimate, however, went unnoticed insofar as geological considerations are concerned.

The establishment, domination, and stagnation of the spontaneous generation school

The Greek philosophers developed the school of thought that later became the mainstream of Western thinking regarding the origin of life, namely, the *spontaneous generation theory*. According to this theory all living forms were formed by the generation power of nature. Moreover, according to this school, many of these living forms are still being formed today under certain favorable conditions, by spontaneous generation, not by means of seeds or parents.

The spontaneous generation theory was adopted and modified by the Stoic philosophers and accepted by the scholars of Rome and Alexandria. It was also adopted by the Christian church in both the East and West and incorporated into its general teachings, becoming, with some ups and downs, a dominant school in the Christian world for almost 2,000 years.

The history of the search into the origin of life will be presented in the next four chapters by focusing on the status of the spontaneous generation school. More details may be found in Farley, 1977, and Magner, 1994.

It is interesting to cite the talmudic Rabbi Shmuel from Kaputkaah as follows: "Birds have scales on their legs, like fish." Hence, birds were created from the mud.

2

Experimental Biology of the Seventeenth Century

Studies serve for delight, for ornament, and for ability.
Francis Bacon, "On studies"

The historian A. Koyre has aptly described the western scientific revolution of the seventeenth century as a transition "from the closed world to the infinite universe"
Ilya Prigogine, *Origin of complexity*

A new era

The beginning of a new era in the search into the origin of life is generally connected to the classical experiments of Francesco Redi. But in order to better evaluate the historical context of these experiments, it is first necessary to understand the world-picture of the scholars of that period. At that time, nobody, including Redi himself, could have imagined that his experiment, based on covering chunks of meat with a piece of fine cloth, would have such far-reaching consequences.

It is difficult for us to fully appreciate the difficulties the scholars of that time faced. Even when they had the resources needed to carry out observations and experiments, they could not order instruments and chemicals from commercial agents. Moreover, the modern scientific methodology had only started to emerge from the mixture of science and alchemy, scholastic philosophy, superstitions, mysticism, and authority of the ancient Greek philosophers, especially Aristotle, which dominated the thinking of most scholars. Taking into account this background, the curiosity and intuition of some scholars of that time should be highly appreciated. Several examples of the world-picture of known scholars of the seventeenth century are discussed here.

First scientific hypotheses on the solar system

In 1644 the great philosopher René Descartes (1596–1650) suggested that the planets and their satellites had formed from a large rotating disk of gas that surrounded the primordial sun. The estimated age of planet Earth, however, conflicted with the then prevailing interpretation of the Bible by the Church.

In the year 1650, the Archbishop James Ussher of Trinity College in Dublin published his calculations of the age of our planet according to his interpretation of the book of Genesis and other ancient texts. These calculations indicated that the earth was created in the year 4004 B.C., on October 22, at 2 P.M. John Lightfoot, Ussher's contemporary at Cambridge University, found a similar age—3928 B.C. These figures were published in various editions of the Bible and were presumably accepted by believers. Moreover, "we should not forget that Boyle, Hooke, and Newton, all towering figures of the scientific revolution, also were ardent votaries of natural theology and the argument from design" (Haynes, 1987, pp. 4–5). Isaac Newton (1643–1727) calculated the age of Earth using a scientific approach and came up with a figure of 50,000 years. Being a religious person, he could not trust this figure and probably assumed he had made a mistake.

The first scholar who understood the real meaning of fossils and who tried to relate them to the ancient history of our planet was Leonardo da Vinci (1452–1519). However, like many of his ideas, this modern hypothesis about the origin of fossils did not have a direct effect on the process of scientific progress in either geology or the search for the origin of life. His suggestion was not accepted because there was nobody to accept it. His explanation was forgotten and was rediscovered only much later, serving as a testimonial to his ingenuity. Indeed, novel ideas, brilliant as they may be, are not enough to change the thinking of human beings when they are not part of a paradigm that can be tested experimentally.

Fossils of ancient creatures had been already noted and explained by Xenophanes (see chapter 1). In the seventeenth century, and as late as the second half of the nineteenth century, it was a widespread belief that fossils were not the remains of ancient plants or animals but had been placed in the rocks by God or Satan to test humans' faith.

The status of the spontaneous generation school

The spontaneous generation theory was considered almost natural in Europe because of the long tradition of belief in this school. Still, in spite of the popular belief in the spontaneous generation teachings, learned men from several disciplines had some difficulties accepting it as is. One argument against this hypothesis was strictly religious. Thus, God created all living creatures during the first days of creation, and since then no new species had been created. How could this belief be reconciled with the spontaneous generation school, according to which living creatures continue to be formed today?

Other kinds of doubts came from scientists. Ample observations supported again and again the dictum of William Harvey (1578–1657), the discoverer of the mechanical principles governing blood circulation: "Omne vivum ex ovo" (All living come from the egg). But at the same time, the mechanisms of generation of many creatures, such as small insects, were not known. Harvey himself did not deny the validity of spontaneous generation, and it seems as if he considered its applicability to certain small animals.

Jan Baptista van Helmont (1577–1644)

With regard to spontaneous generation, the following quotation from van Helmont may sound ridiculous to some: "If one stuffs a soiled shirt into the mouth of a jar containing grains of wheat, the ferment released from the soiled shirt, combining with the odor of the grain, transmutes the mixture into mice in about 21 days" (cited in Deamer and Fleischaker 1994, p. 3).

It should be remembered that van Helmont, a talented scientist, was, among other things, the discoverer of the gas carbon dioxide, the very substance responsible for most of the weight gain of his willow tree. He also coined the term "gas" (from the Latin word *chaos*). But at that time the concept of a chemical element hardly existed (the first discussion and definition of chemical elements was provided by Robert Boyle in his book *Sceptical Chemist,* a short time after van Helmont and Redi published their experiments). His ideas about the spontaneous generation theory reflected the contemporary thinking.

Van Helmont's experiment

In this experiment (which was published only after his death), van Helmont planted a small willow tree in a container of soil, adding only rainwater to irrigate it for five years. By the end of this period, the weight added to the tree was 169 pounds, whereas the weight of the soil in which the tree grew was unchanged. Van Helmont concluded, wrongly but not unreasonably, that plants are formed from water alone (in accordance with Thales!). In his words, "I have proved that plants are produced directly and materially from the single element water" (Pledge, 1959, p. 104). The reader is referred to Wächtershäuser, 1996, 1997) for additional discussion on van Helmont's experiment.

It is interesting to note that van Helmont was not the first scientist to carry out quantitative measurements in biology. He was preceded by Santorio Santorio (1561–1636). Santorio, who is considered the founder of quantitative physiology, designed and constructed a special balance in which he weighed himself before and after eating, drinking, sleeping, resting, and exercising. His experiments were published in 1614 (Magner, 1994).

In another experiment, van Helmont burned 62 pounds of charcoal, of which only one pound of ash remained after the burning. His assumption was that the rest of the material had been driven off as a form of air. He named this air, which is known today as carbon dioxide, *gas sylvestre* (Magner, 1994, p. 239).

Francesco Redi (1627–1697): The first experimental challenge

The initiation of critical experiments connected to the spontaneous generation theory was carried out by Redi. A Tuscan physician with great curiosity, Redi was interested in the problem of the deterioration of meat, which at that time had a very short shelf life. He was also aware of the practice of butchers of his time to cover fresh meat with a cloth or muslin in order to protect it from flies. To study the effect of such a cover, he conducted experiments consisting of a series of treatments, including controls, using a variety of meats and cloths, and observed the appearance of maggots in the meat (see also chapter 4). The presence of maggots in the uncovered meats only was connected, correctly, to the source of these maggots—flies. Thus, it was flies that caused the appearance of worms in the meat, not spontaneous generation!

Redi's experiments were published in 1668 in his book entitled *Experiments on the Generation of Insects*. It is one of the most important experiments in the history of science; moreover, it was a conceptual breakthrough, if one takes into account the long tradition of scholastic philosophy that prevailed in Europe at that time. Redi was one of those first experimentalists who thought that without experimental demonstration, a belief is useless. And even though he was not the first to understand the importance of experiments in the study of nature, it was probably the first time that a simple and elegant experiment has given such a clearcut answer with regard to the question under consideration. It is interesting to note, however, that Redi himself probably continued to believe that the spontaneous generation theory was still applicable to certain small creatures. Indeed, at his time it was believed that after a long period of fertility, during which many generations of creatures were created, Mother Earth lost its ability to generate humans and the big animals, but not small ones. Two centuries were to elapse before this belief would be categorically denied, as we shall see further on, but Redi's experiment initiated its decline.

It is thus not surprising that in a *Lexicon Technicum* published in 1704, spontaneous generation (also called equivocal generation) was rejected in the following words (cited in Mendelsohn, 1976): "Equivocal Generation, is the Production of Plants without Seed, Insects or Animals without Parents, in the Natural way of Coition between Male and Female. . . . The Learned World begins now to be satisfied, that there is nothing like this in Nature."

A technological breakthrough: The optical microscope

The optical microscope was invented in 1590 by the Dutch brothers Francis and Zachary Janssen. The development of the microscope from the stage of a magnifying lens to the stage of a microscope with a magnification power of one order of magnitude larger than a magnifying lens had a decisive impact on the assessment of the validity of the spontaneous generation theory. The microscope developed and used by Antoni van Leeuwenhoek (1632–1723), a haberdasher and civic official of Delft, had a magnification power of 275, just enough to discover microorganisms (1676) and to explore the generation of small animals such as acarites. Apparently, both Robert Hooke (1635–1703) and van Leeuwenhoek, the pioneer microscopists, doubted the validity of the spontaneous generation theory; the latter scholar eventually denied it.

Thus, during the seventeenth century, the spontaneous generation theory was reexamined more critically than in earlier generations. But even though this theory ceased to be universally accepted by biologists, a new experimental approach was needed in order to clarify its validity.

3

Systematic Biology, Doubts, and Uncertainties

The Eighteenth Century

> Although *concepts* and *ideas* occupy a central place in the grand sweep of our understanding of the nature of the world around us, it is a mistake to imagine that they play a greater role than *tools* and *techniques* in achieving scientific progress. Few scientific revolutions are concept-driven.
>
> John Meuring Thomas, *Turning points in catalysis*

Are species permanent?

Carl Linnaeus (1707–1778)

The cataloguing of plants and animals during the eighteenth century culminated in the Linnaeus method, according to which a species is characterized by two central features, namely, *immutability* and *distinctiveness*. In other words, since all living species had been created simultaneously (according to the teachings of the Church), they were not related to each other genetically. One species could not descend from another, nor could different species merge with one another. It should be noted that this was in contrast to the common belief during pre-Linnaean times, according to which transmutations among different species was possible.

The Linnaean method, which fits the ancient belief in the constancy of species, seemingly enhanced the rationale of discarding the problem of the origin of life. But there remained various problematic subjects. One of these followed the development of the optical microscope by van Leeuwenhoek. New forms of life were discovered, such as the ones found upon infusing hay with water (infusoria). These creatures, with their diverse and surprising forms, looked different from the known ones, and their incorporation into the Linnaean classification posed some problems. According to the classification prevailing then, every living entity was thought to be either an animal or a plant. The newly discovered unicellular creatures posed a classification problem: Were these organisms plants or animals?

Another problem came from geology. More and more fossils of ancient creatures were being found in quarries, mines, and roads. These extinct creatures, which were different from the present ones, should have been reconciled with the contemporary scientific dogma of the constancy of species of plants and animals.

These problems eventually brought about changes in the thinking of two of the leading scientists of the time—Linnaeus and George-Louis Leclerc de Buffon. Linnaeus, who believed in the permanence of species at the beginning of his career, became more and more doubtful about this belief, and he expressed himself more and more clearly as his scientific prestige increased. In 1766, about 31 years after the publication of his famous book *Systema Naturae,* he discarded his own teachings, according to which there are no new species.

George-Louis Leclerc de Buffon (1707–1788)

Buffon was even more daring, expressing his opinion that nature is not constant. Moreover, he added, had it not contradicted the Old Testament, it could be suggested that humans and monkeys have a common origin! However, in view of the potential danger of such a statement, he withdrew his view on that matter (Mayr, 1982). Interestingly, Buffon was contemptuous of the study of the lower organisms already known in his time (Farley, 1977, p. 3). Buffon's calculation of the age of Earth is discussed later.

Vitalism: Supernatural forces and their declining role in scientific hypotheses

An integral part of the debate surrounding the spontaneous generation theory was the characterization of life itself. And an inseparable component of this debate was the argument that it is impossible to explain what life is without resorting to agents beyond the known laws of nature. These "vital" forces are the basis of the difference between living and nonliving matter. This approach was very old and could be found in the teachings of some Greek philosophers, and it was the physician Georg Ernst Stahl (1660–1734) who revived it in 1707, when he suggested that life is controlled by nonphysical forces. Thus, every living creature possesses a *vital force,* which is an integral constituent of its existence. This point of view is called "vitalism," though, in view of its initiation by Greek philosophers, it should more adequately be called "neovitalism" (Girvetz et al., 1966).

Stahl was the main author of the *phlogiston theory* of combustion, according to which a pneuma-like principle, phlogiston, escapes from bodies during combustion, calcination, respiration, or fermentation (see Mason, 1991). This theory was destined to be totally discredited by Antoine Lavoisier (1743–1794) by the end of the eighteenth century.

An example of the belief in the vital force may be provided by a citation of Alexander von Humboldt (1769–1859), who wrote in 1793 (Ritterbush, 1964, p. 189): "That internal fire which breaks the bonds of chemical affinity and prevents the elements from being joined to one another at random, we call the vital force."

The need for a vital force reflects the formidable difficulties in understanding what life is in the framework of the world-picture of that time. A citation by Immanuel Kant (1724–1804) best exemplifies this problem. In his treatise "Critique of Judgment" he wrote: "It is absurd for men . . . to hope that another Newton will arise in the future, who shall make comprehensible by us the production of a blade of grass according to natural laws which no design has ordered. We must absolutely deny this insight to men."

Thus, "Not ignoramus—we do not know, but ignorabimus—we never shall know" (Girvetz et al., 1966, p. 334).

The solar system

Theories of the formation of the solar system

In contrast to the study of the complex and incomprehensible natural world that today we call biology, physical laws were discovered by Copernicus, Galileo, Newton, and others; a variety of phenomena in physics and astronomy were explained, predicted, and phrased mathematically. The origin of the solar system was treated first by René Descartes (1596–1650) in 1644 in his cosmic vortex theory, according to which the sun and the planets were formed from a single body of matter. A century later, Buffon hypothesized that the solar system had been formed as a result of the passage of a massive heavenly body near the sun, ripping away part of its material, which subsequently formed the solar system. This collision hypothesis thus challenged Descartes's nebular hypothesis. Immanuel Kant continued with the latter hypothesis and in 1755 came up with his nebular hypothesis. Pierre Simon Laplace (1749–1827) followed Kant and in 1796 presented his planetary ring hypothesis. The acceptance of Laplace's hypothesis by modern physicists had to wait about 130 years.

In this context it is worthwhile mentioning the legendary discussion between Laplace and Napoleon. When asked by the latter (on the boat on their way to conquer Egypt) why he had not mentioned God in his scientific book, Laplace retorted, "I do not need this hypothesis!"

The importance of understanding the origin of the solar system in the context of the search into the origin of life was tenuous in Laplace's time. The relevance of this understanding has yet to be developed and reach its climax in the second half of the twentieth century.

Buffon's calculation of the earth's age

Based on experiments on the cooling rate of metal spheres, Buffon calculated the age of our planet to be about 75,000 years. He also calculated that after about 93,000 more years, our planet would be too cold to be inhabited by living creatures. Though he used an incorrect method of calculation, he managed to free himself from the literal interpretation of the biblical teaching of Creation by suggesting that the days at that time were not necessarily of the same length as the days of his time. It should be noted that the latter argument had already been used by Maimonides (1135–1204).

It is interesting to note that the world-picture of Buffon, who was undoubtedly a prominent scientist, is quite different from ours in many fields. Thus, it is said that Buffon thought that the body of mammals (including humans) native to the Western Hemisphere were smaller than those from the Eastern Hemisphere (Kissane, 1994).

4

Demise and Resurrection of the Spontaneous Generation School

Pasteur and Darwin

> This principle of preservation, I have called, for the sake of brevity, Natural Selection.
>
> Charles Darwin, *On the origin of species*

> It strikes me as almost miraculous that Darwin in 1859 came so close to what would be considered valid 125 years later.
>
> Ernst Mayr, *Toward a new philosophy of biology*

The emergence of the evolution theory

The term "biology" was coined in 1802 by Gottfried R. Treviranus (1776–1837), at the beginning of an era of decisive discoveries in biology, as well as intense controversies, and a decline in the influence of the conservative teachings of the Church. To a large extent, this term was accepted by scientists due to the work of Lamarck, discussed later.

In the history of the search for the origin of life, the nineteenth century is remembered because of Charles Darwin's evolution theory. However, Darwin, who was influenced by scholars from other disciplines, as we shall see later, was preceded by various biologists. One of these was Darwin's grandfather, Erasmus Darwin. But his best-known biologist predecessor was Jean-Baptiste de Lamarck (1744–1829), who published his best-known book, *Philosophie Zoologique,* in 1809, 50 years before Darwin published his book *On the Origin of Species.*

In the history of biology, there are very few examples in which a great scientist is not remembered as such and where his contribution is not only largely underestimated but sometimes even ridiculed.

According to Lamarck's theory for the origin of species and their variations (see, for instance, the textbook by Kutter, 1987):

1. All organisms are characterized by a built-in drive to strive toward perfection.

2. Organisms are characterized by their ability to adapt to their environment.
3. Simple organisms are created spontaneously even today and are able to further develop and change, according to 1 and 2.
4. Traits acquired by organisms during their lifetime are inherited by their descendants.

Lamarck's theory is often explained (and ridiculed) by the famous example of the giraffe neck. According to this example, giraffes gradually developed long necks by eating the upper leaves of trees, thus acquiring this helpful trait, which was transmitted from one generation to the next.

It was Lamarck, in the ninteenth century, who was the first to suggest a consistent theory of gradual transformation of living creatures (Mayr, 1988, p. 199). The central feature of his theory was the "chain of beings," according to which all animals are arranged in an ascending order, starting with the lowest (infusoria) and ending in humankind. This theory deals with the origin of organisms but not with the origin of species (Ruse, 1979, p. 9–10).

Lamarck thought that simple forms of life are constantly being formed from the inorganic world by the action of agents such as heat, moisture, or electricity. Thus, extant mammals and fish do not share a common origin; they are at different stages on the scale of being.

Yet Lamarck, although a Deist, helped to establish an understanding of the gradual nature of the changes in the organization of living creatures and their parts (see Lovtrup, 1987; Jablonka and Lamb, 1995; Küppers, 1990).

Narrowing the gap between biology and the inanimate world

One aspect of the great discoveries of the nineteenth century was the borderline between living and nonliving substances. Thus, in 1827 the botanist Robert Brown (1773–1858) published his observations on the movement of microscopic particles suspended in a liquid. Both organic and inorganic particles were observed to demonstrate this movement (now known as Brownian movement), thus expropriating movement from its former "monopolistic" status as reserved only for living creatures. Another monopoly was broken in 1828, when Friedrich Wöhler (1800–1882) showed that urea, $(NH_2)_2CO$, an apparent organic compound, can be synthesized from inorganic compounds.

In 1835 the great chemist Baron Jöns Jakob Berzelius had already postulated that catalytic processes occur in living organisms. He named the substances involved in these activities "catalysts." It is said that at first even Justus von Liebig and Louis Pasteur did not believe in the ability of these substances to affect chemical reactions without participating in them. The role of enzymes in biochemistry was appreciated only much later, toward the beginning of the twentieth century. In 1897 Eduard Buchner (1860–1917) discovered that yeast extracts can ferment sugar to alcohol, thus proving that the enzymes involved in fermentation can function when removed from the living cell.

It is interesting to note that it was Berzelius who coined the names of the amino acids cysteine and glycine and the term "protein" (from the Greek *protos,* "first; primary substance").

Yet with regard to an understanding of the beginning of life, the scientific community was still discussing the validity of the spontaneous generation theory until the middle of the nineteenth century. An understanding of the cellular organization pattern of living entities, as hypothesized by Theodor Schwann (1810–1882) in his cell theory (1839), did not help bridge this gap either. Moreover, the discovery of chirality in 1848 by Louis Pasteur apparently complicated the problem of the origin of life. Chirality was considered a fundamental property of biogenic molecules and, in line with Pasteur's postulate, any hypothesis dealing with the beginning of life must also explain the origin of chirality (discussed below). The synthesis of chiral metal coordination compounds containing no carbon by Alfred Werner (1866–1919) would have to wait until 1914.

Pasteur: Demise of the cellular spontaneous generation theory

During the 200 years between the classical experiments of Redi (1668) and Pasteur (1862), the spontaneous generation theory was tested experimentally and gradually shifted its focus to smaller and smaller creatures. Following the establishment of the discovery of microorganisms by Leeuwenhoek (1675), and in view of the negation of the spontaneous generation of maggots by Redi seven years earlier, the discussion now focused on microorganisms. At first the spontaneous generation of these creatures could not easily be negated experimentally.

Starting with Louis Joblot, who in 1718 carried out the first sterilization of solutions by boiling, experiments were centered around this technique: Solutions containing the microorganisms under study were heated to the desired temperatures for a known period of time, and the viability of the microorganism was then assessed. Both the conditions needed for sterilization and the high thermal resistance of bacterial spores were not known at first, and seemingly conflicting results were published.

The most famous controversy, between Lazzaro Spallanzani (1729–1799) and John Needham (1713–1781), was the result of the inadequate sterilization methods applied by the latter. This debate, which started in the mid-eighteenth century, was to continue for a hundred years, until it was shown experimentally and decisively that microorganisms do not generate spontaneously.

As a young person, Pasteur (1822–1885) proved himself as a talented artist. At age 19, however, he decided to devote himself strictly to science (see Magner, 1994). He was already a chemist of considerable repute when he became involved in the spontaneous generation controversy. At the height of this controversy, he carried out his brilliant experiments that showed unequivocally that bacteria are not formed by spontaneous generation. His experimental system was based on glass flasks of special design in which nutrient medium was placed. Entrance of bacteria into the bottle was prevented by either using a cotton plug in the flask's neck or preparing a "swan neck" (fig. 4.1).

The flasks and their contents were boiled for some time to achieve sterilization and then left to cool. In the treatments in which airborne bacteria were not able to enter the flasks, no bacteria developed after sterilization. In control treatments with neither sterilization or a cotton plug or a swan neck, bacteria did develop. Conceptually, Pasteur's experiment is an extension of Redi's methodology of physically separating living creatures and their nutrient medium. The separation method is determined by the type of creature whose entrance is being prevented.

The idea that small creatures could enter a flask of aqueous nutrient solutions from the air had already been understood by Spallanzani. Air filtering by means of a long tube of cotton wool was first used by Schröder and by von Dusch (Magner, 1994,

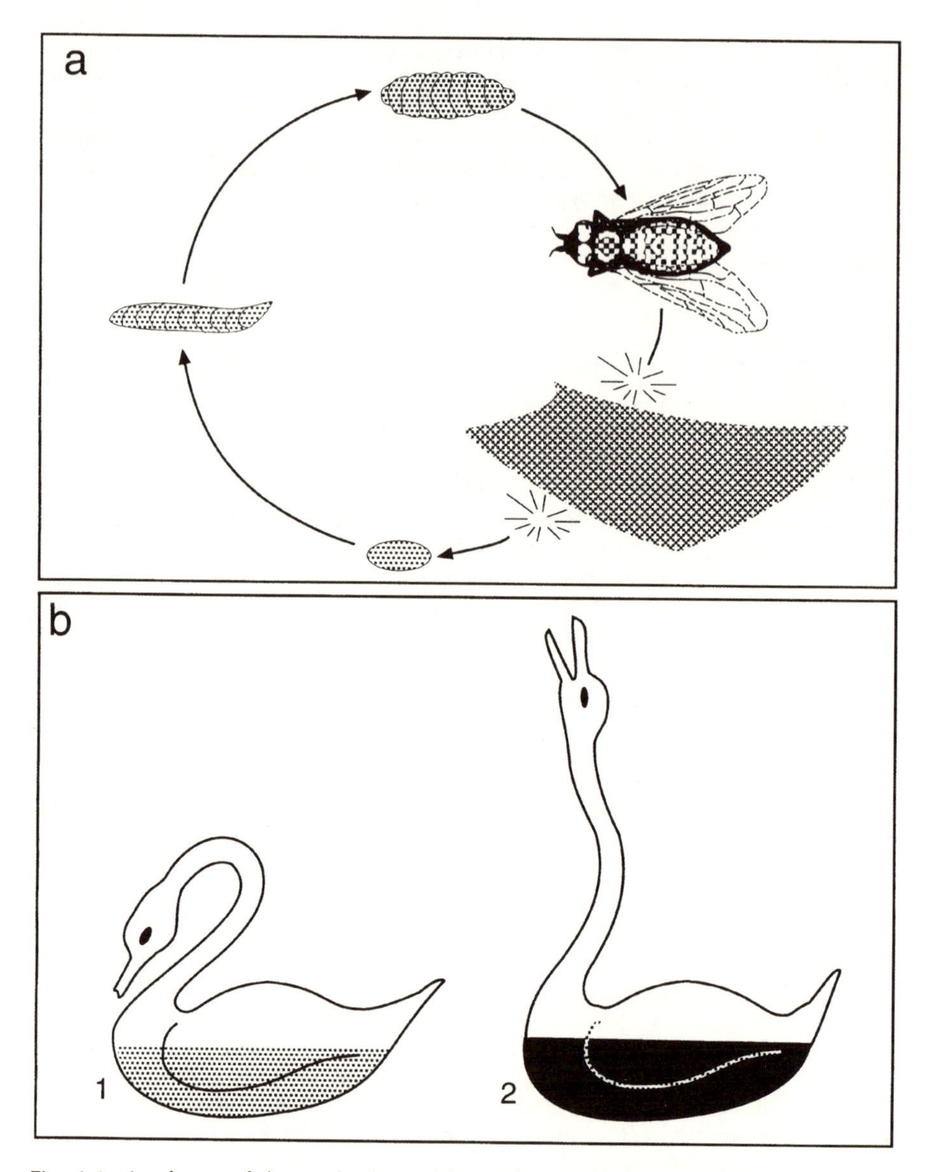

Fig. 4.1. A scheme of the methods used by Redi (a) and Pasteur (b) to exclude flies and bacteria, respectively, from their nutrient substrates. **a.** The life cycle of the fly was interrupted by covering the meat with a piece of muslin. **b.** A paraphrase on the "swan neck." Boiled swan-neck flask (1) remains sterile after boiling. Open flask (2) is contaminated by bacteria.

pp. 269–270). Pasteur's ingenuity seems to lie more in the way he planned and carried out his experiments than in the technological means he used.

Thus, two centuries after the first challenge to the spontaneous generation theory, it was proved that the then smallest known creatures—bacteria—are not formed spontaneously. For a moment it seemed that one must accept Harvey's dictum from

1651, "Omni vivum ex ovo," and Pasteur's triumphant declaration, "Life is bacteria and bacteria is life. Never will the doctrine of Spontaneous Generation recover from the mortal blow struck by this simple experiment" (cited by Hartman et al., 1985). But beyond the immediate dramatic effect of this declaration (too theatrical according to present behavioral patterns), there was an acknowledgment of the failure to explain the origin of life by scientific hypothesis. It should be noted that like Newton, Pasteur was a devoted Christian. His conclusions, however, which on the face of it seemed to contradict the prevailing verbal interpretation of the Holy Scriptures, were based on experimental results. Newton, on the other hand, based his calculations of the age of planet Earth on theoretical considerations, and presumably because of that—suspecting that they may be wrong—he tended to distrust them.

Although Pasteur's work ended 200 years of controversy, it raised new questions without answers regarding the origin of life. Pasteur's great authority and his determination with regard to the death of the spontaneous generation theory may have helped to create the standstill into which the search for the origin of life entered during the subsequent decades. For a brief time it seemed that old ideas about the eternity of life were the only way to deal with these problems.

But almost at the same time that Pasteur started his famous experiments regarding the spontaneous generation of bacteria, Darwin's book *On the Origin of Species* (1859) was issued. As noted by the historian Farley (1977, p. 7), "Before Darwin, the controversy was over the origin of present day organisms. . . . After Darwin, the controversy also involved the origin of life." At that stage, the implications of Darwin's theory were only hinted at; a few more scientific generations would be needed before the accumulating knowledge of various scientific disciplines would reach the critical mass needed to reach the next stage of understanding. Thus, the seeds of the next stage of the search into the origin of life had been sown. It would take a long time, however, before they could sprout.

Another discovery by Pasteur

In 1848, Pasteur discovered optically active organic substances (fig. 4.2). It was immediately obvious to Pasteur that this surprising discovery had to do with a fundamental attribute of life. As such it was also related to the origin of life and destined to remain on the agenda of origin-of-life researchers in the future, as will be discussed later.

Pasteur manually separated the two kinds of crystals and showed that solutions of these separated crystals each rotated polarized light to the same extent but in opposite directions. A racemic solution (of the two crystal forms) did not rotate polarized light (see Lehninger et al., 1993; Mason, 1991).

By-products of basic research

Two of the most obvious results of the scientific debate over the spontaneous generation theory have been of great importance since their development (see Horowitz, 1986):

Canning industry

As a result of France's involvement in wars in Europe, there was a need for a new technology to preserve large quantities of food. The lucrative prize promised by the

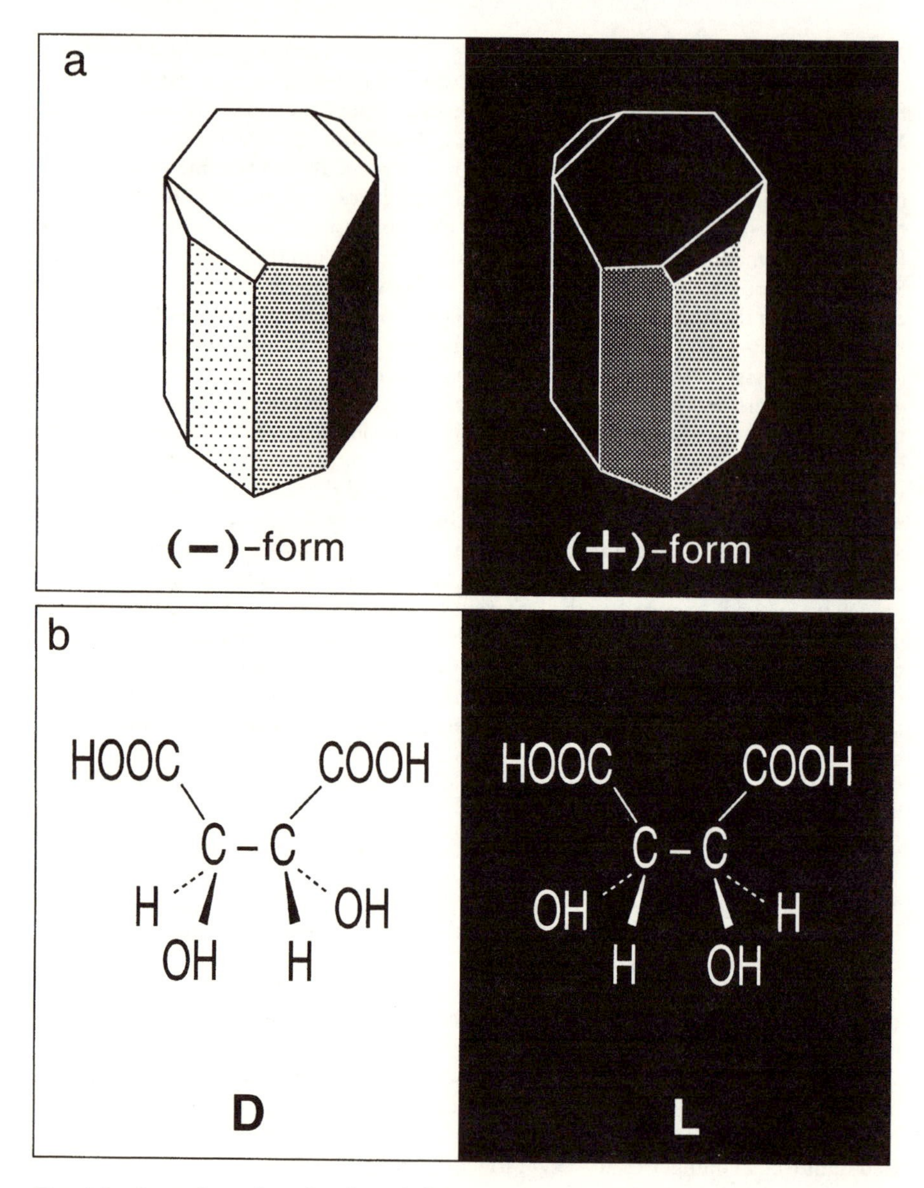

Fig. 4.2. Crystals and molecules of the two stereoisomers discovered by Pasteur: **a.** (−) and (+) forms of sodium ammonium tartarate crystals (Adapted from Mason, 1991, by permission of Oxford University Press); **b.** D and L tartaric acid molecules. (Following Lehninger et al., 1993)

French government for the inventor of such a technology in 1795 was won 14 years later by a resourceful Parisian confectioner, Nicolas Appert, who knew about Spallanzani's experiments. By filling large glass bottles with food, then boiling them, he sterilized it, thus applying the latest scientific discoveries of the time and founding the huge canning industry of the modern world.

Medical disinfection

Though this achievement was realized after Pasteur's classical work, it is a direct result of the spontaneous generation controversy. Joseph Lister, who was the first to develop medical disinfection, was a physician who worked with Pasteur for some time and was familiar with the recent sterilization methods of his time. His first experiment on the disinfection of open wounds was carried out in 1865. Two years later his own method—bandages presoaked with carbolic acid—became a huge success.

Charles Darwin (1809–1882) and Alfred Russell Wallace (1823–1913): The background of the evolution theory

> If I were to give an award for the single best idea anyone has ever had, I'd give it to Darwin, ahead of Newton and Einstein and everyone else. In a single stroke, the idea of evolution by natural selection unifies the realm of life, meaning, and purpose with the realm of space and time, cause and effect, mechanism and physical law.
>
> Daniel C. Dennett, *Darwin's dangerous idea*

Darwin did not work in a vacuum. For instance, the "struggle for existence" between species and evolutionism had been discussed much earlier by several scientists (Ruse, 1979). As I mentioned earlier, he was preceded by early hypotheses of naturalists such as Buffon, Lamarck, and the comparative anatomist Georges Cuvier (1769–1832). In the 1790s the latter, a staunch opponent of the evolution theory, made the first suggestion about the extinction of animals in certain geological eras (Ruse, 1979). This subject is destined to be brought again to the limelight 90 years later by Alvarez. Still, the idea of gradual evolution was accepted by various scientists even before Darwin's time, but it was Darwin (1809–1882) who developed it into a consistent theory supported by many observations (Mayr, 1988, p. 199).

But there were two more scholars whose books had a decisive influence on the development of Darwin's evolution theory. Charles Lyell's (1797–1875) book *Principles of Geology,* which is considered the beginning of modern geology, was published in 1830. In it, Lyell developed his uniformitarianism hypothesis, according to which the earth evolved slowly and rather uniformly, without catastrophes.

Darwin not only read the book, but also took it with him on his famous five-year voyage on the *Beagle* (1831–1836). He did this following the advice of his teacher, Professor Henslow, who arranged for him to be appointed chief naturalist on the boat. It turned out (Fifty years ago, 1996, p. ix) that "it was only by a series of chances that Darwin found himself aboard the *Beagle* at all, the final hazard being Fitzroy's objection to the shape of Darwin's nose, which he as a disciple of Lavater, considered as an indicator of insufficient determination."

But in spite of additional advice from Henslow—not to accept Lyell's uniformitarianistic point of view—there is no doubt that his evolution theory was a continuation of Lyell's teachings in geology (Mayr, 1982). Darwin was not influenced, however, by Lyell's teachings about organic evolution. It is interesting to note that when Darwin returned to England after his five-year voyage, he considered himself a geologist, not a biologist (Ruse, 1979).

The second source of influence on Darwin came from Thomas Robert Malthus's (1766–1834) essay entitled "Essay on the Principles of Populations," published first in 1798 and read by Darwin in 1838.

In this essay Malthus suggested that whereas populations of plants, animals, and human beings increase geometrically, in a doubling fashion, food production and natural resources available for the population under study increase linearly. Consequently, any population will either lose large numbers of individuals or level off at a stable level adjusted to the available resources.

The evolution theory was developed quite independently by another naturalist, Alfred Arthur Wallace (1823–1913), at about the same time that Darwin had conceived his own theory. Darwin's work was much broader and better founded than that of Wallace; in any case, it was Darwin who got the credit for the theory of evolution.

Alfred Russell Wallace, a penniless field naturalist who spent some time in the Malay penninsula, came to the idea of natural selection quite independently.

> He wrote it up and innocently sent his paper off to England, thinking that if anyone was interested in this topic it would probably be that chap in Kent, Charles Darwin, who was rumored to hold novel ideas about the problem. Wallace was quite unaware that the letter would come as a bombshell to his recipient, who had been carefully honing his natural selection theory for years. (Gamlin, 1992; see Ruse, 1979)

The manuscript sent by Wallace prompted Darwin to complete the writing of his version. The two of them read their manuscripts at the meeting of the Linnaean Society of London on July 1, 1858. The reports of the two scientists appeared in 1858 in the *Journal of the Linnaean Society,* and the president of the society wrote in his annual report that no revolutionary scientific idea was discovered during this year (Mayr, 1982). One year later Darwin published his famous book *On the Origin of Species by Means of Natural Selection.*

The social impact of Darwin-Wallace's evolution theory has been tremendous. A glimpse of this impact can be exemplified by the following anecdote (Keynes, 1995): "When Charles Darwin published *On the Origin of Species* in 1859, the Bishop of Worcester's wife was most distressed. 'Let us hope it is not true' she is said to have remarked, 'But if it is, let us pray that it does not become generally known!'"

The priority of Darwin's contribution to the development of the evolution theory over that of Wallace has not been accepted by some historians and scientists, who claim that Wallace's conclusions, which were known to Darwin beforehand, helped the latter to phrase his theory. However, most historians do not accept this argument, and Wallace himself never attempted to cast doubt on Darwin's integrity. Among those who discredit Darwin of his priority is Jacques Ninio, who argued (Ninio, 1983) that the idea of evolution has been expressed by various researchers and philosophers starting from Lucretius. Ninio maintained that this idea

> was adopted again in the last century by Chambers and by Malthus, and then developed officially by Wallace and Darwin. It is the name of Darwin, above all, with which this theory is linked in our collective consciousness. The skill with which Darwin created his own public image, adding his name alone to the idea of natural selection, serves as an example and inspiration source for many scientists even today.

Ninio's harsh criticism should not be accepted without question. For instance, Darwin's modesty is reflected in his writing. Moreover, when the Scottish naturalist Patrick Matthew announced that he had published ideas similar to those of Darwin many years before the *Origin of Species,* Darwin apologized, acknowledging his own ignorance for not being aware of Matthew's early work (Dennett, 1996, p. 49). Titus Lucretius Carus (55–96 B.C.) was a Roman poet; his didactic poem entitled "On Nature and Universe" describes the atomic doctrine of Epicurus (who followed the teachings of Democritus). Most historians do not consider his teachings to be related to evolution.

Though nothing like the evolution theory had been suggested by scholars of other cultures, it is interesting to cite Chuang Tzu (399–295 B.C.), a Chinese philosopher, as follows (Chan, 1963, p. 204):

> All species have originative or moving power (*chi*) [*chi* is a vital force which can be activated and transformed, for instance, through specific breathing (Sattler, 1986)]. . . . When they obtain water, they become small organisms like silk. In a place bordering water and land, they become lichen. Thriving on the bank, they become moss. On the fertile soil they become weeds. The roots of these weeds become worms, and their leaves become butterflies.

Some scholars maintain that this concept is related to evolution.

Overcoming Pasteur's crisis

The approaches taken by scientists in the wake of Pasteur's experiments can be divided into three categories, as follows (Kamminga, 1982).

Denial

Several scholars denied the validity of Pasteur's experiment, while attempting to produce counterexamples (Bastian, 1911). It should be noted that implied in the spontaneous generation theory is the spontaneous generation of certain diseases, therefore it entails more than just curiosity. However, due to the lack of experimental evidence, its adherers soon became an unimportant minority and hardly contributed to the continuing discussion on the origin of life. This line of thought will not be further discussed here.

Circumvention

Some scholars doubted the relevance of Pasteur's experiments by postulating that life is eternal in the universe. The problem of the origin of life is thus a problem not of origin per se, but of transplanetary transfer.

Bridging

Circumventing Pasteur's results by attempting to bridge between the inorganic and organic worlds. According to this approach, which later became mainstreaming the search into the origin of life, Pasteur's experiments are irrelevant to this search because bacteria do not represent a primordial stage in the beginning of life. The origin of life process had been a series of gradual transformations of simple carbon compounds, culminating in the emergence of the first primitive cells. Bacteria are the product of such processes, not their primordial stage.

Transfer rather than origin of life

According to this school of thought, life is a cosmic phenomenon and is eternal like the universe itself; it has existed somewhere in the universe since forever and has been transferred from one planet to another. The subject under discussion is thus meaningful only in the context of the mechanisms of that transference; it is no longer a biological problem but, rather, a physical one. As a matter of fact, the major contribution to the establishment of this school of thought was made by physicists.

From Anaxagoras (fifth century B.C.) to Richter (nineteenth century A.D.)

In its most rudimentary form, the concept of life as a cosmic phenomenon may be related to Anaxagoras. In its new form, it was first phrased by Hermann Richter in 1865 (Kamminga, 1982). Accepting Darwin's evolution theory, Richter looked for the starting point of evolution. As a paraphrase on Harvey's dictum of 1651—"Omne vivum ex ovo"—he now suggested, "Omne vivum ab aeternitate e cellula" (All life has always come from cells). Accordingly, life is eternal, propagating itself uninterruptedly from time immemorial. He coined the term "cosmozoa," the germs of the cosmos, which transfer life from one place to the next in the universe.

The eternity of life was conceived by Justus von Liebig (1803–1873) in a wider context, namely, the eternity of matter. According to his 1868 theory (see Engels, 1940), it can only be assumed that life is as old and eternal as matter itself and thus cannot be formed. The implication of this approach is a lack of connection between the appearance of life, which is eternal, and the conditions that prevailed on Earth prior to the arrival of life on this planet. This implication is of special importance in view of the hypotheses on the formation of our planet, according to which the high temperatures on primordial Earth would have burned away any organic compound.

The physicist Lord Kelvin (1824–1907) (William Thomson before becoming lord) was also involved in the controversy on the eternity of life and the origin of life on Earth (Kamminga, 1982). He totally rejected the spontaneous generation theory, as well as the possibility that life could have been formed from inorganic matter "without coming under the influence of matter previously alive." This to him was as valid "as the law of gravitation." In 1871, he suggested as a possible source of life on Earth fragments of heavenly bodies, such as meteorites, containing the seeds of life. Moreover, he claimed that "strong proofs of intelligent and benevolent design lie around us," showing "that all living beings depend on one acting creator and ruler." Twenty-six years later, he was still suggesting the scattering over Earth of "seeds of all species of the present day" (see Kamminga, 1982), thus rejecting Darwin's evolution theory and attempting to graft his firm religious belief to science.

Many scientists objected to Kelvin's ideas because of the incorporation of strong religious fundamentalism into his scientific reasoning. The transfer mechanism he suggested raised objections because of the high temperatures expected during the movement of the meteorite fragments through the earth's atmosphere. Such temperatures would burn up any living entity.

Hermann von Helmholtz (1821–1894) tended to support Kelvin's view that "life on Earth have been derived from another inhabitable planet" (Kamminga, 1982). The mechanism he suggested for the transfer of life from one planet to another was air currents sweeping bacteria up into space. He was more careful than Liebig, however: Considering the eternity of life as a scientific hypothesis, he stipulated the acceptance of this hypothesis on one's failure to form living organisms from inorganic substances. Thus, as implied in his approach, the solution of the problem of the eternity of life and its origin should be postponed for future generations.

Transition from search into the origin of cellular organisms to search into the origin of molecules

The spontaneous generation controversy was exacerbated in France in 1862, following the translation of Darwin's *Origin of Species* into French, three years after it had been published in England. Pasteur's first response, in 1864, was that the beginning of the evolutionary process was necessarily preceded by "an original act of spontaneous generation" (see Mason, 1991). Based on his own experiments, then, this idea was doomed to fail. It is interesting to note, however, that Pasteur himself was not happy with the total negation of the spontaneous generation theory. As a matter of fact, in 1878, 15 years after his dramatic declaration, he said (see Mason, 1991): "Spontaneous Generation? I have been looking for it for 20 years, but I have not found it, although I do not think that this is an impossibility." Obviously, novel ideas about the physical basis of life were needed in order to overcome the crisis induced by Pasteur's experiment.

Darwin's evolution theory does not deal specifically with the origin of life, even though he considered the possibility that all living forms on Earth had descended from one "primordial form." "Although the title of Darwin's great book (1859) promised a solution to the problem of the origins of biochemical diversity, it is agreed that his work far better explained the maintenance of species by natural selection than their first appearance." (Margulis, 1993a).

The implications of the evolution theory on the beginning of life, as we understand it today, were not clear at first; they were implied, however, in the central idea of common descent (fig. 4.3), which goes back to an unknown, remote past.

Darwin was also interested in the implications of his theory on the understanding of the origin of life. He was a religious person before the development of the evolution theory (Mayr, 1988), and this may explain his refraining from taking part in the discussions on the spontaneous generation theory. His firm religious belief was changed, however, as a result of the development of his theory. He did entertain the idea of an ancient ancestor, but at least at first, he thought this remote entity was created by a divine force. Apparently he preferred to express his ideas on this matter privately rather than publicly. Thus, in a letter to his friend Hooker in 1863 he said: "I have long regretted that I truckled to public opinion, and used the Pentateuchal term of creation, by which I meant 'appeared' by some wholly unknown process. It is mere rubbish, thinking at present of the origin of life; one might as well think of the origin of matter."

But in 1871 he may have been more mature to reflect:

> If (and oh! what a big if!) we could conceive in some warm little pond, with all sorts of ammonia and phosphorus salts, light, heat, electricity etc., present, that a protein compound was chemically formed ready to undergo still more complex changes, at the present day such matter would be instantly devoured or absorbed, which would not have been the case before living creatures were formed.

(Note that in Darwin's time, following the discovery of the electric current emanating from frog's muscles by Luigi Galvani in 1786, electricity was considered a characteristic sign of life.)

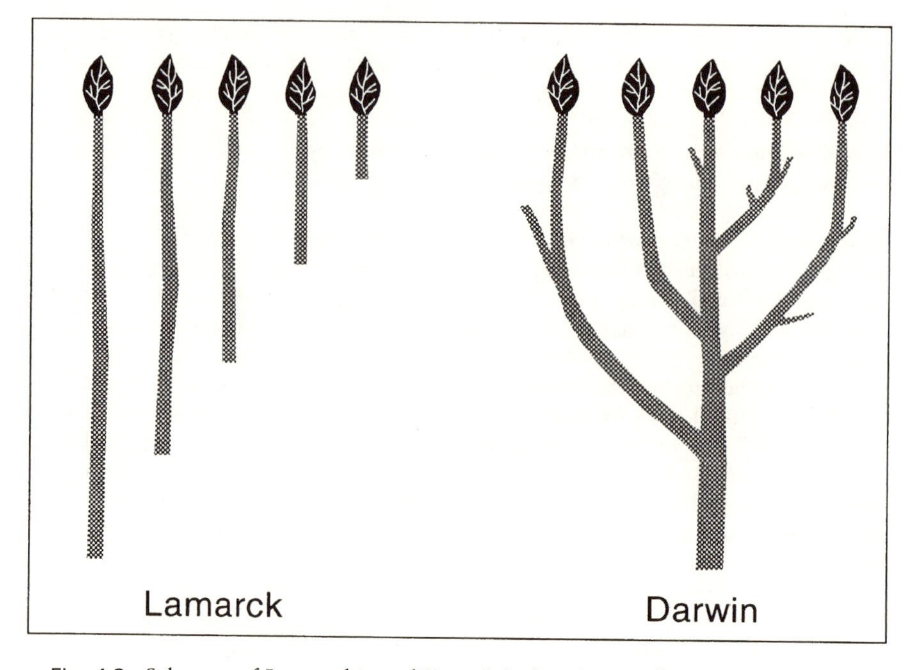

Fig. 4.3. Schemes of Lamarck's and Darwin's theories. (Following Ruse, 1979)

Simpson (cited by Yockey, 1992) recently noted that in his concept of the "warm little pond" Darwin was preceded by other scientists who expressed similar ideas. But even if Darwin had used ideas expressed earlier by others, it is also the combination of the "warm little pond" with the first glimpse into the common origin of the entire living world that make such a realization so interesting.

Darwin's vision of the "warm little pond" may be considered not only as a back-extrapolation along the evolutionary pathway, but also as the beginning of the concept of *chemical evolution*. Indeed, for a contemporary reader, Darwin's vision of a "warm little pond" is equivalent to the central thesis of Oparin and Haldane, published more than 50 years later.

It is still unclear to what extent this astounding passage of Darwin's vision affected the thinking of those prominent scientists who later developed the theoretical conceptual basis of the study of the origin of life. Moreover, it is not at all certain that Darwin's intention was in line with the approach suggested by Oparin and Haldane. Thus, according to Woese (1987), the "warm little pond" image was never intended to be "a prescription for life's beginning." Moreover, as noted by Mason (1991), Darwin's vision became significant only after the amino acids, polypeptides, and sugars were studied by Emil Fisher and his contemporaries. Only then could the "warm little pond" conjecture become a viable theory.

Darwin's idea of a gradual transition between living and nonliving states was adopted by several contemporaries. Indeed, in 1874 the physicist John Tyndall (1820–1893), a strong supporter of Darwin, described the origin and development of life as a gradual transition between natural forms, thus rejecting supernatural creative forces, as follows (cited by Bernal, 1967): "Trace the line of life backwards, and see it approaching more and more to what we call the purely physical condition."

First ideas about the necessary chemical mechanisms

Soon after the publication of Darwin's evolution theory, there were many attempts to account for protein structure, which was believed to be central to the question of the origin of life. Eduard Pflügger (1829–1910) was the first to suggest a chemical mechanism for the formation of the components of a living cell. In 1875, he suggested that simple organic compounds, such as cyanide (CN^-), as well as its polymers, could have been formed on the primordial earth, under the conditions prevailing then. His suggestion was not accepted by other scientists in the newly formed scientific discipline that later became known as *biochemistry*. But the vague idea that evolution at the molecular level preceded evolution at the cellular level was introduced into the thinking of scientists.

Nucleotides were discovered by Friedrich Miechner in 1869, but their role in genetics was destined to wait until 1943, when Oswald T. Avery first demonstrated it.

Phylogenetic trees

The phylogeny of living forms, both extinct and contemporary, is represented by a phylogenetic tree; the beginning of this tree is the common ancestor, from which all later forms were derived (fig. 4.4).

In 1866, Ernst H. Haeckel (1834–1919), a staunch Darwinist, suggested a new kingdom of microorganisms, which he named "protista" (from the Greek *protistos*, "primary." In phylogenesis, protists are single-celled eukarya). Thus, rather than the ancient classification of all living forms into plants and animals, Haeckel divided biology into three kingdoms, namely, plants, animals, and protista, which had diversified from one root. The first creatures were autotrophic microorganisms thriving on CO_2, which was abundant in the early atmosphere; heterotrophism evolved later.

Haeckel considered Darwin the Newton of biology. According to him (Emmeche, 1994, p. 9), "Darwin achieved the fundamental physical explanation for the multiplicity of life. . . . The divine design explanation was now unnecessary."

The geological picture

The importance of geology in the search for the origin of life was recognized rather early, as shown by the following statement, made by Haeckel in 1866 (Farley, 1977): "Any detailed hypothesis whatever concerning the origin of life must, as yet, be considered worthless, because, up till now, we have not satisfactory information concerning the extremely peculiar conditions which prevailed on the surface of the earth at the time when the first organisms developed." At the end of the nineteenth century and the beginning of the twentieth, many more geological findings enriched the knowledge of our planet's past. However, the oldest known geological fossils did not go back beyond a few hundred million years; the geological record of the earth's remote past had yet to be discovered.

The controversy surrounding Earth's age is finally settled

The importance of time in the search for the origin of life and the evolution theory was considered critical both by Darwin and his opponents. Reliable calculations of the age of planet Earth or, at least, the time period available for the evolutionary

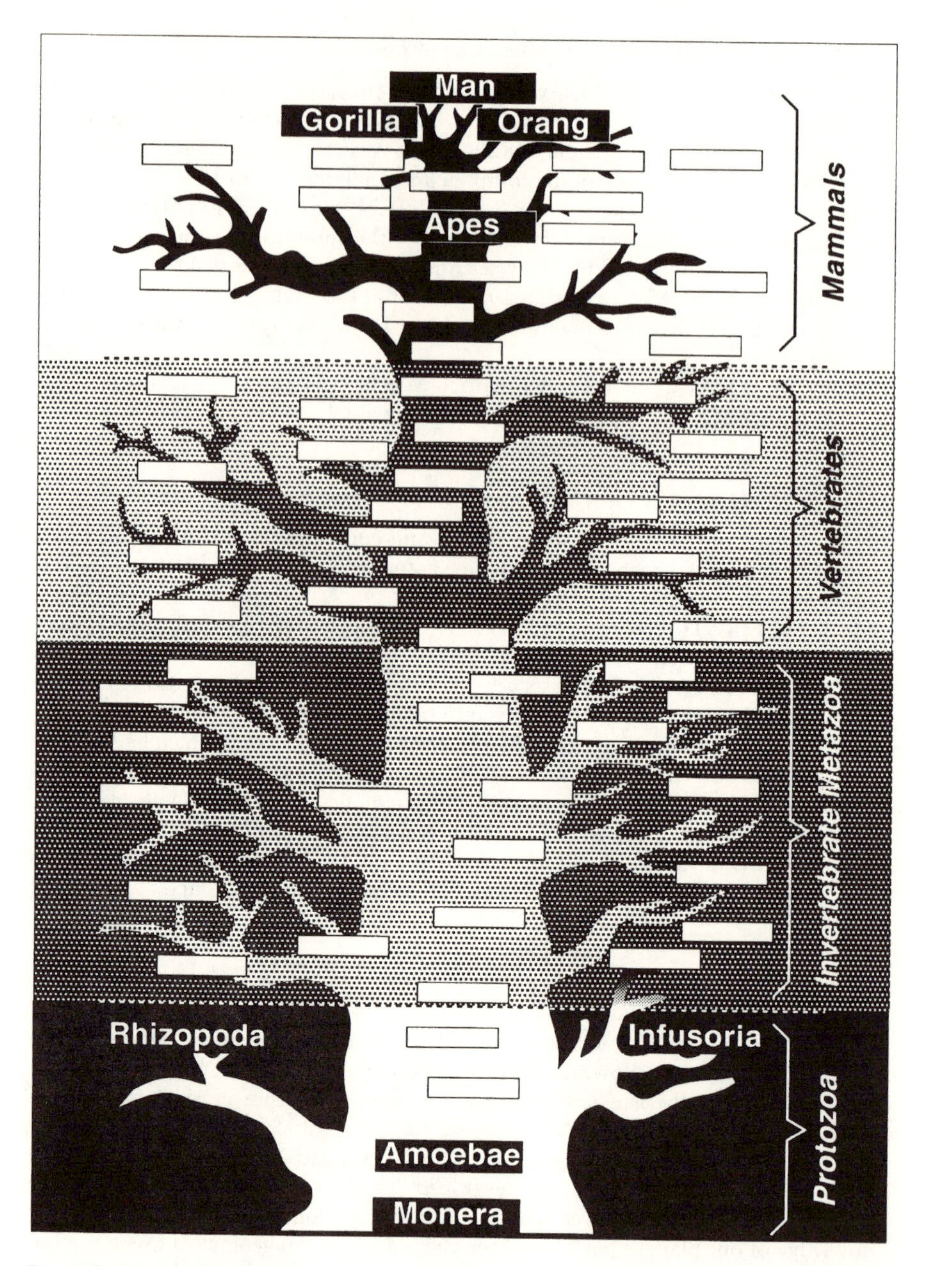

Fig. 4.4. Haeckel's phylogenetic tree. For simplicity only a few names at either end of the tree are given. (Adapted from Haeckel's evolutionary tree of life [1910], reproduced in Sheldrake, 1988)

process became a critical factor in the search for the beginning of life on this planet. Obviously, the old estimates by Newton and Buffon had to be replaced by more reliable ones. It was Darwin who, in line with the new geology of Lyell, made a daring estimate of the geological time available for evolution. His calculation was a rough estimate of the time needed for the denudation of the chalk cliffs of the Weald (Sussex, southern England), assuming a constant encroachment rate of one inch per hundred years. According to these calculations, which were published in the first edition of *Origin,* the age of the Weald was about 300 million years; thus, the age of Earth must be much older!

This was one of the few calculations that Darwin published, and he later regretted it; in the second edition of *Origin* he reduced these age estimates, and he did not include them at all in the third edition.

It is said that Galileo once remarked that the book of nature was written in the language of mathematics. Elegant and succinct as this saying may seem, it should be noted that Darwin, one of the most prominent men to have understood the book of nature, did not have the mathematical skills needed, according to Galileo, to read this book.

Toward the end of the nineteenth century, the uniformitarianists came under heavy attack by their opponents, who stressed the importance of catastrophes in the history of the earth, as evidenced by the lack of continuity in the geological record. Among the latter was the physicist Kelvin, who was also involved in other scientific controversies.

Kelvin presented some important arguments against the evolution theory, as summarized in the following three items (Burchfield, 1975):

1. Natural selection could not, in his opinion, explain the origin of life.
2. The time needed for natural selection was much larger than the actual time available for this process.
3. Chance cannot bring about the observed planning in nature.

Of these three arguments, the second one can be quantified. Thus, Kelvin applied Helmholtz's hypothesis of heat generation by way of nebular gravitational condensation to these calculations and came up with an age of about 20 million years. Later he revised this number, stressing that according to his "irrefutable" calculations, Earth was 10 million years old. Obviously, such a period of time is too short for an evolutionary process to explain biology. Wallace, the co-inventor of the evolution theory, tried to adjust this theory to the newly imposed limitations introduced by Kelvin's calculations, but there was no way to reconcile the two approaches.

The objection of the geologist T. C. Chamberlin, a contemporary of Kelvin's, to the scientific carelessness of the latter was described by Burchfield (1975, p. 141): "His theme . . . was the air of certainty which Kelvin gave to speculations that could by no means be considered certain. Courteously but firmly he suggested that the use of such phrases as 'certain truth,' 'very sure assumptions,' and 'half an hour after solidification' were so misleading as to convey untruth."

Less than ten years after the publication of Kelvin's calculated value, Henry Becquerel discovered radioactivity (1886), and two years later Marie and Pierre Curie discovered radium. Soon it became clear that estimates of heat balance and the calculated age of the planet would have to be changed dramatically in view of the radioactive disintegrations inside the earth. This is just another example of the importance of the assumptions upon which all calculations are based. Moreover, the discovery of radioactivity not only refuted Kelvin's calculations but also resulted in the development of very accurate methods for estimating the earth's age (Badash, 1989). The most recent measurements of this kind show this age to be 4.6 billion years.

5

The Modern Era

Spontaneous Generation at the Molecular Level

> The common assumption is that the earth and its atmosphere have always been as they are now, but if this is assumed it is necessary to account for the present highly oxidized conditions by some processes taking place early in the earth's history. Briefly, the highly oxidized conditions is rare in the cosmos.
>
> Harold Clayton Urey, "On the early chemical history of the earth and the origin of life"

The second rise and fall of Panspermia

In 1908, Svante Arrhenius (1859–1927) published a new version of Richter's Cosmozoa theory and gave it the ancient name Panspermia, which had been coined more than 22 centuries earlier by Anaxagoras. Arrhenius accepted the thesis of his predecessors Richter, Kelvin, and Helmholtz with regard to an alternative to abiogenesis; his main innovation was a newly introduced theory concerning the pressure exerted by solar radiation. He suggested that this pressure was the mechanism by which living germs are transported in space.

Arrhenius assumed that bacterial cells had been formed in the universe on planets like ours. He was aware of the problem of the heat generated by meteoritic fragments falling in the gravitational field through the atmosphere, and he also assumed that collisions between large bodies such as planets were rare in the universe. His solution to the problems involved in cosmic transport of living entities was to assume that the traveling units were *bacterial spores*. These spores were first transported by upward air currents to heights of 100 km or so. There, strong electrical discharges would push particles against gravity. Travel through interplanetary space was made possible by *radiation pressure,* which could carry the spores over very large distances. Upon entering the gravitational field of another planet, the spores would fall down, probably by being attached to grains of interstellar dust big enough not to be pushed away by the solar radiation pressure. Due to the size of the spores and their carriers, the problem of frictional heat would be negligible, and a soft landing would take place. The living spores could then germinate under appropriate conditions. It was noted by Bernal (1967, p. 109) that Panspermia "is really a doctrine of the effective eternity of life."

A central consideration of Arrhenius's scenario is the ability of the bacterial spores to withstand the harsh conditions prevailing in space. Experiments on this subject were begun after the publication of Arrhenius's book, and it soon became obvious that exposure of spores to the strong UV irradiation expected in space brought about their death. Following the publication of these experiments, the popularity of the Panspermia theory decreased dramatically.

But there was an additional reason for the demise of the Panspermia theory: Some time after its publication, another theory on the origin of life was suggested, quite independently, by Oparin and by Haldane. The popularity of this newer theory soon eclipsed the Panspermia theory and became a dominant paradigm for many decades. Interestingly, the Panspermia theory was destined to return to the scientific agenda in different forms, unimaginable by Arrhenius.

Organic synthesis as the first step of life's origins

Life as we know it should have been preceded by organic synthesis. Meteorites containing organic compounds have been known since the time of Berzelius and Wohler (chapter 4), therefore organic syntheses could have started either extraterrestrially or terrestrially. Most of the early attempts to explain the origin of life focused on terrestrial synthesis of organic compounds. Extraterrestrial sources of organic compounds have only recently become more popular among scientists.

The idea of organic synthesis as the first stage of life's origins was also reflected by attempts to characterize the sites of such syntheses on primordial Earth's surface. Thus, in 1908 Chamberlin and Chamberlin (cited by Deamer and Fleischaker, 1994) wrote: "The primitive organic synthesis may have taken place (1) in the ocean, (2) in some body of fresh water, or (3) on the land, or, more specifically, (4) in the soil." And later they added: "May we not take it for granted that the higher presumption will lie in favor of that localization which brings into closest interaction the requisite material in unstable states, attended by the maximum range of concentrations, condensations, catalytic, electrical, nascent and other favoring conditions?"

We shall see later that these presumptions are as valid today as they were in 1908, serving as general guidelines for the possible sites of the very first stages of chemical evolution. The term "chemical evolution" was coined by Benjamin Moore in 1913 (Kamminga, 1988b), but its implications for the origin of life were destined to become clearer only decades later.

Subcellular state of matter

The second half of the nineteenth century and the beginning of the twentieth constitute the era of *colloid chemistry* (from the Greek *colla,* meaning "glue") established during the mid-1860s by Thomas Graham (1805–1869) and applied to both organic and inorganic systems. At the beginning of the twentieth century, Wilhelm Ostwald (1853–1932) developed the theory of the colloidal state of matter. Inorganic colloids were classified in the same category as proteins because all of them were considered colloids. The proteins, as *biocolloids,* were thus distinguished from the inorganic colloids. It therefore became natural to focus on the colloidal state as the most significant subcellular state of matter. In the words of Duclaux (1904, cited by Laszlo, 1986, p. 139): "All living things are colloids and colloids, in many respects, are life."

The first suggestion of a relation between colloids, viruses (discovered in 1896), and the beginning of life was made in 1908 by Macallum (see Farley, 1977, p. 159), who argued that the protoplasm represents a colloidal state, the particles of which are "alive," similar to viruses. And he went on to state: "When we seek to explain the origin of life, we do not require to postulate a highly complex organism . . . as being the primal parent of all, but rather one which consists of a few molecules only and of such a size that it is beyond the limits of vision with the highest powers of the microscope." The discoveries of bacterial viruses in 1915 by Fredrick W. Twort (1877–1950) and of phages in 1917 only strengthened this point of view. And in this spirit, it was Félix d'Hérelle (1873–1949), the discoverer of the phage, who said, as late as 1926 (cited by Farley, 1977, p. 160): "Life does not require a cellular organization; . . . it results from a special physico-chemical state of matter, that is, the protein micella."

A new beginning to the phylogenetic tree

Back-extrapolation of this line of thought toward the beginning of life is thus obvious: The beginning of the phylogenetic tree of life is what d'Hérelle called "unimicellar entities" or "protobes" (fig. 5.1).

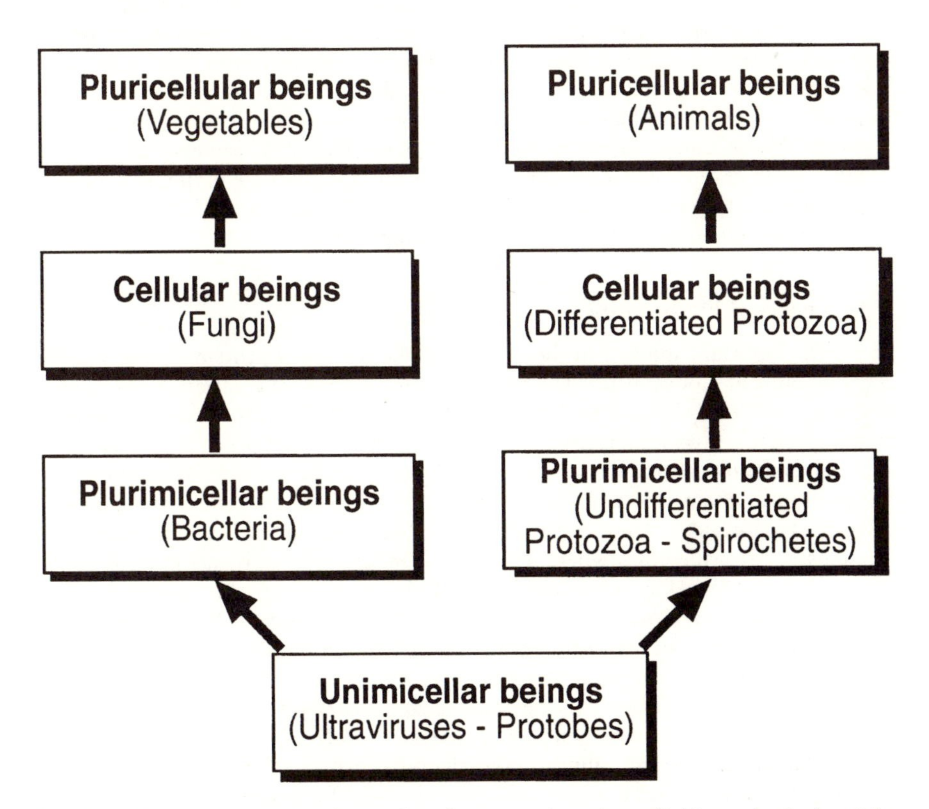

Fig. 5.1. The phylogenetic relationships between the primordial hypothetical entities (unimicellar beings) and the known organisms according to d'Hérelle. (Following Farley, 1977)

Centrality of the protoplasm

An important feature of the study of chemistry of life at the end of the nineteenth century and the beginning of the twentieth was the centrality of the **protoplasm** (see Kamminga, 1988a, 1988b). The protoplasm was viewed as one substance, complex as it may be, which cannot be separated into its components without destroying its ability to function in the central processes of life, such as growth and metabolism. Accordingly, the beginning of life should have been the origin of a primordial and less sophisticated protoplasm. Progress in cytological and biochemical research, such as the isolation of active enzymes from the cell, resulted, already at the beginning of the twentieth century, in the recognition that the protoplasm is a heterogeneous substance made up of many subsystems. The origin of life should thus be looked for in much simpler systems. Moreover, these obscure primordial entities cannot be explained in terms of single molecules but only by the combination, collaboration and coordination of several components.

At the molecular level, the characterization of life was still unclear. For instance, in 1927, at the 73rd meeting of the American Chemical Society, the chemist Vaughan wrote:

> I can say with much confidence that the conversion of non-living into living matter is accompanied by increased molecular liability. By this I mean that the atoms and electrons within the molecules are energized. Their orbits are enlarged. Within their orbits they move with greater speed. Their chemism is intensified so greatly that they are now able to drag into their orbits atoms and possibly molecules which have hitherto been beyond their grasp. In other words the molecules begin feeding on outside matter. . . . This means that metabolism or trading in energy begins. Such is the first evidence of life. Have we any idea on the nature of these primitive living molecules? Yes. They were and are protein molecules. There is no life save in proteins.

Emergence of biochemistry

The most important impact on the search for the nature of protein came from the new discipline of biochemistry. The name "biochemistry"—a translation of the term *biochemie,* coined by Hoppe-Seyler in 1877 (Florkin, 1972, p. 1)—covers all the molecular approaches to biology. Important progress in this field had already been made in the nineteenth century. Thus, various examples of catalysts were already known. The prominent chemist Jöns Jakob Berzelius (1779–1848) used the term "enzyme" (from the Greek *en* "in," and *zumé,* "leaven," "down-loosening") in decomposition reactions, to describe cases where an active agent in a reaction was not consumed (Mason, 1991). Berzelius also suggested, in 1838, the term *protein* (from the Greek *protos,* meaning "first; primary substance") for the basic unit of all albuminous materials. Soon afterward, this term became popular with the modified meaning of a class of albuminous materials in general.

Early in the twentieth century, the first enzymic theory was incorporated into biochemistry. The enzymes were considered *colloidal catalysts* by Hofmeister in 1901, in accordance with the prevailing colloid-chemistry world. The first crystallized enzyme (urease) was prepared by James B. Sumner in 1926, but the full significance of the difference between what we now call "macromolecules" and colloidal particles (made of small molecules bound to each other by intermolecular forces) took some years to establish. Gradually it became clear that micelles are irrelevant to biochemistry; but the first electron microscope, and with it the main structural features of proteins, had to wait until 1936. The biochemistry of the heredity material was still unknown, even though the term "gene" had been introduced to science by Wilhelm I. Johannsen already in 1909.

First ideas about "synthetic biology"

Optimism surrounding the possibility of synthesizing living entities in the laboratory was reflected in the works of several researchers. Some of the first publications about the experimental synthesis of living forms, however, were based on the erroneous interpretation of various colloidal structures as living entities (Oparin, 1957).

One of the first pioneers of synthetic biology was Jacques Loeb (Laszlo, 1986; Turney, 1988), an eminent physiologist who emigrated in 1891 from Germany to the United States because of the anti-Semitism in German universities. As early as the beginning of the twentieth century he conceived biological manipulations as a means to synthesize novel living organisms in a test tube. The following statement of Loeb was made in 1911 (cited by Cassirer, 1950) and is of special interest in this context: "I have the feeling that the technical limitations of our young science are to be blamed for not being able to synthesize life."

Another approach to the synthesis of living entities is reflected in these words by Ostwald, who in 1919 was cited by Gaskell as follows: "Like the chemistry, so should the physics of organized bodies to be analyzed for separate processes, the combination of which would create the synthetic biology."

The methodology suggested by Ostwald is the conventional scientific methodology based on Descartes's dictate that one must divide the problems under study into as many parts as possible and then study them separately—the simplest first and the more complex later. The suggested analysis of the foundation of biology through a chemical-physical approach seems to echo the belief of a group of European physiologists who in 1847 were already vowing to "constitute physiology on chemico-physical foundation and give it equal scientific rank with physics" (cited by Smith and Morowitz, 1982). More than a century later, this idea seems almost natural to students, even those in high school.

Obviously, the strategy delineated by Ostwald was not enough to serve as a new paradigm. The latter should include an understanding, prioritization, and simplification of the central processes in the living cell, on the one hand, and a general principle for the role of the environment in such simplified processes and scenarios, on the other. Moreover, a scientific paradigm is a prerequisite not only for the planning of an experiment but also for the interpretation of the observations, as already stated at the end of the nineteenth century by "the father of physiology," Claude Bernard (cited by Farley, 1977, p. 3): "The experimentalist who does not know what he is looking for, will never understand what he finds."

It would be hard to find a better confirmation for Bernard's statement than the cases of W. Loeb and Fritz Haber. In 1913, Loeb found that silent electrical discharges in certain mixtures of gases gave, among other things, the amino acid glycine (cited by Miller, 1955). At about the same time, Haber carried out similar experiments in which he passed electrical discharges through various carbon-containing gases. His conclusion was that by applying such discharges it is possible to synthesize "any substance known to organic chemistry." Obviously, Haber could not realize the significance of his finding for the study of the origin of life (Chang et al., 1983, p. 62).

As we shall see, a similar finding four decades later, with the background of a general hypothesis on the origin of life, brought about the formation of the novel scientific discipline "chemical evolution." The observations of the two scientists, Loeb and Haber, were not connected to any hypothesis on the origin of life and therefore did not play a role in the development of the first modern paradigm in this scientific discipline.

Emulation of biological forms

Several scientists at the beginning of the nineteenth century were misled by the similarity between living cells and various inorganic, mostly colloidal, forms produced in the laboratory. Erroneously, they interpreted these forms as living entities. In view of the ideas about the transition between inorganic compounds and living organisms, this interpretation may not be very surprising. It should be also noted that in 1919 the prominent biologist Ernst Haeckel published a book about crystals in which he ascribed attributes of living organisms to these inorganic forms.

The biogeochemical spontaneous generation theories of Oparin and Haldane

The biochemist Alexandr Ivanovitch Oparin (1894–1980) published his first book, entitled *Vozniknovenie Zhizny na Zemle* (The Origin of Life on Earth), in 1924. In this small book, Oparin presented his main ideas very succinctly in the Russian language. The enlarged and revised version of this book appeared in 1936, was translated into English in 1938, and soon became very influential, recognized as a landmark in the search for the origin of life. The main elements of his teaching were the adoption of evolution as a central theme and the integration of various scientific disciplines, mainly organic chemistry, biochemistry, geochemistry, and astrophysics, in a coherent and partially testable scenario (Oparin, 1961).

Based on available studies of his time, Oparin suggested that the prebiotic earth had a reducing atmosphere. Under the hypothesized reducing conditions, the synthesis of organic molecules—including the building blocks of biochemistry or their predecessors, as well as their further chemical transformation—would be feasible. Gradually, the organic compounds thus formed became more and more complex. In line with the then fashionable colloid-chemistry world-picture, the organic molecules were suggested to have formed colloidal aggregates in the aqueous environment in which they were formed. These colloidal aggregates—"coacervates" in the jargon of colloid chemists—were able to absorb and assimilate organic molecules from their environment in a way reminiscent of metabolism; presumably they served as the first chemical entities capable of undergoing evolutionary processes, which eventually led to the emergence of primitive living forms. Oparin coined the name "eubiont" (from the Greek *eu,* "well," implying "true," and *bio,* "life") for this hypothetical creature, which today would probably be considered an extreme anaerobic heterotrophic prokaryote.

The first organisms, according to Oparin's scenario, were anaerobic and heterotrophic, or dependent on preformed nutrients. Chemotrophism and phototrophism emerged later in evolution, as a result of exhaustion of the organic nutrients synthesized and provided by the prebiotic environment. Moreover, the evolution of photosynthetic organisms slowly enriched the atmosphere with oxygen, giving rise to the present-day biosphere.

As noted in 1974 by Oparin (cited by Lazcano, 1994a), the question of the origin of life at the time he became interested in the topic was in the realm of imagination:

> It appeared as if it was a forbidden subject in the world of science. The problem was generally felt insoluble in principle using objective research methods. It was felt that it belonged more to the sphere of faith than knowledge, and that, for this reason, serious scientists should not waste their time and efforts on hopeless attempts to solve the problem.

Quite independently, in 1929 (before Oparin's first book was translated into English), J. B. S. Haldane (1892–1964), a biochemist and geneticist, published a succinct review of his own ideas about the origin of life, which were very similar to those of Oparin. Haldane proposed that the primordial sea served as a vast chemical laboratory powered by solar energy. The early atmosphere was oxygen-free, and the combination of dissolved carbon dioxide and ammonia with UV radiation uninhibited by ozone (as is the case today) gave rise to a host of organic compounds in the upper layer of the primordial sea. Consequently, the primordial sea became a dilute solution of large populations of organic monomers and polymers, which he likened to a "hot diluted soup." The first cell emerged after the acquisition of a lipid membrane, as well as additional systems. Following many failures, the first cells emerged, out of which a single cell, the common ancestor of the entire living world, was selected. (For a discussion about the universal ancestor, see chapter 10).

The name "prebiotic soup" coined by Haldane soon became a buzzword in the jargon of both scientists and laypersons. More than any other concept, the prebiotic soup has become a symbol of the new world-picture hypothesized by Oparin and Haldane in connection with to the origin of life.

Heterotrophy replaces autotrophy

Both Oparin and Haldane replaced the hypothetical *autotrophy* suggested by Haeckel almost 70 years earlier with *heterotrophy*. The former trophic method, however, was destined to be revived several decades later, and reenter again the vocabulary of origin-of-lifers, as discussed further on.

Schrödinger's aperiodic crystal

One of the most influential books on the study of the origin of life was based on a series of lectures by Schrödinger, given in Dublin in 1944, called "What Is Life?" (Schrödinger, 1944). It is considered "*one of the most talked about books in twentieth-century science*" (Welch, 1995; emphasis in original). One of the most cited ideas in this book has to do with information storage in aperiodic structures. For example: "Let me anticipate . . . that the most essential part of a living cell—the chromosome fiber—may suitably be called an *aperiodic crystal*" (Schrödinger, 1944, p. 5; emphasis in original). This expression was coined eight years before the discovery of the DNA structure by Watson and Crick, so it may be considered the forerunner of molecular biology.

Other expressions that have had a great effect on the biophysical thinking of the second half of the twentieth century are "Order from order," "order from disorder," and "negative entropy." For more information the reader is referred to Elitzur, 1995, and Welch, 1995.

Norman Horowitz's "backward evolution" and the reflection of evolution in extant biosynthetic pathways

An interesting hypothetical feature starting to emerge from the Oparin-Haldane scenario was added in 1945 by Horowitz as follows: The prebiotic soup contained simple molecules, as well as complex ones to which the simpler molecules served as precursors. Since the heterotrophic creatures that evolved first utilized the complex chemical entities and brought about their depletion, the biochemical synthesis of organic "nutrients" by those chemical entities had to evolve backward, from complex products to their simpler precursors. Thus, complex molecules, such as amino acids, were progressively depleted from the prebiotic soup as a result of their incorporation into the early organisms. As a result of this depletion, only organisms that "learned" how to synthesize those complex molecules out of the still-abundant simpler intermediates survived. Upon further gradual depletion, only those organisms that had "learned" how to synthesize other precursors would survive by utilizing simpler intermediates and developing new biosynthetic pathways (Mason, 1991, p. 235). Moreover, according to Horowitz the biosynthetic pathways represent an evolutionary development (Granick, 1957; see also Keefe, Lazcano, and Miller, 1995).

Bernal's involvement of clays

In 1951 still another feature was added to the Oparin-Haldane scenario, this one by the biophysicist John Desmond Bernal (1901–1971). Bernal (1951) suggested that a dilute prebiotic soup would have been ineffective in the first hypothetical reactions, in which the dissolved building blocks of the earliest "living" entities would have needed to react at a reasonable rate. In order to overcome the obstacle of a dilute prebiotic soup, Bernal resorted to the minerals suspended in the waters or sedimenting on the bottom of the prebiotic sea. These minerals, predominantly aluminosilicates known as *clay minerals,* could have served as adsorbents for organic molecules such as amino acids and sugars. Enhancing the concentration of biochemical building blocks on the surfaces of clay minerals was considered to be the first step leading to condensation and increased complexity, which would be essential for the evolution of living forms. Moreover, according to Bernal's suggestion, minerals, which are known for their catalytic properties in a large number of organic reactions, would act as catalysts in these hypothetical reactions under study.

Bernal's hypothesis is in line with Chamberlin and Chamberlin's 1908 view about the possible environments in which the first stages of the origin-of-life processes could have taken place. In retrospect, it seems surprising that the Oparin-Haldane scenario, which is partially testable, was not tested experimentally until the 1950s, even though the essence of their approach was probably known to many scientists.

Oparin, Haldane, and Bernal were all confirmed Marxists, defenders of dialectic materialism. As noted by de Duve (1991), it is hard to evaluate the influence of their political and social ideologies on their scientific thinking. Like some other intellectuals in Britain during the 1930s and 1940s of this century, Bernal was a Marxist to the extent that "he supported the Stalinist regime long after it was prudent or sensible so to do" (Harris, 1994). Bernal also became known as an art expert who attempted to reunite art and science—two disciplines that had a common origin in our civilization and only became separate disciplines relatively recently. In the words of Harris (1994): "He argued that it was interest in the visual art in the past which had led to the accurate observation of nature, and that the requirements of architecture had given rise to the discipline of mechanics. He suggested that the social responsibilities of the artist were similar to those of the scientist."

Lorch (1975), in a paper entitled "The charisma of crystals in biology," noted that scholars throughout the history of science had been fascinated by the possibility of finding similarities between crystals and living things. Bernal's hypothesis does not invoke such similarities. Moreover, attributing qualities of living creatures to crystals was suggested in the 1960s by Cairns-Smith (1966, 1982) and resulted in considerable publications during the early 1980s. This is discussed in chapter 20. No less interesting is the semantics used by Kauffman (1993) in his book on the origin of life, in the title of his second chapter "The crystallization of life." The third sentence of this chapter, a cross-breeding of science and poetry, reads as follows: "Life, in a deep sense, crystallized as a collective self-reproducing metabolism in a space of possible organic reactions."

Calvin, Urey, and Miller: The advent of the experimental era

Prompted by Oparin's model, Melvin Calvin and his associates made the first attempt, in 1950, to synthesize organic molecules from inorganic compounds by applying a beam of 40 MeV He^{2+} ions generated from helium in the then new cyclotron at Berkeley. They bombarded a gas mixture of hydrogen (H_2) and carbon dioxide (CO_2) equilibrated with an aqueous solution of ferrous ion (Fe^{2+}) by these helium ions in an effort to simulate natural radioactivity as a source of energy for prebiotic reactions. They found that more than 20% of the dissolved carbon dioxide was transformed into formic acid (HCOOH) and less than 0.1% into formaldehyde (H_2CO). Acetic acid (CH_3COOH), multicarbon acids (lactic $CH_3CHOHCOOH$, succinic $COOHCH_2CH_2COOH$, fumaric COOHCHCHCOOH, etc.) were also detected. The report of this experiment (Garrison et al., 1951; Calvin, 1951), however, was hardly noticed by the scientific community.

At about the same time, Harold Urey (1893–1981) tried to quantify the properties of the putative primitive atmosphere of early Earth, addressing three separate questions: the synthesis of the chemical compounds characterizing living organisms, the evolution of complex chemical reaction pathways and dynamics characterizing living organisms, and the availability of free energy sources needed to maintain the central chemical reactions of the primordial "living" entities (Urey, 1952). According to his calculations, this primordial atmosphere consisted of methane (CH_4), ammonia (NH_3), hydrogen (H_2), and water (H_2O). Based on his model of the early atmosphere, he suggested that his graduate student, Stanley Miller, carry out a simulation experiment for the synthesis of organic compounds in such an atmosphere. The experimental system was constructed and tested by Miller in 1953. It consisted of a large glass flask, which served as a reaction chamber, and a subsidiary glass flask (fig. 5.2). After a few days of continuous spark discharges of 60,000 V, the solution became darker. Chemical analyses carried out after one week showed that the main accumulated products in the gaseous phase were carbon monoxide (CO) and nitrogen (N_2). Most of the ammonia was consumed, and a little methane remained. Most of the dark matter in the solution could not be identified; it was a mess of organic polymers. About 10% of the carbon in the experimental system was converted into iden-

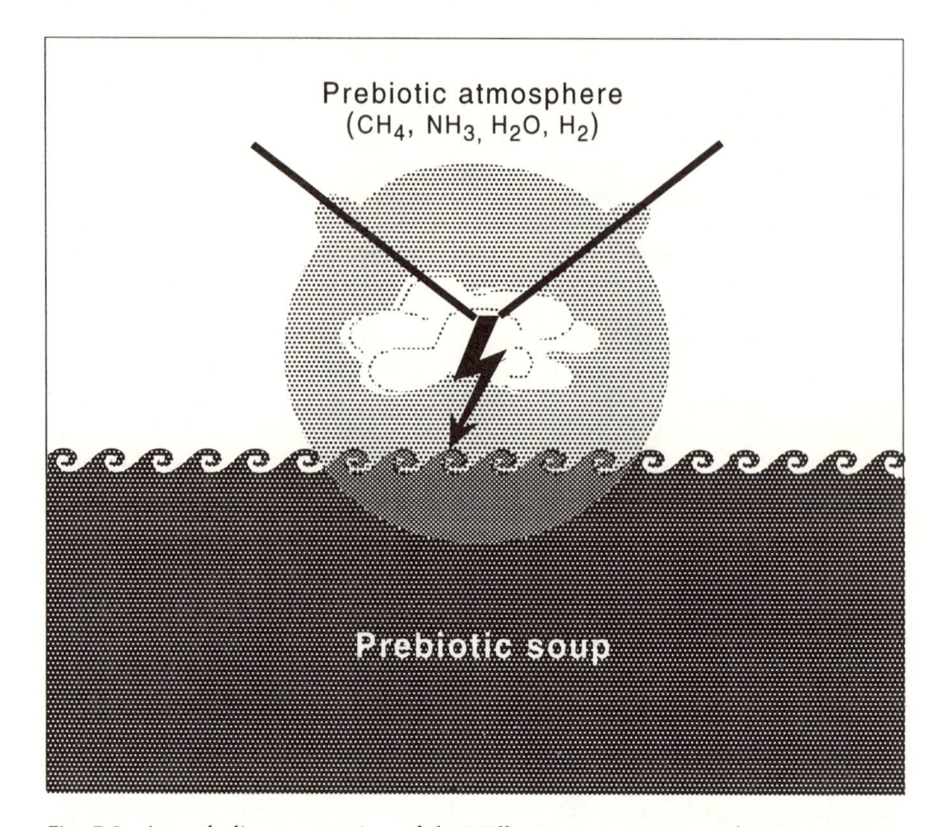

Fig. 5.2. A symbolic presentation of the Miller-Urey experiment of 1953 as a simulation of a prebiotic environment. The reaction cell is a large flask equipped with tungsten electrodes connected to a Telsa coil of the type used in chemical laboratories to detect leaks in glass apparatus; the energy source was electrical discharges, simulating the putative lightning discharges of the primitive atmosphere. The main water reservoir is the primordial ocean. The two are connected, thus simulating fluxes of water and soluble molecules from the primitive atmosphere and back into the ocean.

tifiable organic compounds. Analysis of the solution showed that about 2% of the carbon was recovered in some 25 amino acids synthesized during the experiment. In addition, several fatty acids, hydroxy acids, and amide products were also identified in the mock ocean. The main amino acids produced were glycine ($HCHNH_2COOH$), alanine (CH_3CHNH_2COOH), and aspartic acid ($COOHCH_2CHNH_2COOH$). Several nonbiological amino acids were also identified (see Orgel, 1994).

Unlike the experiment of Calvin's group in 1950, which went practically unnoticed in the scientific community, Miller-Urey's work was immediately recognized as an important breakthrough in the study of the origin of life. By including a nitrogen source in the reaction chamber, the Miller-Urey experiment became relevant to the origin of life: The products of the reaction included the building blocks of proteins. A novel scientific discipline, "prebiotic chemistry," was born. Within a short time, many scientists all over the world were trying to repeat the Miller experiment, using a variety of energy sources and gaseous compositions. At first the research was focused on the synthesis under "prebiotic conditions" of amino acids and, later, on other central building blocks of biochemistry: nucleotides, sugars, and lipids. But very soon scientists began to realize that even if all these building blocks could be synthesized under prebiotic conditions, they still faced the problem of condensing the building blocks into polymers and organizing them in assemblies capable of carrying out the functions characterizing life. This last problem was gradually to be acknowledged as an unimaginably high hurdle. Thus, as a result of the Miller-Urey experiment, more questions were brought up, and new problems had to be resolved.

The year 1953: The introduction of two novel pillars

It was noted by Erik Mosekilde and Lis Mosekilde (1991, p. vii) that "from time to time, perhaps a few times each century, a revolution occurs that questions some of the basic beliefs and sweeps across otherwise well guarded disciplinary boundaries. These are the periods when science is fun, when new paradigms have to be formulated, and young scientists can do serious work without first having to acquire all the knowledge of their teachers." One such period started in 1953. The most important discovery of 1953 was undoubtedly the structure of the DNA molecule by James D. Watson and Francis H. C. Crick. The Miller-Urey experiment was published three weeks later; however, unlike this contribution, that of Watson-Crick was a theoretical one—a model of the structure of the DNA molecule, based upon X-ray analysis. No less important, they recognized that DNA is the molecule of heredity, the structure that is the key to understanding of the mechanism of heredity. By this the deep connection between chemistry and biology has been established (Eigen, 1992).

The importance of this discovery was also recognized as soon as it was published. And as in the case of the Miller-Urey experiment, it brought about the establishment of a novel scientific discipline—molecular biology. At that time, the direct connection between the origin of life and molecular biology was rather vague. But within a few years, it became accepted by most researchers that the two novel disciplines were among the central pillars on which the search for the origin of life was destined to be based.

According to Judson (1979, p. 47), Watson "has often said that the decisive influence on him was the book 'What Is Life?' by the physicist Erwin Schrödinger." Yet with regard to the details of Schrödinger's science, Crick noted (Judson, 1979, p. 245), "I wasn't conscious of any influence of what he called the aperiodic crystal—I don't suppose the man had ever heard of a polymer!" Orgel (1995), on the other hand, noted, "I believe that he had in mind repeating and random copolymers."

Darwinian evolution: A central pillar in the study of the origin of life

The actual mechanism of evolution—"evolution by natural selection," with its slogan "survival of the fittest"—was discovered by Darwin and Wallace. However, with the development of the modern population genetics, the Darwinian concept had to be extended, and its extension resulted in the *synthetic evolution theory*. In its new form, the evolution theory treated the development of a population in which several variants of the same gene compete with each other. A succinct description of the basic ideas of this theory can be found in Küppers 1990.

It should be noted that the first and central pillar of the modern search into the origin of life had already been laid by Darwin. The centrality of evolution in biology was best phrased by the geneticist Dobzhansky (1973; see also Sober, 1993) as follows: "Nothing in biology makes sense except in the light of evolution." Whereas this slogan was coined in the context of biology, it is also applicable in the context of the origin of biology, namely, the origin of life. This can be visualized as applying the evolution theory to populations of molecules.

The spontaneous generation theory in retrospect

Max Planck (cited by Mason, 1991, p. 161) stated in 1948 that "a new scientific truth does not triumph by convincing its opponents and making them see the light, but rather because its opponents eventually die, and a new generation grows up that is familiar with it." In contrast to Planck, the adoption of a new theory may be explained mainly by its usefulness, according to Mason (1991).

The historical-philosophical view of the transformations of the spontaneous generation theory from its primordial to its modern form was summarized by Mendelsohn (1976, p. 40). Reviewing the apparent overthrow of the spontaneous generation theory three times—by Redi and Leeuwenhoek, Spallanzani, and Pasteur—he concluded that "spontaneous generation, as a paradigmatic concept, did not die each time, but rather in descending order of size of organism its usefulness as an explanatory model for the generation of organisms was replaced." The new paradigm initiated by Oparin, Haldane, Urey, and Miller seems to fit very nicely into its historical place according to Kuhn's philosophical teaching, as discussed by Prigogine (1984, p. 308).

The post–Miller-Urey era

Soon after the Miller-Urey experiment, many scientists entertained the belief that the main obstacles in the problem of the origin of life would be overcome within the foreseeable future. But as the search in this young scientific field went on and diversified, it became more and more evident that the problem of the origin of life is far from trivial. Various fundamental problems facing workers in this search gradually emerged, and new questions came into focus, as the new scientific field of chemical evolution became established. Despite intensive research, most of these problems have remained unsolved.

Indeed, during the long history of the search into the origin of life, controversy is probably the most characteristic attribute of this interdisciplinary field. There is hardly a model or scenario or fashion in this discipline that is not controversial. This is the subject matter of part IV. But before going into the complex and fascinating puzzle that is the study of the origin of life, it is necessary to briefly review the relevant foundations of biology, geology, and cosmology, on which part IV stands. For additional points of view see Misra, 1992, and Ruse, 1997.

Summary of part I

The scientific search into the origin of life is summarized schematically in fig. 5.3. It was started by the Greek philosophers, with observations and characterizations of

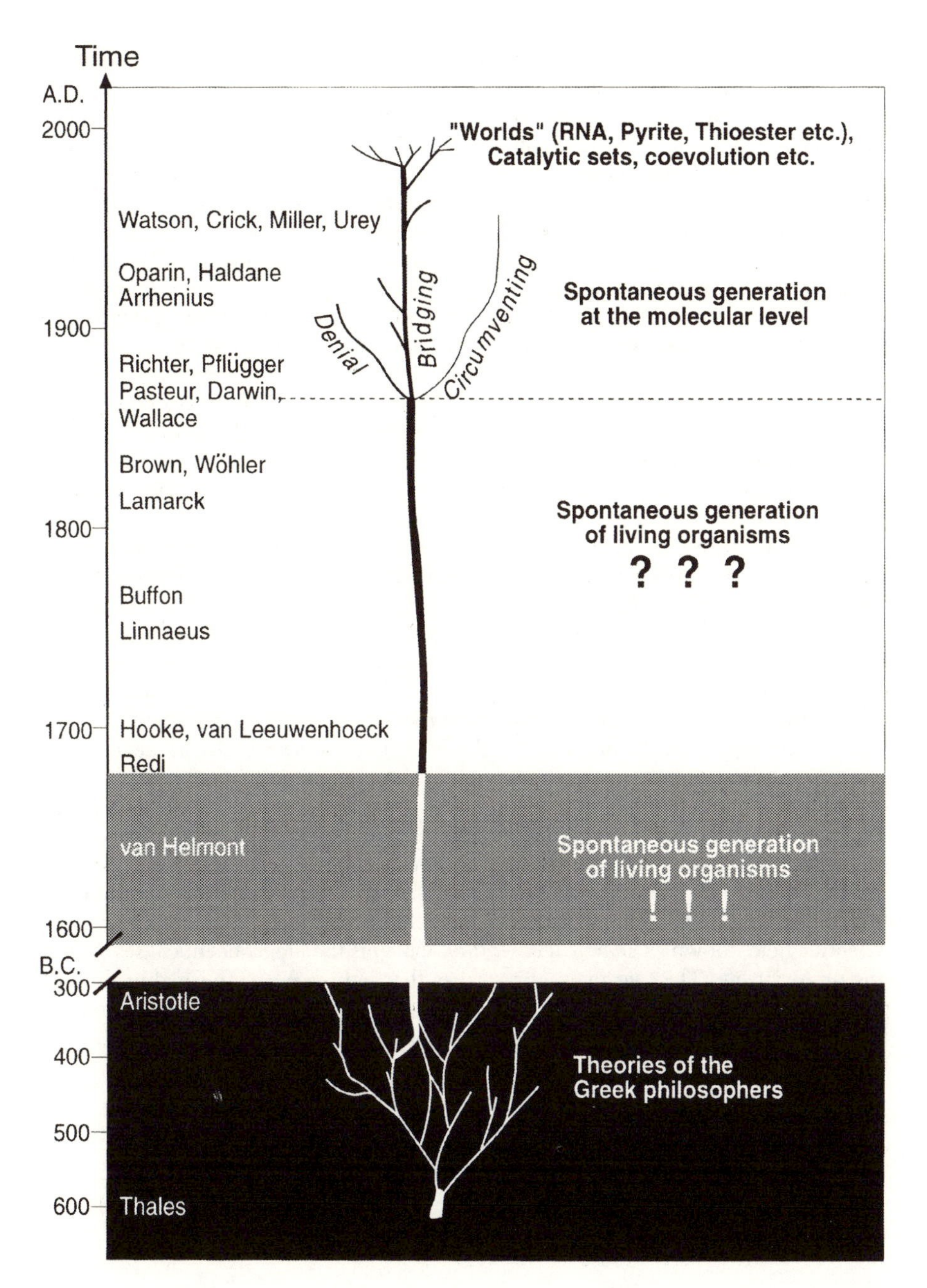

Fig. 5.3. A scheme of the major landmarks in the history of the search into the origin of life, starting with Thales and ending with some of the names of the most recent theories.

what was considered "natural" phenomena. This was followed by logical explanations, without implying unnatural agencies. Of the various hypotheses proposed, the spontaneous generation proposition was selected by the post-Aristotelian scholars and, in collaboration with the teachings of the Church, gradually became the dominant school in the explanation of the origin of life. Until the seventeenth century there was little doubt about the validity of the established spontaneous generation theory.

The era of scientific testing of the spontaneous generation theory began with Redi, who showed experimentally that flies are not generated spontaneously. Redi's experiment coincided with the development of the first microscopes, and soon a new generation of experimentalists began testing the validity of spontaneous generation in microscopic creatures. Over the next 200 years, it was gradually established that this proposition, if valid at all, is applicable only to microscopic creatures. At the end of this time period, Pasteur showed that the then smallest known creatures, bacteria, are not generated spontaneously. Pasteur's experiments coincided almost exactly with the publication of Darwin-Wallace's evolution theory, the theory that was destined to serve as the most important basis for the modern search into the origin of life in the twentieth century.

The crisis caused by Pasteur's experiment with regard to the origin of life was circumvented by two approaches, namely, resorting to the infinity of the universe on the one hand, and shifting the origin problem to the molecular level, on the other. The first approach resulted, at the turn of the century, in the Panspermia theory, established by Richter and Arrhenius. The latter approach was suggested during the 1920s, independently by Oparin and Haldane, in their similar biogeochemical scenarios.

The first successful experiment based on the Oparin-Haldane paradigm was carried out in 1953 by Miller and Urey. In this experiment electrical discharges were applied to a reducing atmosphere consisting of methane, ammonia, hydrogen, and water, thus simulating a prebiotic environment of the kind suggested by Oparin and Haldane. The amino acids and other organic compounds produced in the reaction vessel were interpreted as experimental support for the Oparin-Haldane hypothesis.

The Miller-Urey experiment, together with the discovery of the DNA structure by Watson and Crick in 1953, are the landmarks assigned to the establishment of the scientific field known as *chemical evolution,* which is the biogeochemical search into the origin of life. The two main approaches that have been established in this old-new scientific field are prebiotic synthesis of the building blocks of biochemistry and the organization of these building blocks into the first living entities.

Part II

CENTRAL FEATURES OF LIFE AS WE KNOW IT IN OUR PHYLOGENETIC TREE

> Biology was referred to as a "dirty science," an activity, according to the physicist Ernest Rutherford, not much better than "postage stamp collecting." At best it was a second-class, "provincial" science.
>
> Ernst Mayr, *Toward a new philosophy of biology*

> Biology belongs to one of the surprising sciences, where each rule must always be supplemented with several exceptions (except this rule, of course).
>
> Claus Emmeche, *The garden in the machine*

Extant living entities carry over many clues to earlier stages of our phylogenetic tree. The following chapters address various aspects of life that have been studied in the context of the origin of life. The topics discussed in these six chapters were selected by origin-of-life researchers during the last decades because of their centrality in biology. They cover a wide spectrum of observations, from morphology to biochemistry and molecular biology, and also include a discussion of the characterization of life. Their relevance to the origin of life is summarized at the end of each chapter.

6

A General Morphological-Functional Characterization of the Cell

The cell: The minimal unit of extant life

A fundamental organizational characteristic of life is its being based on discrete units—*cells*. Living creatures on our planet vary enormously in the number of cells they are made of, in the size, shape, and properties of those cells, and in the extent of their dependence on each other. However, life as we know it (with the exception of viruses, which are discussed later) can exist only in the form of cells.

Organelles are not dependable living entities

Organelles are subcellular entities that can be isolated by centrifugation. No parts of the cell, even organelles such as mitochondria and chloroplasts, which replicate independently of the whole cell (and were probably acquired by an old ancestor billions of years ago), can sustain life in a nonliving medium.

Cell size is always limited

In order to be effective, solute molecules, such as metabolites, must be able to move through the cell in a matter of seconds. Molecule movement in the cell is carried out by diffusion, and the size of a cell is therefore limited by the rate of this process. Indeed, the upper size of most cells is in the range of 30 to 50 μm. In some special cases of larger cells, such as in certain algae, special mechanisms for enhanced movement by diffusion have developed during the process of biological evolution.

Cells are derived from other cells

Whatever the definition of life is, one of its major characteristics is that it is transferred from one generation to the next. In other words, all cells are derived from previously existing cells. The method of this transfer is always the same, namely, by division into two cells.

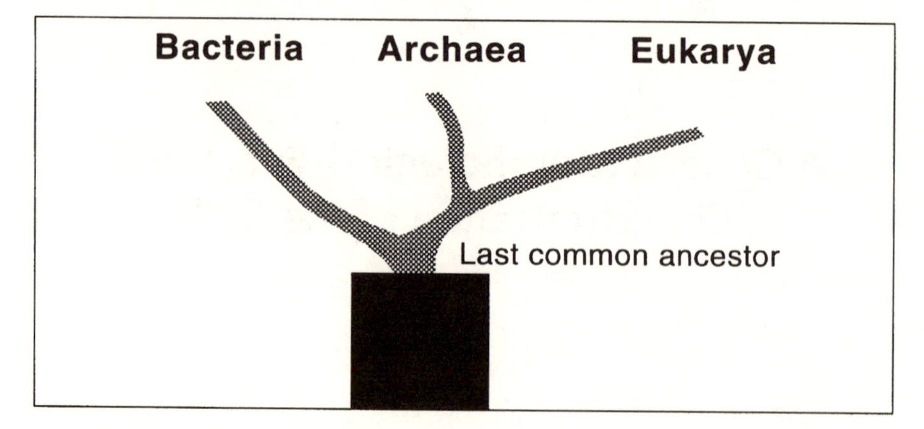

Fig. 6.1 The upper part of this figure illustrates the three domains of living entities (Eukarya = Eukaryotes) and their common ancestor (Adapted from Woese et al., 1990). The lower part of the figure used to be, and still is, a kind of a "black box" for many people. It is the subject matter of part IV.

Three domains of living forms

In an attempt to organize all life forms into a phylogenetic system that relates their origin and order of emergence, three "domains" have been identified, as follows: (1) Archaea—formerly named archaebacteria (from the Greek *archaios,* meaning "ancient"); (2) Bacteria—formerly named eubacteria (from the Greek *eu,* meaning "well," implying "true"); (3) Eukarya—formerly named eukaryotes (from the Greek *karyon,* meaning "kernel, nucleus," denoting cells with a true nucleus).

The phylogenetic relationships between these three domains are shown in fig. 6.1 and discussed in chapter 10. The three domains have a common origin (Woese et al., 1990), the "last common ancestor" which, presumably, no longer exists today. This hypothetical creature was preceded by more ancient chemical entities, as discussed in part IV.

Two kinds of cell organization

In view of the immense diversity in size, shape, and function of cells, even in one pluricellular organism, it is comforting to realize that there are only two kinds of organization in all known living forms. These bear the names of the two "kingdoms" of all life forms, according to the old classification and nomenclature. Thus, the following groups denote cell organization rather than phylogenetic classification, as follows: (1) Prokaryotes (from the Greek *pro,* meaning "before," and *karyon,* meaning "kernel") encompass archaea and bacteria, which consist of the archaebacteria and the eubacteria, including cyanobacteria, mycoplasma, and rickettsiae. (2) Eukaryotes comprises *eucarya* (eukarya), which include fungi, algae, protozoa, slime molds, and all plants and animals.

Prokaryotes

Prokaryotic cells have no membrane-bound nuclei, nor do they have any internal system surrounded by a membrane inside the cell (with a few exceptions). Implied in their name is the understanding prevalent in the time when it was coined, that prokaryotic cells emerged before the more complex nucleated cells, the eukaryotes. This view has recently changed (Woese et al., 1990).

All prokaryotes are unicellular microorganisms. They are divided into two groups, namely, archaea (formerly known as archaebacteria) and bacteria (formerly known as eubacteria). These names are somewhat misleading; they were given when it was believed that the archaebacteria originated before the eubacteria, thus representing the phylogenetic relationships between the two. However, according to the modern phylogenetic trees, these two groups are the descendants of a common ancestor.

It is generally held that the size of prokaryotic cells is controlled by their surface-to-volume ratio because of the effectiveness of the diffusion of metabolites inside the cell, which decreases with cell size. An outstanding exception to this generalization is a recently discovered giant prokaryote, the largest bacterium yet described. With its cells measuring 600×80 μm, this bacterium is more than a million times larger in volume than a typical bacterium such as *E. coli* (Angert et al., 1993). The identification of this creature as a bacterium was based not on its size but, rather, on the its RNA sequence. Nevertheless, it should be noted that most prokaryotic cells are smaller than eukaryotic ones.

Ubiquity of archaea

Traditionally these prokaryotes were thought to predominate only in environments characterized by extreme conditions, such as high temperatures. Archaea have been recently shown to be an abundant component of Earth's biota (DeLong et al., 1994; see also Olsen, 1994). Moreover, previously unrecognized members of the Archaea domain are adapted to many niches in the biosphere, including symbiotic partnerships with eukaryotic hosts (Preston et al., 1996; Stein and Simon, 1996).

Some central cellular features

This description focuses on those features that can be traced back in time and are thus either directly related to the origin-of-life process or are considered continuous evolutionwise with those features.

The plasma membrane (cell membrane) The plasma membrane encloses the cytoplasm, thereby separating the cell wall from the main cell volume. It is the true boundary between the cytoplasm and the chemical environment, due to its selective permeability. Organisms may differ in the composition of their plasma membranes, but this membrane is found in all living forms and is a vital component of the cell. Thus, the cell wall of bacteria can be removed by certain enzymes that hydrolyze it or may even be absent, if its synthesis is prevented; however, the naked cells with their plasma membrane may remain viable. It is interesting to note that in such cases, special measures must be taken to avoid high osmotic pressures in the internal cell volume—the cytosol—since the plasma membrane has a very little mechanical strength.

The plasma membrane is typified by selective permeability, conferred to it by its protein components, and thus controls the chemical composition of the cytosol. It also contains various kinds of proteins that are involved in the electron-transfer reactions taking place in the cell. Thus, photochemical reactions of photosynthesis in bacteria take place in the cell membrane in pigment-protein complexes known as reaction centers.

A more detailed picture of membranes has emerged recently as a result of the application of new research methods. The novel picture is characterized by dynamic activities in which components of the membrane can not only diffuse randomly but can also be confined transiently to certain domains and can even experience highly directed movement (Jacobson et al., 1995).

The cytosol The cytosol is the aqueous, viscous sap in which the organelles and other insoluble components of the cytoplasm are dispersed. The cytosol is rich in proteins, in the form of hundreds of kinds of enzymes, and is the main seat of metabolism.

The ribosomes The ribosomes are the sites of protein synthesis. A typical cell contains about 15,000 ribosomes, each of which synthesizes proteins according to the genetic properties of the cell. The ribosome measures about 25 nm and is made of two parts of unequal size. The detailed structure of the ribosome of *E. coli* has been recently studied by cryoelectron microscopy (Frank et al., 1995). This reconstruction revealed details that, together with biochemical details of its function, will help establish a higher level of understanding of protein synthesis. Revealed by this technique is an irregular structure replete with bulges, bridges, and even tunnels, totally unknown until now. As Zimmermann (1995) notes, "The structure of the ribosome as derived from cryo-electron microscopy is likely to become the new pattern for modelling the ribosome and ribosome-ligand interactions."

P. B. Moore (1993, pp. 122–123) points out, "The ribosome is not an organelle, a term one still sees attached to it in the literature; it is not comparable in any way to intracellular entities like the mitochondrion or the lysozome, which are organelles. It is an enzyme. It would make sense to rename the ribosome to reflect this fact; 'polypeptide polymerase' would be appropriate." An additional role for the ribosome was recently suggested by Siegel (1997).

The chromosome Bacteria contain a single, densely packed circular chromosome (also known as genophore; see Margulis, 1993b), embedded in the cytosol, without any surrounding membrane to separate it from other components of this environment. Its total length is about 1 mm, approximately 1,000 times the cell diameter.

The chromosome is a double-stranded polynucleotide, DNA, which contains the information needed for the cell to grow and replicate. During the replication process, the DNA double strand separates into two strands, each carrying the same information required for the growth and reproduction of the daughter cells. Upon completion of cell division into two daughter cells, each new cell possesses one DNA strand, which serves as a template in directing the synthesis of its own complementary strand from available building blocks in the cytosol.

Plasmids Plasmids are small circles of DNA, located in the cytosol of most bacteria. Though dispensable, these small genes affect the properties of their host cells; under natural conditions they can be exchanged between cells upon conjugation. Moreover, it is possible to introduce them artificially into the bacterial cells, using genetic engineering techniques. The advantages of these techniques have to do with the properties conferred by the plasmids to their host cells: the ability to metabolize or synthesize certain compounds.

Cell division Bacterial cell division involves the formation of a *plasma membrane furrow*. This structure grows between the duplicated chromosomes such that each daughter cell contains one chromosome. The chromosomes are attached to the cell membrane and remain attached during cell division.

Additional cellular features A major component of bacterial cells included in this group is the cell wall. It is not considered relevant to the origin-of-life process.

Eukaryotes

The eukaryotes are a very diverse group, comprising the unicellular algae, fungi, and protozoa on the one hand, and the pluricellular fungi, plants, and animals, on the other. The eukaryotic cells are characterized by a nucleus, mitochondrion, and other organelles. They are larger than the prokaryotic cells and also more complex and sophisticated, containing many subcellular structures. According to the endosymbiotic theory (Margulis, 1993b, p. xxv; see also Gupta and Golding, 1996), these organisms are the result of symbiosis between two kinds of cells, which began some 3.5 Ga ago.

Following Morowitz (1992), it seems that the smaller and simpler prokaryotes exhibit the necessary aspects of living organisms that are dealt with in this book. Standard textbooks supply details on this group.

Metabolism and the trophic methods of using carbon source

> Cells are little black holes for energy.
> David Mauzerall, "Light, iron, Sam Granick, and the origin of life"

The totality of all the chemical processes taking place in the cell is known as *metabolism* (from the Greek *metabole*, "change"). This term aptly describes the dynamic characteristics of the living cell. The general pattern of this dynamics is the import of chemical elements in various forms (nutrients) from the environment, and their processing. The process by which the nutrients are transformed into the integral components of the cell is called *anabolism* (from the Greek *ana*, "up," and *bole*, "throwing"), or *biosynthesis*. Biosynthesis is an energy-requiring process, and the cell obtains the needed energy from the environment. There are three kinds of energy sources that cells can use, namely, light, inorganic chemicals, and organic chemicals. The process by which chemicals are broken down and their energy released is known as *catabolism* (from the Greek *cata*, "down, away, against, fully, used to form derivatives").

Living forms are divided into two groups in terms of the carbon source for the synthesis of all their organic molecules, as follows:

Autotrophs Autotrophs (from the Greek *autos*, meaning "same, self," and *trophe*, meaning "nourishment"; *autotrophos* would mean "supplying one's own food") utilize mineral carbon, CO_2, as a carbon source together with mineral compounds such as H_2O, NO_3^- and SO_4^{2-} to synthesize their organic compounds. The energy needed for their activities is derived from either the sun or chemical reactions.

Heterotrophs Heterotrophs (from the Greek *hetero*, "akin to, different, other than usual, other than oneself") utilize preformed organic molecules that have been synthesized, either directly or indirectly, by autotrophs. These are then modified and combined in various ways, resulting in the same macromolecules that characterize autotrophs.

Organisms can also be divided according to their energy sources, as follows:

Phototrophs Phototrophs (light-feeders) use solar energy to form their organic compounds.

Chemotrophs Chemotrophs (chemical-feeders) require the intake of oxidizable chemical compounds.

The combinations between the sources of energy and carbon give four kinds of metabolism, namely, photoautotrophy, chemoautotrophy, photoheterotrophy, and chemoheterotrophy.

Range of important environmental parameters

Upper and lower limits of temperature

The temperature range in which extant organisms can live covers a rather wide range, from below 0°C to above 100°C (Cowan, 1995). The five categories representing thermal conditions needed by organisms are psychrophiles, mesophiles, thermophiles, extreme thermophiles, and hyperthermophiles. The hyperthermophiles are generally characterized by a $T_{minimum}$ lower than 70°C, a $T_{optimum}$ higher than 90°C, and a $T_{maximum}$ lower than 115°C (Cowan, 1995). Recent reports, which are still controversial, suggest that for certain hyperthermophilic creatures, the maximum temperature is as high as 150°C.

The cellular functions normally stop at temperatures close to 0°C, but in many cells these activities are restored upon increasing the temperature. Various creatures can withstand very low temperatures. The possible hyperthermophilic origin of life is discussed in chapter 16.

Lowest limit of pH

The ability of living creatures to thrive in extreme environments seems to push time and again toward new records. The most acidophilic creatures have been identified recently (Schleper et al., 1995) as a species of archaea that can grow at and slightly below pH0. Similarly, a few eukaryotes can grow at pHs slightly above 0.

Clues to the study of life's origin

The features of life discussed in this chapter reflect a very high level of complexity and sophistication, a result of long evolutionary processes. In order to explore the origin of extant organisms and the environments in which such processes could have taken place, one should use the old assumption that the earlier the living entities are, the simpler is their structure and internal organization. Moreover, such a study should focus on the molecular level of the cell and look for traces and fingerprints of earlier evolutionary stages, as well as indications pertaining to their environments.

7

General Chemical, Biochemical, and Molecular-Biological Characterization

> When we attempt to define life, or living, we immediately come up against a fundamental and apparently irreducible paradox: living organisms are composed of inanimate molecules. . . . Must we then say that "life" is the interaction of all the inanimate components of this whole? In other words, that nothing is alive in a cell except the whole of it?
>
> Martin Olomucki, *The chemistry of life*

The chemical status of the cell is far from static, even when it seems to change only slightly with time. Chemical processes, that is, the synthesis and breakdown of molecules, take place in the cell continuously. These processes create fluxes of matter and energy between the cell and its environment that are characteristic of living organisms. The scientific disciplines that study these processes are thermodynamics, biochemistry, and molecular biology.

Universal constituents

Water

All cell functions are carried out in aqueous solutions or at water interfaces, where water serve as both solvent and metabolite. As a solvent, aqueous solutions have a decisive effect on the structure of all biomolecules. As a metabolite, water is the source of protons, as well as hydrogen and oxygen, in membrane photochemistry and in photosynthesis, respectively. Water is an essential component of the environment in which living creatures function. Moreover, it is also essential at the cellular level, where the internal environment is basically aqueous, as well as at the molecular level of the cell constituents, where it is attracted to organic and inorganic molecules and ions, thus forming hydration layers on certain chemical moieties. Due to their size and chemical properties, water molecules can diffuse into the cell from outside and from the cell to its immediate vicinity, depending on the chemical potential difference between the outer and inner environments. Upon removal of water from the environment in which living cells are found, water molecules from the cell tend to diffuse outside. Under natural conditions, such a dehydration process is caused, for instance, by evaporation, freezing, or increasing solute concentration.

Severe dehydration stops the living activities of the cell, and, unless special precautions are taken, this is a permanent and irreversible process. Under certain conditions, however, water may be removed from living cells without impairing their ability to function as living entities upon rehydration. Indeed, the evaporation of water under high vacuum and freezing temperatures has been shown to be an efficient technique for the removal of most of the water from prokaryotic cells, without causing them to die out; upon rehydration they recover their original life activities. Whether this results from the fixation of the cell's "blueprint" by the freezing process or the involvement of additional factors, it demonstrates the essential role of water in cellular function (Morowitz, 1992).

Elemental composition

The elemental composition of living cells reflects some general feature of the chemical compounds that constitute them, as well as the geochemistry of the environment in which life evolved. The principal atomic components in biology may be divided into three groups according to their percent body weight (table 7.1).

Interestingly, the atomic components carbon, hydrogen, oxygen, nitrogen, phosphorus, and sulfur are among the most abundant elements in the universe; in the cell, they normally constitute between about 1% (with the exception of S) and 60% of the body weight of living creatures, forming most of the covalent bonds in the cell. Based on a multitude of observations, it is safe to conclude that the chemistry of the major constituents of all known living creatures is the chemistry of these elements; they are the major atoms of biochemistry.

The other two groups of atomic constituents—the secondary elements and the microconstituents—are vital to the proper functioning of living cells. However, their chemistry in the cell is different from that of the primary elements. The secondary group, which normally constitutes between 0.05% and 1% of the body weight of living creatures, is characterized by the ionic bonds it forms with the biochemical mol-

Table 7.1 Principal atomic components of biology and their range of percentage of body weight of living systems.

Primary constituents (1–60)	Secondary constituents (0.05–1)	Microconstituents (trace elements) (0.05 or less)
C	Na	B
H	Mg	Fe
N	S	Si
O	Cl	Mn
P	K	Cu
	Ca	I
		Mo
		Zn

Morowitz, 1992. Reprinted with permission by Yale University Press.

ecules. Moreover, due to their chemical properties, the elements of this group are the major constituents of the cell's liquid phase and are present there mainly in the form of ions, and to some extent also in the form of soluble complexes. In certain cases, these elements form less soluble salts and precipitate in the cell's liquid phase to form organic or inorganic crystals.

The microconstituents group consists of elements that normally comprise less than 0.05% of the body weight of living organisms. The elements of this group are involved in a variety of enzymatic reactions, as well as in the composition of various biochemical compounds.

Metal clusters

In some iron-containing proteins (the iron-sulfur proteins), the iron is associated with inorganic sulfur atoms, forming structures known as clusters. Clusters of two to four iron atoms have been observed, and there seems to be evidence of six iron atoms. We shall return to these clusters in part IV.

Universal set of small organic molecules and their polymers

The ability of the major atomic components of the cell to combine into molecules of considerable complexity in the size range common in biology is enormous. However, the actual number of compounds that are used in biology is relatively small, comprising only hundreds of compounds. The major biochemical groups to which this generalization applies are biopolymers of three kinds, namely, proteins, nucleic acids, and sugars (table 7.2).

It should be noted that the α carbon of all amino acids, except glycine, is asymmetric. Thus, all amino acids except glycine can adopt either of two stereoisomeric structures, called by convention the D and L forms. With rare exceptions, only the L form of amino acids is found in proteins. The sugars and sugar moieties of nucleotides are only of the D form. This subject is discussed in the context of the origin of life in part IV.

Following Morowitz (1992), the group of compounds occurring in all living cells is designated as a universal subset of biological metabolites. These compounds are made of the six primary elements of biology, namely, C, H, N, O, P, and S. Because this universal set may change to some extent in the future, as a result of further study, it is a tentative list based on our present know-how. Even so, it serves as one of the

Table 7.2 Universal compounds of biology.

1. Amino acids—the standard 20 building blocks of proteins
2. Purines, pyrimidines, and derivatives, including nucleotides and nucleosides
3. Lipids and precursors, including glycerol, fatty acids, and isoprenoids
4. Sugars and derivatives, including intermediates of the citric acid cycle
5. Vitamins, coenzymes, and precursors
6. Miscellaneous, including water, carbon dioxide,ammonia, and phosphoric acid

Source: Adapted from Morowitz, 1992; reprinted with permission by Yale University Press.

most important generalizations of biology and as a common denominator for all known forms of life (Morowitz, 1992).

The major biopolymers in the cell are proteins, nucleic acids, and polysaccharides and their derivatives. It has long been established that upon hydrolysis, each of these three kinds of biopolymers gives rise to a surprisingly small number of molecular species, the building blocks (monomeric subunits) of that polymer. These are the amino acids (the building blocks of proteins), nucleotides (the building blocks of nucleic acids), and sugars (the building blocks of polysaccharides and other related biopolymers).

I shall now focus briefly on selected features of those molecules and processes that are of special importance in the study of the origin of life, as discussed in part IV. These are amino acids and proteins, nucleotides and nucleic acids, and biological information transfer on the one hand, and biomembranes and compartmentation, on the other.

Amino acids, peptide-bond formation, peptides, and proteins

Based on their structure and properties, the 20 common amino acids can be divided into groups according to attributes such as hydrophilicity and hydrophobicity, polarity and acidity.

Peptide-bond formation

The structure of amino acids is shown in equation 7.1. The a carbon in amino acids is bonded to an amino group (—NH2) (or imino group —NH, in proline), a carboxyl group (—COOH), a hydrogen atom (H), and a side-chain R group, which is specific for each amino acid. At pH values typical to cells, the amino and carboxyl groups are ionized to form $—NH_3^+$ and $—COO^—$.

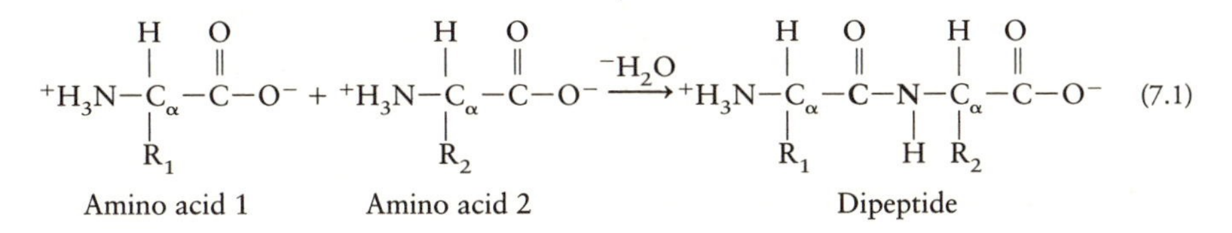

$$ {}^{+}H_3N-C_\alpha H(R_1)-C(=O)-O^- + {}^{+}H_3N-C_\alpha H(R_2)-C(=O)-O^- \xrightarrow{-H_2O} {}^{+}H_3N-C_\alpha H(R_1)-C(=O)-N(H)-C_\alpha H(R_2)-C(=O)-O^- \quad (7.1) $$

Amino acid 1 Amino acid 2 Dipeptide

The condensation of two amino acids into a dipeptide is shown in equation 7.1, where the amino acids are covalently bonded via a substituted amide linkage, termed a *peptide bond* (the bond between the bold C and N atoms in equation 7.1). Amino acids thus connected into a linear, unbranched chain form a peptide. A random sequence of amino acids linked by peptide bonds into a single chain is called a polyaminoacid (Lodish et al., 1995) or a random peptide. Longer peptides are referred to as polypeptides. Peptides generally contain 20–30 amino acid residues, whereas polypeptides contain as many as 4,000 residues. The term "protein" is used for a polypeptide that has a three-dimensional structure.

Nucleotides, phosphodiester-bond formation, and nucleic acids

A nucleotide consists of three parts, namely, a phosphate group, a pentose (a five-carbon sugar molecule), and an organic base (fig. 7.1). The pentose is ribose in RNA and deoxyribose in DNA. The organic bases differ between the two biopolymers: The

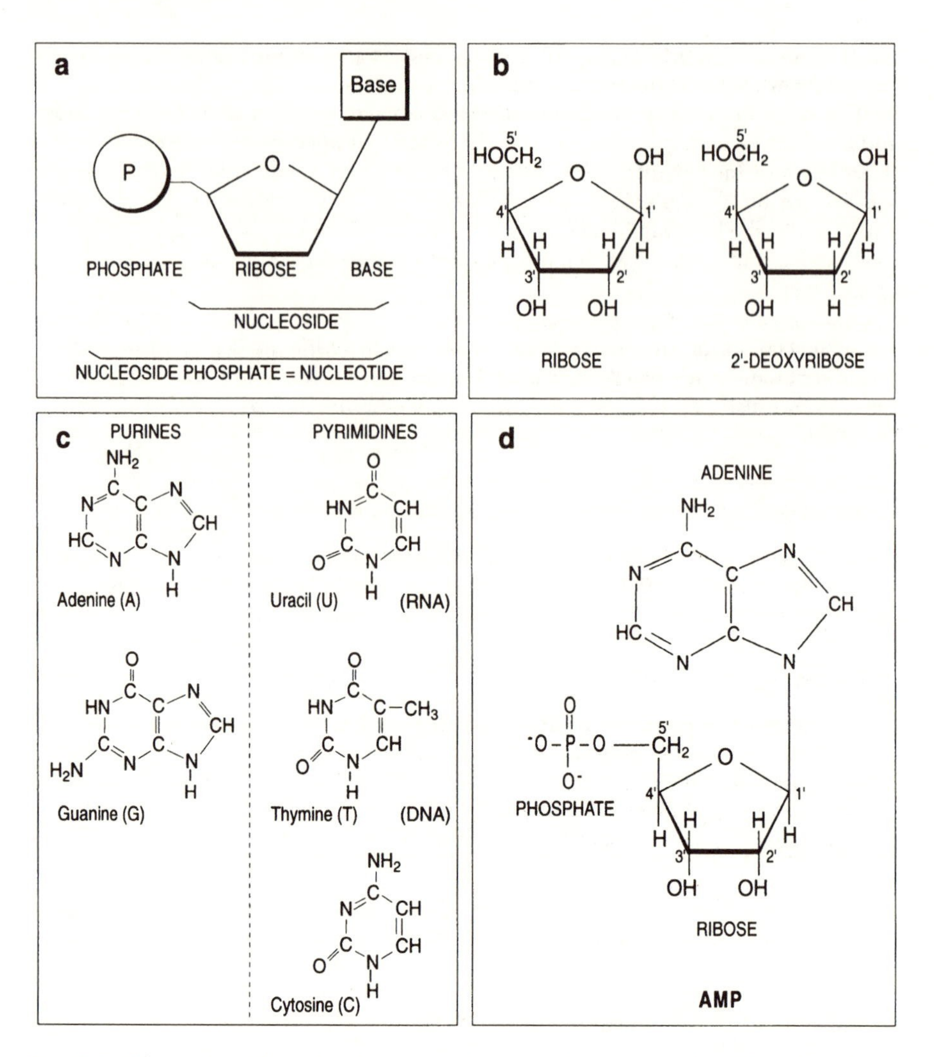

Fig. 7.1. The nucleotide molecules. a. A schematic view of a nucleotide and its three constituents. b. Haworth projection of the structures of ribose and deoxyribose, with the conventional numbering (with primes) of the carbon atoms. c. The chemical structures of the five nucleic acid bases. d. The chemical structure of the nucleotide adenosine 5′-monophosphate (AMP). Note that the C—N linkage between the sugar and the base can adopt either the α form (the base lies below the plane of the ribose) or the β form (the base lies above the plane of the ribose); only the β linkage is found in extant cellular nucleotides. An additional phosphate group (P—O—P) in ester linkage would give adenosine 5′-diphosphate (ADP) and two additional phosphates (P—O—P—O—P)—adenosine 5′-triphosphate (ATP). (Following Stryer, 1995; Lehninger et al., 1993)

bases adenine (A), guanine (G), and cytosine (C) are found in both DNA and RNA. Thymine (T) is found only in DNA; uracil (U) is found only in RNA. Chemically, the base components A and G are named purines; C, T, and U are named pyrimidines.

Phosphodiester-bond formation between two nucleotides

The condensation reaction between two nucleotides occurs when the hydroxyl group at the 3′ carbon of a sugar of one nucleotide forms an ester bond to the phosphate of another nucleotide, where one water molecule is eliminated (equation 7.2, following Lodish et al., 1995):

$$\begin{array}{l}(\text{base})_1 \\ | \\ (\text{sugar})-\mathbf{OH}\end{array} + \mathbf{HO}-\overset{\overset{\displaystyle O}{\|}}{\underset{\underset{\displaystyle O^-}{|}}{P}}-O-\begin{array}{l}(\text{base})_2 \\ | \\ (\text{sugar})\end{array} \xrightarrow{-H_2O} \begin{array}{l}(\text{base})_1 \\ | \\ (\text{sugar})\end{array}-O-\overset{\overset{\displaystyle O}{\|}}{\underset{\underset{\displaystyle O^-}{|}}{P}}-O-\begin{array}{l}(\text{base})_2 \\ | \\ (\text{sugar})\end{array} + \mathbf{H_2O} \qquad (7.2)$$

DNA and RNA

The information carriers of the cell are biopolymers called nucleic acids. Cells have two chemically similar information carrying molecules, namely, deoxyribonucleic acid (DNA) and ribonucleic acid (RNA). The monomers (building blocks) of both DNA and RNA are nucleotides with only four different nucleotides in each of these biopolymers. The length of cellular RNA molecules ranges from less than 100 to many thousands of nucleotides. The number of nucleotides in a cellular DNA molecule is much larger and can exceed 100 million.

The structural information of proteins and nucleic acids

Each of the cell proteins and nucleic acids has a precisely defined sequence of building blocks. This sequence is determined by a specific portion of the cell's genetic system, the gene. The total number of proteins and nucleic acids in the biological world is very small relative to the huge number of possible polymers of the same size and of the same number of building blocks. Still, this relatively small number of cellular polymers perform all the functions needed for living organisms. Obviously, during the evolution of living creatures, only a small number of molecules made from the building block candidates was selected to construct those universal compounds.

Flow of biological information

Generally speaking, any specific binding or "recognition" between two molecules involves information. This includes the binding of substrates by catalysts (enzymes or ribozymes), the recognition of antigens by antibodies, and a host of processes involved in biological regulation, communication, sorting, dispatching, and defense (de Duve, 1991, p. 21). Typically, the recognition is very sensitive to even slight changes in the properties of the biopolymers under consideration. Indeed, in the final analysis, this recognition depends, under ordinary conditions, on the exact sequence of the relevant biopolymers. Thus, biological information, which is stored in biopolymers in the form of sequences of building blocks, must be transferred with a high degree of fidelity. It should be noted, however, that 100% fidelity means lack of variability and, therefore, lack of evolution.

It is recalled that in any molecule there is continual turnover of the atoms. Thus, the atoms of the informational molecules under study are continuously being replaced by other, identical atoms. It is the molecular structure, however, that is of interest to us, and the atomic replacements do not have structural implications.

Transfer of genetic information

Information transfer is the process by which the sequence of a biopolymer serving as a template orderly affects the sequence of other biopolymers. The structural information is thus transferred from one biomolecule to the other, always with the help of catalysts. No wonder this process is very accurate, with a number of "proofreading" mechanisms and other means to correct mistakes. Biological information transfer may be divided into the replication of DNA molecules and protein synthesis.

Replication of DNA

This process is sometimes called "self-replication." However, this name is somewhat misleading, since the replication process is carried out with the involvement of many enzymes; no DNA molecule can replicate itself. In this process, the information stored in one DNA molecule is transferred to another DNA molecule in two stages, namely, formation of a strand complementary to the first DNA strand, and formation of a strand complementary to the latter complementary strand, thus bringing about replication of the original DNA molecule. The resulting strand is a copy of the original strand (equation 7.3):

$$\text{DNA strand} \rightleftharpoons \text{complementary DNA strand} \qquad (7.3)$$

The formation of a complementary strand on a DNA strand is shown in fig. 7.2.

Transcription

Transcription is the synthesis of RNA molecule directed by a DNA template, with the corresponding sequence of building blocks of the RNA molecule. In the case of information transfer in protein synthesis, the structural information thus transferred is delivered in the form of messenger RNA (mRNA) to the site of protein synthesis, namely, the ribosome. Some of the RNA molecules have different functions. The overall process of RNA synthesis on a DNA template, which is catalyzed by a specific enzyme, can be divided into three stages: *initiation*, *elongation*, and *termination*. By the end of this process the newly formed RNA molecule is released as a result of the breakup of the complex of DNA, RNA, and the enzyme.

Translation to a protein

The information stored in the DNA molecule is first transcribed to an mRNA molecule, where DNA serves as a template on which mRNA is formed, and then translated into a protein molecule, as illustrated in fig. 7.3. The mRNA molecule serves as the template on which the adaptor molecules of tRNA, each carrying a specific amino

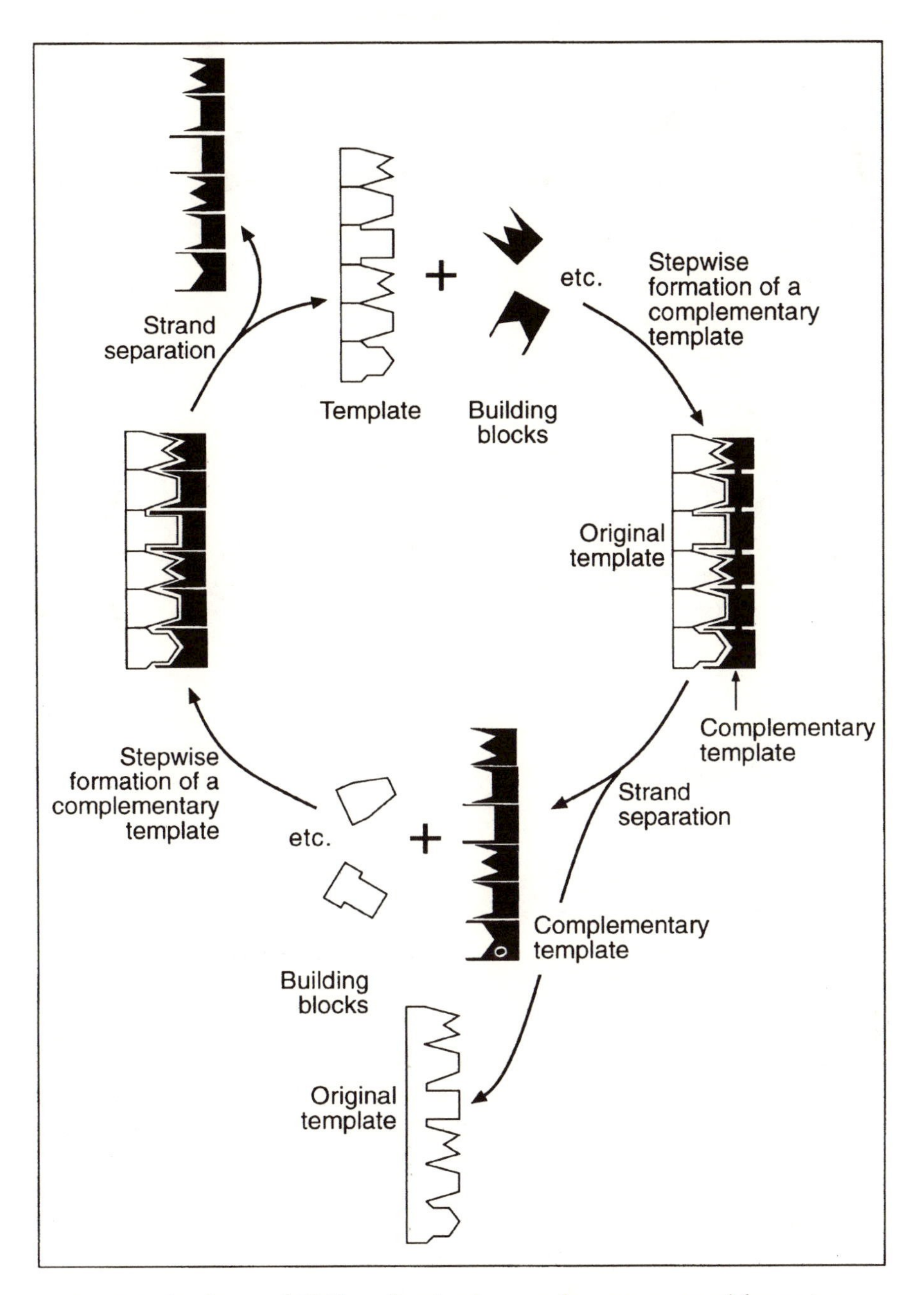

Fig. 7.2. A scheme of DNA replication by complementary strand formation.

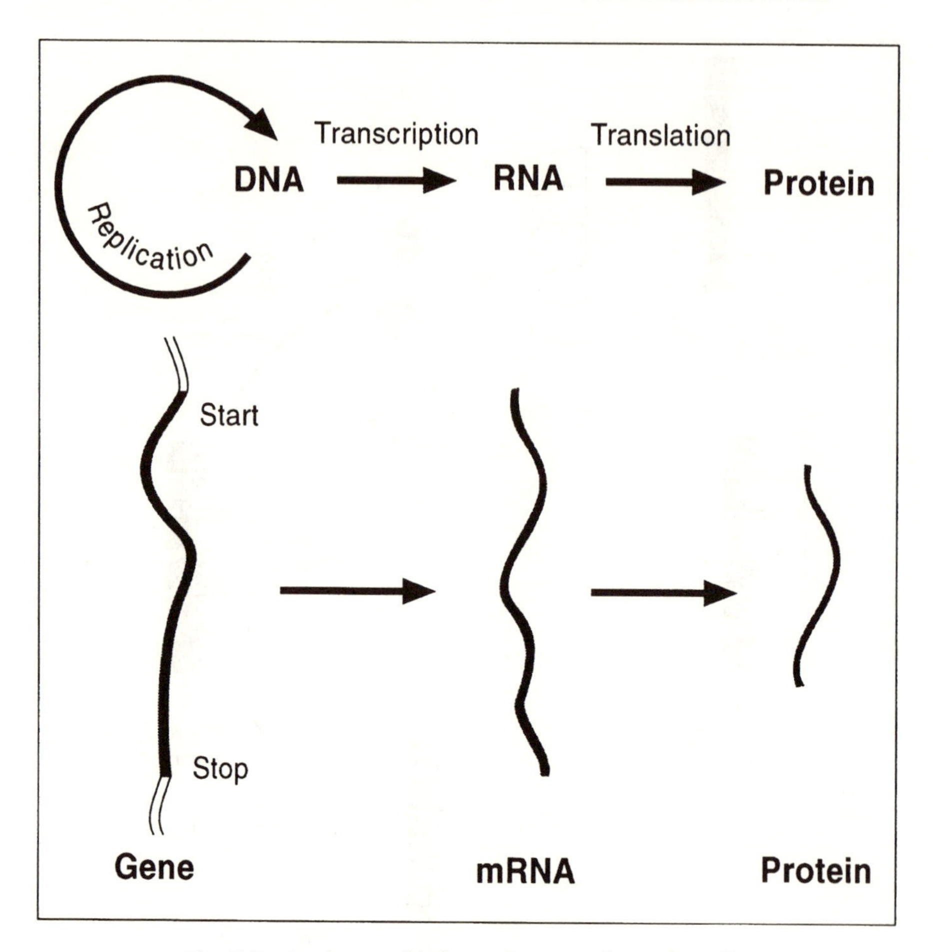

Fig. 7.3. A scheme of information transfer in the cell.

acid, are directed to establish the sequence of the ensuing protein. All these processes are mediated by a host of enzymes.

Because of the importance of the translation reaction in some of the models dealing with the origin of the translation reaction (part IV), it is necessary to focus for a while on some central aspects in this reaction.

Loading of amino acid on tRNA

The formation of a peptide bond between two amino acids (equation 7.1) is thermodynamically unfavorable. The energy barrier is overcome by activating the carboxyl group of the amino acid to be peptide-bonded to the growing peptide. The activated intermediates in protein synthesis are amino acid esters between the carboxyl group of the amino acid and either the 2′- or 3′-hydroxyl group of the ribose moiety at the 3′ end of tRNA; in some cases, the activated amino acid migrates very rapidly between the 2′- and the 3′-hydroxyl groups of this ribose. The amino acid ester is called an aminoacyl-tRNA, or charged (loaded) tRNA.

In addition to the activation problem, the linkage to tRNA is needed because amino acids by themselves cannot recognize the codons on the mRNA. The codons can be recognized, however, by the anticodon domain on tRNAs, which function as adaptor molecules (between the amino acid and the mRNA). The activation of amino acids and their subsequent linkage to tRNAs are catalyzed by specific enzymes called aminoacyl-tRNA synthetases (aaRSs; activating enzymes). The first step of the reaction is the formation of an aminoacyl-adenylate from amino acid and ATP:

$$^{+}H_3N-\underset{H}{\overset{R}{C}}-\underset{O^-}{\overset{O}{\overset{\|}{C}}} + ATP \rightleftharpoons {}^{+}H_3N-\underset{H}{\overset{R}{C}}-\overset{O}{\overset{\|}{C}}-O-\underset{O^-}{\overset{O}{\overset{\|}{P}}}-O-\text{Ribose}-\text{Adenine} + \text{ppi} \quad (7.4)$$

Aminoacyl adenylate
(Aminoacyl-AMP)

The second step of the reaction is the transfer of the aminoacyl group of aminoacyl-AMP to the tRNA, thus forming aminoacyl-tRNA:

$$\text{Aminoacyl-AMP} + \text{tRNA} \rightleftharpoons \text{aminoacyl-tRNA} + \text{AMP} \quad (7.5)$$

The sum of these reactions and the hydrolysis of pyrophosphate (in which energy is released; see chapter 9) is a highly exergonic reaction, namely:

$$\text{Amino acid} + \text{ATP} + \text{tRNA} + H_2O \rightarrow \text{aa-tRNA} + \text{AMP} + 2\ \text{pi} \quad (7.6)$$

in which aa-tRNA denotes aminoacyl-tRNA.

The translation of the information stored in the mRNA molecule to the sequence of the corresponding protein is governed by the genetic code, which relates the nucleotide sequence of the template with the amino acid sequence of the resulting protein (fig. 7.4). De Duve (1995b, p. 66) "In terms of information, the ribosome is illiterate. It acts blindly and links any two partners that are in the right chemical conformation and are properly aligned in regard to its catalytic center."

The genetic code

The genetic code, which relates amino acids to trinucleotide sequences was established well over a billion years ago and since then has been adopted, together with the translation machinery, by all living organisms. The amino acid is linked to the specific nucleotide triplet

> by virtue of the aminoacylation reaction where an amino acid is joined to a RNA, which contains the anticodon trinucleotide corresponding to the attached amino acid. The two components of the code—the attached amino acid and anticodon trinucleotide—are contained in separate domains of the two-domain L-shaped tRNA structure, where they are separated by a distance of ~76 A. (Schimmel et al., 1993)

Information transfer via the translation process includes recognition of the amino acid and the nucleic acid (the tRNA) by the enzyme *aminoacyl tRNA synthetase*. The two recognition reactions are at the heart of the act of translation. The number of

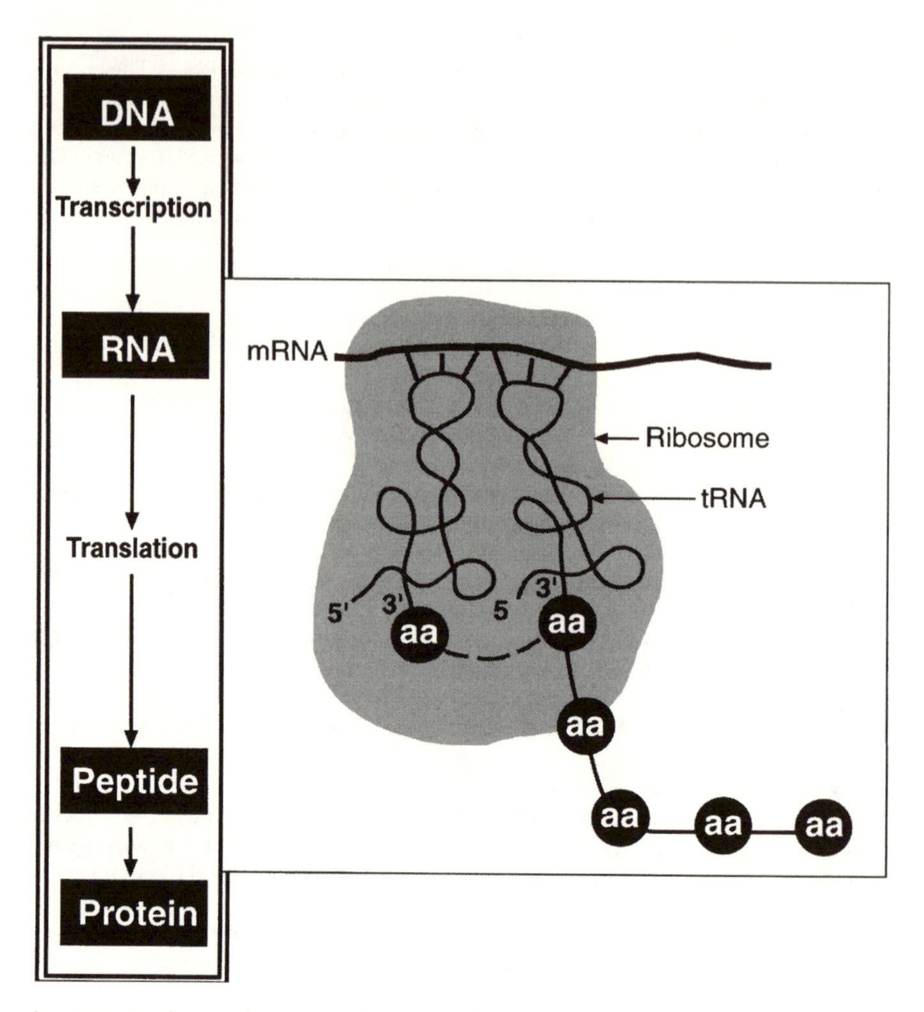

Fig. 7.4. A scheme of the central template-directed reactions in the cell and the main constituents of the translation machinery. Enzymes are not shown.

enzymes equals the number of amino acids (20), whereby each enzyme recognizes all the distinct tRNA molecules (up to six) that are specific to the same amino acid. Thus, the 20 pairs of recognition sites define the primary code of the genetic translation (de Duve, 1991, p. 181). Until recently, the genetic code was considered a universal feature of all living entities. Recently, however, a few exceptions have been found to the universality of the genetic code (fig. 7.5). These are discussed in chapter 17.

The central dogma

This postulate, already phrased by Crick in 1957, states that "the transfer of information from nucleic acid to nucleic acid, or from nucleic acid to protein may be possible, but transfer from protein to protein, or from protein to nucleic acid is impossible" (cited by de Duve, 1995b). It should be noted that recently, Ghadiri and his

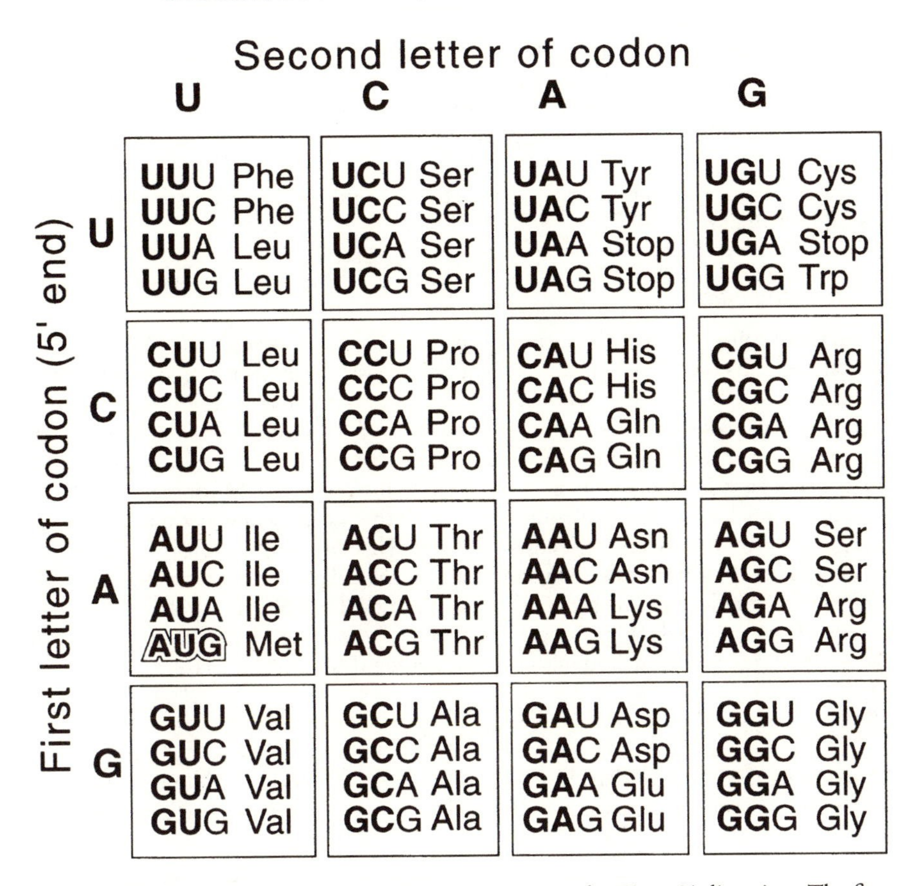

Fig. 7.5. The genetic code. The codons are written in the 5′-to-3′ direction. The first and second bases of each codon play the major role in specifying the amino acid. The initiation codon AUG is typed in bigger letters.

collaborators (D. H. Lee et al., 1996) reported self-replication of a peptide molecule in a test tube. This most interesting observation is discussed in chapter 16, in the context of molecular evolution.

The cell's blueprint

Since the discovery of DNA's structure in 1953, it has been known that this genetic "blueprint" of every cell is handed down from one generation to the next. This is the way in which living entities overcome the inherent chemical instability of the complex structure that is the living state; life thus involves the perpetuation of chemically unstable molecules, whereas evolution takes place as a result of mutation, followed by selection.

During the information-transfer process, defined segments of the DNA sequence are "read" by means of a complex cellular system, and these sequences are then translated into corresponding sequences of amino acids in the form of a protein molecule. Specific segments of DNA that encode a single protein molecule of defined functions are called genes. The DNA sequence is thus expressed in features characterizing life, such as metabolism and replication, performed with the help of proteins (e.g., enzymes). As noted earlier, the structural properties of the DNA molecule—that is, the sequence of its building blocks—represent the information contained in this molecule.

The method by which information is stored in the DNA of an organism is discussed by Küppers (1990).

Exons and introns

In bacterial cells, a protein is normally encoded by single strand of DNA, which is transcribed into an mRNA molecule. The mRNA strand is then translated into the amino acid sequence of the protein. In contrast, coding sequences of most eukaryotes, the so-called exons, are interrupted by noncoding sequences called introns (intervening sequences). The introns are divided into two groups, I and II. In these cells, the entire length of the gene is first transcribed into a large RNA molecule. Before leaving the nucleus, several enzymes remove all the intron sequences, greatly shortening the remaining RNA molecule (fig. 7.6). This processing step is called RNA splicing. The RNA molecule then moves to the cytoplasm and directs the synthesis of the protein molecule. The process of intron removal by splicing is called RNA processing. The intron, which is a part of the *primary transcript* of the RNA (and DNA encoding it), is not included in mature, functioning RNA (mRNA, rRNA, or tRNA).

Exon shuffling

Many researchers believe that the exons are not random molecular domains but represent distinct structures with specific functions. According to Gilbert's "exon shuf-

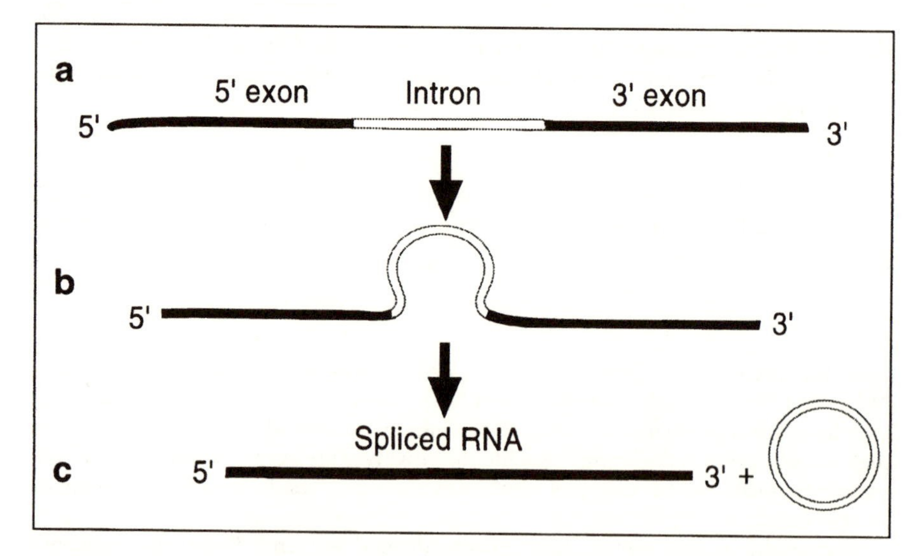

Fig. 7.6. Initial, intermediate, and final stages in the splicing process of group II introns. The splicing mechanism of group I introns is similar but not identical to the present one. (Following Lehninger et al., 1993)

fling" theory (Gilbert, 1987), the split genes have the advantage of being able to make various combinations of the same DNA molecule (see also Long et al., 1995). In this capacity, the exons serve as modular domains capable of forming many combinations of "mosaics" composed of the same building blocks, namely, exons. By this method, the diversity of combinations between the exons greatly increases, since one gene has the potential of coding for more than one protein (de Duve, 1995b, p. 224; Kauffman, 1993).

Interestingly, there seems to be no evidence in metabolic enzymes for such a modular behavior. The latter feature "is observed primarily in proteins involved in advanced regulatory systems in advanced organisms; in particular, proteins involved in the immune system, blood clotting, and other regulatory pathways that emerged only within the last 400 million years" (Benner et al., 1993, p. 44). Thus, in order to evaluate the importance of exon shuffling in the early stages of life's evolution, more phylogenetic data and additional indications are needed.

Biomembranes and compartmentation

Biological membranes are organized, sheetlike structures consisting predominantly of lipids and proteins. Lipids play several roles in the cell; however, in this discussion we focus on lipids as membrane constituents.

Main constituents of biological membranes

Phospholipids are the major class of membrane lipid. Two other classes, namely, glycolipids and sulfolipids, will not be discussed here. Membrane lipids are relatively small amphiphatic (from the Greek *amphi,* meaning "double") molecules that have both a hydrophilic and a hydrophobic moiety (fig. 7.7).

Biological membranes are made of two layers of amphiphilic (amphipathic) molecules joined laterally, as well as tail to tail, by van der Waals interactions between their hydrophobic moieties, which are their hydrocarbon chains. The hydrophilic heads of these molecules are directed toward the water phase in which the bilayers are dispersed (figure 7.8). Biological membranes are very flexible structures, where the lipid molecules can move about freely in the plane of the bilayer. As a result of this flexibility, lipid bilayers are self-sealing systems capable of organizing themselves into closed structures.

The basic constituents of the plasma membrane are lipids and proteins, organized as shown schematically in fig. 7.8. Characteristically, the membrane consists of lipid bilayer in which specific proteins are embedded such that they can mediate between the cytosol, on the one hand, and the aqueous external environment, on the other. All known cells are surrounded by a selectively permeable membrane that separates the cell from its environment. The presence of a membrane isolates the cell to some extent and forms an internal environment that is different from the external one. Cellular biopolymers such as proteins and nucleic acids cannot diffuse through the membrane. In contrast, water, ions, and small organic molecules can diffuse in and out of the cell. Moreover, the external and internal solutions are characterized by different concentrations, giving rise to transmembrane potential.

Biomembranes are not just mechanical barriers; they constitute selective membranes involved in the transport of specific molecules and ions in and out of the cell. They also serve a central function in energy conversion processes, that is, photosyn-

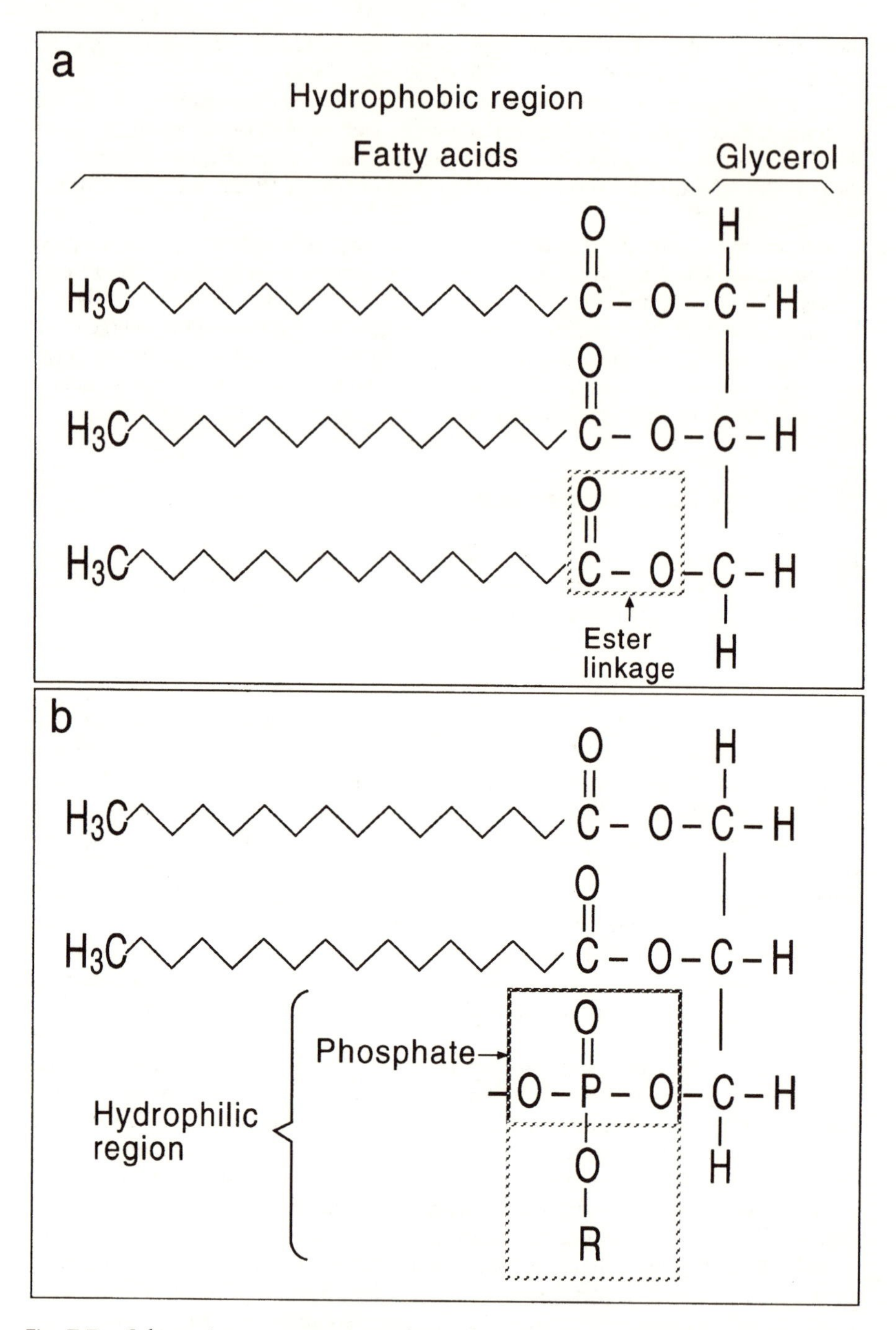

Fig. 7.7. Schematic representation of simple and complex lipids. a. Simple lipid (triglyceride): fatty acids linked to glycerol by ester linkage. b. Complex lipid: phospholipid. (Adapted from Brock and Madigan, 1991)

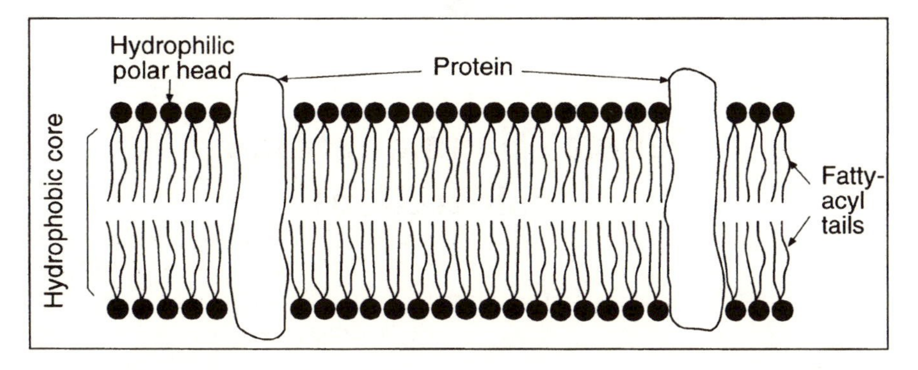

Fig. 7.8. Schematic representation of a cross section through a typical biological membrane. The phospholipid bilayer is the basic structure of all cellular membranes. It consists of two leaflets of oriented phospholipid molecules and embedded proteins, most of which span the bilayer as shown. The polar groups of the phospholipid molecules line both external surfaces, endowing them with hydrophilic properties. The fatty-acyl tails form the hydrophobic interior of the bilayer.

thesis and oxidative phosphorylation. The first occurs in the inner membranes of the chloroplasts, whereas the latter takes place in the inner membranes of the mitochondria. Thus, membranes are an indispensable part of the cell.

Clues to the study of life's origin

A close look at the complex biochemical and molecular-biological processes in extant cells reveals the enormous difficulties in the study of the origin of life and, at the same time, also suggests clues to the latter process. Assuming continuity between extant organisms and the primordial entities of the prebiotic evolution era, these clues include the role of water and the elementary composition and centrality of the universal set of small organic molecules and their polymers.

8

General Thermodynamic Considerations

Everything is driven by motiveless, purposeless decay.
P. W. Atkins, *The creation*

The only definitive statement that thermodynamics makes about evolution is that ultimately it will result largely in CO_2, H_2O, and N_2. Fortunately, the sun allows us to live in a world that is kinetically controlled due to an input of energy. Nevertheless, since many equilibria are reached rapidly compared with the age of the earth, thermodynamic principles can be used to predict what reactions are possible in this transistory world.
Douglas H. Turner and Philip C. Bevilacqua,
Thermodynamic considerations for evolution by RNA

According to the first law of thermodynamics, energy can neither be created nor destroyed (fig. 8.1). It can be changed, however, from one form to another. For instance, chemical energy can be changed into heat or mechanical energy; mechanical energy can be changed into heat or electrical energy, and so on. According to the second law of thermodynamics, each change from one form of energy to another is accompanied by a decrease in the ability of the energy to perform work. For instance, certain protein molecules are capable of converting part of their chemical energy into mechanical energy. But during the process, part of the chemical energy is changed into heat and dissipated to the environment without performing work.

Basic thermodynamic functions and direction of chemical reactions

Free energy

The component of the total energy that is available to perform work is called the Gibbs free energy and denoted by G. The overall change in the free energy in a reaction, ΔG, is defined as:

$$\Delta G = G_{products} - G_{reactants} \tag{8.1}$$

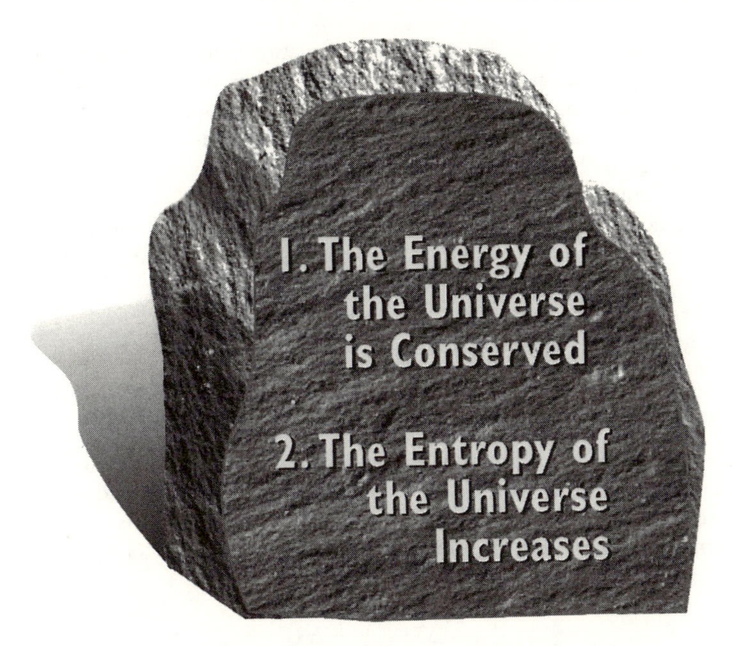

Fig. 8.1. The first and second laws of thermodynamics.

Enthalpy

The heat of reactants or products is equal to their total bond energy, or their enthalpy, denoted H. Enthalpy is absorbed or released in a chemical reaction when bonds are broken or formed. Thus, the overall change in enthalpy in a chemical reaction, ΔH, is equal to the overall change in bond energy:

$$\Delta H = H_{\text{products}} - H_{\text{reactants}} \tag{8.2}$$

Exothermic reaction

In an exothermic reaction, heat is given off, which means that the products contain less energy than the reactants. The difference between the initial and final states, ΔH, is designated by convention as negative.

Endothermic reaction

In endothermic reactions, heat is taken up, which means that the products contain more energy than the reactants. ΔH is thus positive.

Entropy

The second law of thermodynamics also states that all physical and chemical processes in closed systems proceed from an ordered to a random, or disordered, state. In the present context, a system is defined as an entity that does not exchange mass with its surroundings but can exchange heat. A measure of the degree of randomization is represented by the thermodynamic function called entropy (from the Greek *en,* "in, within, inside," and *tropia*—"change"), designated *S*. An ordered system has a low value of entropy, whereas a randomized system is characterized by high entropy. Unlike temperature, pressure, energy, and mass, the entropy of a system cannot be measured directly but is inferred from other quantities.

The term "entropy" was coined by Rudolf Clausius in 1851. Its application to biology is still controversial, as can be seen, for instance, in the controversial book by Yockey (1992). A vivid and didactical description of this thermodynamic function can be found in Lehninger et al. 1993.

The overall change of entropy in a system, ΔS, is the difference between the entropy of the products and that of the reactants:

$$\Delta S = S_{\text{products}} - S_{reactants} \tag{8.3}$$

According to the second law of thermodynamics, a reaction tends to take place spontaneously when the total entropy of the system and its surroundings increases in the reaction. In other words, a reaction tends to occur when ΔS is positive.

The importance of the surroundings in the calculation of ΔS is shown in equation 8.4:

$$\Delta S_{\text{total}} = \Delta S_{\text{system}} + \Delta S_{\text{surroundings}} \tag{8.4}$$

Free energy, enthalpy, and entropy

For a system in equilibrium at constant pressure and temperature T, these functions are related to each other as follows:

$$G = H - TS \tag{8.5}$$

Thus, these three functions are needed to define the state of equilibrium of a system.

When a reaction takes place and a molecule is transformed into another one, a change in the free energy, ΔG, also takes place. This change, rather than the absolute value of G of the molecules, is of interest to us, since it can be used to predict the direction of the reaction.

Predicting the reaction direction

The change in free energy is designated as ΔG and is defined as follows:

$$\Delta G = G_{\text{products}} - G_{\text{reactants}} = \Delta H - T\Delta S \tag{8.6}$$

The ΔG of the reaction is defined for constant temperature and pressure. It is a useful parameter for predicting the direction of a chemical reaction or mechanical process. Thus, if ΔG is negative, the reaction or the process under study can take place spontaneously. This reaction is called exergonic (from the Greek *ergon,* "work"). If ΔG is positive, the reaction or process under study *will not* take place spontaneously. This reaction is called endergonic. If ΔG is zero, the system under study is in equilibrium; any conversion of a reactant into a product will be balanced by conversion of a product into a reactant.

It can also be shown that the change in entropy, ΔS, of a system and its surroundings must be positive for a reaction to take place. This means that the overall degree of disorder in the system and its surroundings must increase. Thus, there are two equivalent ways to predict the direction of a reaction, both of them derived from the second law of thermodynamics; for a reaction to proceed spontaneously, the following equations need to be satisfied:

$$\Delta S_{total} > 0 \tag{8.7}$$

$$\Delta G_{system} < 0 \tag{8.8}$$

It should be noted that ΔG refers only to the *system* under study, whereas ΔS refers to the *system and its surroundings*. Because the measurement of ΔG is easier than that of ΔS, ΔG is a more useful function.

The equilibrium constant of a reaction

Consider this reaction:

$$A + B \rightleftharpoons C + D \tag{8.9}$$

The equilibrium constant of the reaction, K_{eq}, is given by this equation, where parentheses denote activity (if the concentration is relatively low, the activity can be approximated by molar concentration):

$$K_{eq} = (C)(D)/(A)(B) \tag{8.10}$$

The equilibrium constant of a reaction under standard conditions of temperature, pressure, and reactant concentrations is denoted K'_{eq} and is related to the free energy change of the reaction under these standard conditions, $\Delta G^{0\prime}$, as follows:

$$\Delta G^{0\prime} = -RT \log_e K'_{eq} = -2.303\ RT \log_{10} K'_{eq} \tag{8.11}$$

For a lucid discussion of the second law the reader is referred to Atkins, 1994.

Biology and the second law of thermodynamics

Thermodynamics is a substance-independent theory, which applies to *all* physical systems, irrespective of their chemical composition, structure, or specific forms of energy. It thus seems to be applicable to the basic concepts of biology (see Elitzur, 1994a). Indeed, the fact that two of the prominent founders of thermodynamics were biologists (Hermann von Helmholtz, 1821–1894, was trained in medicine, but his scientific contributions were in physiology and physics; Julius Robert Mayer, 1814–1878, was a physician) is not likely to have been just a coincidence. Indeed, the two disciplines address energy exchange, order, and irreversible changes (Elitzur, 1994a).

Entropy in biology

Interestingly, some of the basic tenets of thermodynamics have always been, and still are, controversial with regard to their applicability to biology. Among those scientists who have argued against the applicability of the laws of physics to biology, one can find such prominent names as Helmholtz, Kelvin, Bohr, Wigner, Brillouin, and Schrödinger. This issue has been reviewed recently by Elitzur (1994a).

Entropy is a rather complex concept, and its reflection in phenomena connected to the living state has been controversial. Indeed, several partially overlapping points of view have been adopted to characterize entropy since this term was first suggested by Clausius in 1851 as a basic tenet of the second law of thermodynamics. The central thermodynamic concepts that have been discussed in relation to the application of this discipline to the theory of evolution and the origin of life are "entropy," "order," and "information" (Elitzur, 1994a). For more points of view, the reader is referred to Berry, 1995.

In an attempt to base the origin of life on the foundation of the second law of thermodynamics, Elitzur (1994a) discussed the various characterizations of entropy and order and related them to the corresponding manifestations of life. One argument against the attempt to base biology (including the origin of life) on physical laws was based on an inadequate use of the second law. According to this argument, in any closed system there must be either a conservation of or an increase in entropy, whereas the phenomenon we call life is characterized by a decrease in entropy, and order is formed from disorder. A closer examination reveals, however, that life does not defy the second law of thermodynamics, because a living organism exchanging matter and energy with its environment is an open system, where the entropy of the living organism *and its surrounding* does increase. The use of entropy in biology is still highly controversial, as can be seen, for instance, from the polemical comments of Yockey (1995) on Elitzur's paper (1993; 1994a) and the reply of Elitzur (1996) to Yockey's comments.

More about information

In chapter 7 I stated that the information of DNA is structural. Information, however, is also a thermodynamic property, as was first shown by Claude Shannon (1916–) in 1948. And even though his original work was aimed at and connected to communicating systems, it soon became apparent that his theory was of special importance in biology, where the concept of information is closely related to those of entropy and order. Higher entropy is associated with randomness, whereas lower entropy is associated with nonrandomness. In a given system in a state of thermodynamic equilibrium, where entropy reaches a maximum, information is at a minimum.

One aspect of information deals with the correlation between a certain reality and a sequence of symbols (Elitzur, 1994a). Both of these attributes form an ordered relationship, which is unlikely to be formed by chance. Information and entropy are thus closely related; the former is the opposite of the latter. Still another aspect of information may be defined operationally as "the capacity to store and transmit meaning or knowledge, not the meaning or knowledge itself" (Gatlin, 1972). This definition, to which the concept of *capacity* is central, is analogous to the operational definition of energy as the capacity to do work. The vocabulary relating entropy and order is discussed by Gatlin (1972).

It should be noted that information is meaningful only when there is an entity that can read it, thus establishing a sender-recipient relationship (Küppers, 1990). Much confusion has arisen with the introduction of the concept of "entropy" as a measure of "information," equating "information" with "negentropy" (negative entropy). For more discussions on this topic, the reader is referred to Eigen, 1992; Elitzur, 1994a; Küppers, 1990; and Wicken, 1987.

Information content of a DNA molecule

The information content of a sequence of symbols is "the average number of binary yes/no decisions that are necessary in order to identify unambiguously a particular sequence of symbols" (Eigen, 1992). In other words, the information content of a defined set of symbols is measured by the number of possible sequences that can be produced from its symbols. The information content of a DNA molecule can be measured by the numbers of possible sequences of the same length that can be prepared from the building blocks of this molecule. The numbers of sequences are additive—that is, the total amount of information in the DNA sequence under study is the sum of all possible sequences.

The probability of the appearance of a particular sequence is given by the reciprocal of the information content. Since the number of sequences is additive, whereas the probabilities are multiplicative, the quantity of information is generally expressed not by the number of alternative arrangements but by its logarithm, where logarithms to base 2 are used (see Eigen, 1992). Recall that the logarithm of a product equals the sum of the logarithms of the factors.

The unit in which the information content is given is the binary digit (bit), where the bit number corresponds to the length of the sequence of binary characters (Eigen, 1992).

Nucleic acids are made up of four nucleotides. Therefore, a DNA sequence of length N has 4^N (or 2^{2N}) different sequences. Assuming that all sequences are equally probable, the information content of this sequence is $2N$ bits. For comparison, the number of nucleotides in the genetic system of microorganisms is on the order of 4 million. A printed text containing the genetic information describing a bacterial cell would be about a thousand-pages long. The genetic information needed to describe a human cell is about a thousand times larger than that needed for bacteria.

In order to calculate the number of variants of a specified DNA molecule, consider a strand of 1,000 nucleotides, that is, of the size of one gene. The number of variants, according to the preceding example, is 4^{1000}, or approximately 10^{602}. This number is beyond our intuitive grasp. (For more details see Eigen, 1992.)

Examples of thermodynamic applications

Activation energy

The free energy change of a reaction is calculated for a state of equilibrium and does not contain any information regarding the reaction rate. Thus, many reactions having a negative ΔG do not proceed at a measurable rate under certain conditions. For instance, a mixture of oxygen and hydrogen gases at room temperature seems chemically stable until the gases are exposed to a high enough energy source, such as an electric spark. This initiates an explosive exergonic reaction.

The energy input required for such a reaction to proceed is called activation en-

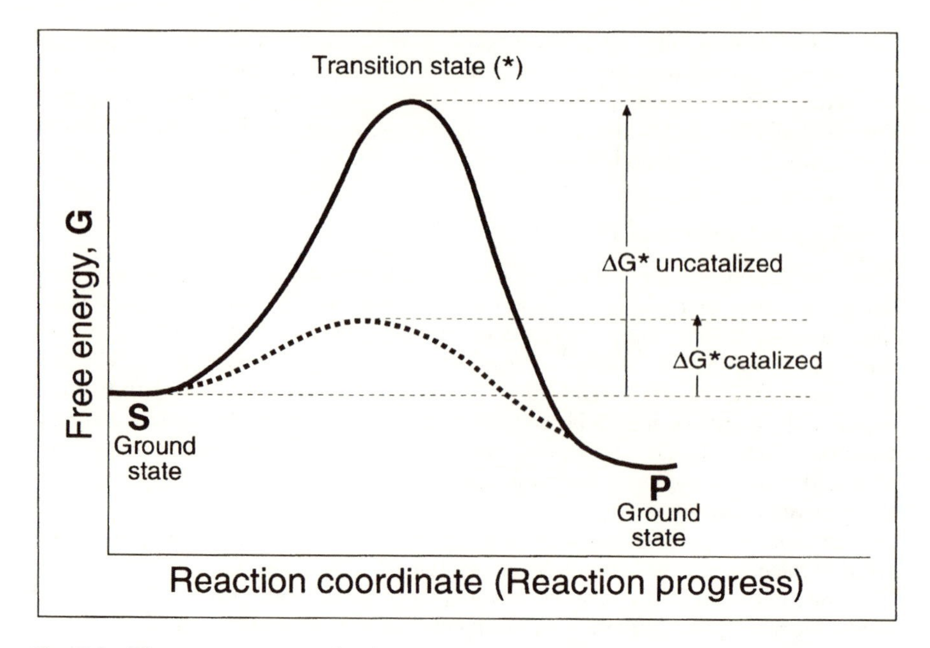

Fig. 8.2. The energy course of a chemical reaction S → P. (Following Lehninger et al., 1993). The solid line represents an uncatalyzed reaction, the broken line a catalyzed reaction.

ergy. The need for energy in order to initiate a reaction may have various causes: In some cases, mutual repulsion between the electron clouds of the reacting molecules needs to be overcome before they can get close enough for a chemical reaction to take place. In other systems, bonds must be broken before new ones can form. In still other cases, electrons must be excited before they can pair up in a covalent bond.

The activation energy is normally represented as free energy versus reaction coordinate (fig. 8.2). The increase in the free energy—the activation barrier—represents the transition state that must be overcome in the conversion of reactants into products (Lehninger et al., 1993). In many cases the reaction has more than one intermediate, where each intermediate is characterized by a trough. The activation energy for each intermediate is the peak just preceding its trough. In a multistep reaction, the highest activation energy must be achieved for the reaction to proceed.

Catalysts

The activation energy of a chemical reaction is not constant; it depends to a considerable extent on molecules with which the reactant is associated. Thus, a certain class of compounds can reduce the activation energy (fig. 8.2). The chemical entities that, by interacting with reactants of a chemical reaction (substrates), bring about a reduction of the activation energy of this reaction are called catalysts.

A catalyst is a substance that increases the reaction rate but is not itself permanently changed by the reaction. It does not change the free energy of the reaction—

that is, it accelerates the rates of forward and reverse reactions by the same factors. Thus, the equilibrium constant of the reaction, K'_{eq}, is not affected by the catalyst.

Many inorganic catalysts are naturally occurring; some of them are used in the chemical industry. Many others have been synthesized for various purposes.

Autocatalysis

Autocatalysis can be described as "the enhancement of the (absolute) rate of growth of a population as that population increases" (Epstein, 1995). This process is ubiquitous in the reproductive process of biological systems.

> As the Reverend Thomas Malthus recognized when he suggested that population tends to grow geometrically, the number of individuals born is, on average, proportional to the number in the population. This perceptive clergyman was also aware that there are limitations to population growth; he proposed that the food supply grows only arithmetically.

In chemical terms, autocatalysis is a catalytic process in which a catalytic molecule helps in the synthesis of another, identical molecule. This is represented by a reaction of the type (Elitzur, 1994a):

$$R \xrightarrow{p} p + b \qquad (8.12)$$

R stands for the reactant, *p* the reaction-product, and b the by-product. *p* plays the function of both catalyst and product. (For a discussion see Elitzur, 1994a; Lifson, 1997). The importance of autocatalysis in the origin of life is probably connected to primitive molecular *systems* rather than to *single molecules* (chapter 23).

Clues to the study of life's origin

Both biology and the origin of biology, namely the origin of life, are assumed to obey the fundamental laws of thermodynamics. This assumption, however, does not suggest specific clues regarding the entities and environments involved in the transition from inanimate to animate.

9

Central Biochemical Molecules and Processes

A biochemical system can be regarded as a system of catalysts regulating the transformation of other compounds so as to make available the system energy and matter for its further increase and maintenance.

Martynas Yčas, "A note on the origin of life"

Biocatalysts

Catalysis is especially important in biology, because living organisms normally cannot tolerate the high temperatures that may be needed to overcome the activation energy of a reaction. Biology's solution to this problem has been to use catalysts (fig. 8.2), where in addition to biocatalysts, various metals such as magnesium, iron, zinc, and molybdenum can also catalyze biological reactions. Most of the processes in the cell depend on complex biocatalysts, each of which is specific to just one reaction. Moreover, there is no known natural way by which contemporary biocatalysts could have formed, except through evolution. This is a central issue in the study of the origin of life and will be discussed in part IV.

Biocatalysts: Enzymes and ribozymes

The known biocatalysts are divided into two groups, namely, enzymes and ribozymes. The first group, which consists of the great majority of known biocatalysts, is made of proteins, whereas the second is made of ribonucleic acids (RNAs). Some enzymes are made up of protein molecules associated with nucleic acids; these will be discussed later.

Reaction rate

In biological systems at constant temperature, the reaction rate is determined predominantly by reactant concentration, pH, and the biocatalyst that catalyzes this reaction. In a reaction involving two or more reactants, the reaction rate depends on their concentration; at a high concentration molecules are more likely to encounter one another; pH affects the reaction rate because it influences the physicochemical properties of both the reactants and the catalyst. The reaction rates can be measured by the number of molecules "processed" (transformed) by an enzyme molecule per unit time. For most enzymes this rate ranges from about 1,000 to more than 500,000 molecules per minute. The fastest known enzyme is catalase, which can transform 5 million peroxide molecules per minute.

Since a catalyst increases the reaction rate by lowering the activation energy, this does not affect the free energy change, ΔG, or the equilibrium constant, K'_{eq}, of the reaction (chapter 8). Thus, the reaction rates in both directions are enhanced by the same proportion. The mechanism by which activation energy is lowered involves binding the substrate (the reactant that the catalyst catalyzes). Binding strains the substrate bonds and makes them more reactive. Some enzymes bind several substrate molecules, juxtaposing them in a way that facilitates their interaction. In all cases, the overall effect is a decrease in the activation energy of the reaction. The three steps of enzymatically catalyzed reactions, namely, binding, catalysis, and release, are given in the following scheme, where E is the enzyme, S is the substrate, and P is the product:

$$E + S \underset{}{\overset{\text{binding}}{\rightleftharpoons}} ES \xrightarrow{\text{catalysis}} EP \xrightarrow{\text{release}} P + E \qquad (9.1)$$

Specificity

A most important feature of biocatalysts is their specificity to just one reaction. In order to be so specific, the biocatalyst molecules must be large and complex, so as to accomodate the exact size and geometry of the substrate. Moreover, biocatalysts must be able to select their specific substrate molecules in a medium that contains other similar molecules. Thus, these large molecules must always assume the same three-dimensional configuration. This last feature highlights the importance of pH, which affects, among other things, the configuration of the biocatalyst, as well as the binding substrate.

For a discussion on the dynamics of enzymes, see Williams, 1993.

Metabolic channeling

The concerted function of the cell's enzymes is an important feature of their activities. Rather than the cell as a "bag of enzymes," according to the view accepted until recently, the enzymes and other components of the metabolic system form complexes that are frequently associated with the cell membrane or the cytoskeleton. The metabolic pathways are thus channels, where intermediates are transferred directly between the enzymes without being mixed with the cell fluid (Welch and Easterby, 1994; Edwards, 1996).

Processes requiring free energy input

All living organisms require a continuous supply of free energy for both the biosynthesis of molecules and the active transport of ions and molecules across the cell membrane. The cell of autotrophic organisms is characterized by energy-capturing, energy-converting, and energy-consuming mechanisms. The cell of heterotrophic organisms is characterized by the two latter functions only.

Energy-rich molecules

ATP, ADP, and AMP

As a general rule, the molecule used in biological processes as a coupling agent for the driving of reactions with positive ΔG is adenosine triphosphate (ATP) (fig. 9.1). The hydrolysis reaction of ATP has a negative ΔG, and the energy required by reactions such as the reduction of CO_2 to glucose or dehydration-condensation is provided, directly or indirectly, by this reaction.

The ATP contains one phosphoester bond and two pyrophosphate linkages, sometimes called phosphoanhydride bonds, which are formed by the removal of one water molecule and the condensation of two phosphate molecules. Upon hydrolysis, these pyrophosphate bonds release their energy, which becomes available for chemical reactions. The ATP molecule thus possesses a high free energy of hydrolysis, and the phosphate bond is sometimes called a high-energy phosphate bond.

ATP can be hydrolyzed in two ways, namely, removal of the terminal phosphate to form adenosine diphosphate (ADP) and inorganic phosphate (Pi) or removal of terminal diphosphate to yield adenosine monophosphate (AMP) and pyrophosphate (PPi) (equations 9.2, 9.3). These reactions, which are enzymatically catalyzed, can be described as follows:

$$\mathrm{ATP} + 2\mathrm{H_2O} \rightarrow \mathrm{AMP} + \mathrm{PPi} + \mathrm{H^+} \Delta G^{0\prime} = -7.3 \text{ kcal/mol} \tag{9.2}$$

$$\mathrm{ATP} + \mathrm{H_2O} \rightarrow \mathrm{ADP} + \mathrm{Pi} + \mathrm{H^+} \Delta G^{0\prime} = -7.3 \text{ kcal/mol} \tag{9.3}$$

In these reactions $\Delta G^{0\prime}$ is the standard free energy change of the reaction. The standard conditions are: pressure of 1 Atmosphere; an initial concentration of all reactants and products of 1 M (except for protons: the pH is 7.0).

It should be noted that if the hydrolysis of ATP takes place freely, the energy it releases is dissipated to the environment in the form of heat, rendering it unavailable for the performance of chemical reactions. It is the gradual, enzymatic energy transfer that enables utilization of the phosphate bond. Note also that the synthesis of ADP from AMP and phosphate, as well as that of ATP from ADP and phosphate (or from AMP and pyrophosphate), is involved in the investment of free energy from an external source. Mechanistically, this need for energy is explained as the electrostatic energy necessary to rearrange the negatively charged oxygen atoms of the phosphate group. This energy is released upon hydrolysis.

The hydrolysis of ATP to ADP and Pi, and the regeneration of ATP from ADP and Pi, constitute the *ATP cycle* (fig. 9.1b). The source of energy for the formation of ATP from ADP and Pi is either solar radiation (photosynthesis) or energy-rich compounds (e.g., glucose). The hydrolysis of ATP is enzymatically coupled to the performance of many cellular reactions.

Because it is a mediator in both exergonic and endergonic reactions, ATP is sometimes viewed as an energy "shuttle" (see, for instance, Swanson and Webster, 1985): ATP hydrolysis is the exergonic reaction, with negative ΔG; ATP formation is the endergonic reaction, with positive ΔG. Some people call ATP the fuel of life. This phrase is misleading, since ATP is not found in the environment as a source of energy ready for use. It is present in the cell in minute amounts and is continually regenerated from ADP and Pi (for a succinct discussion see also de Duve, 1991).

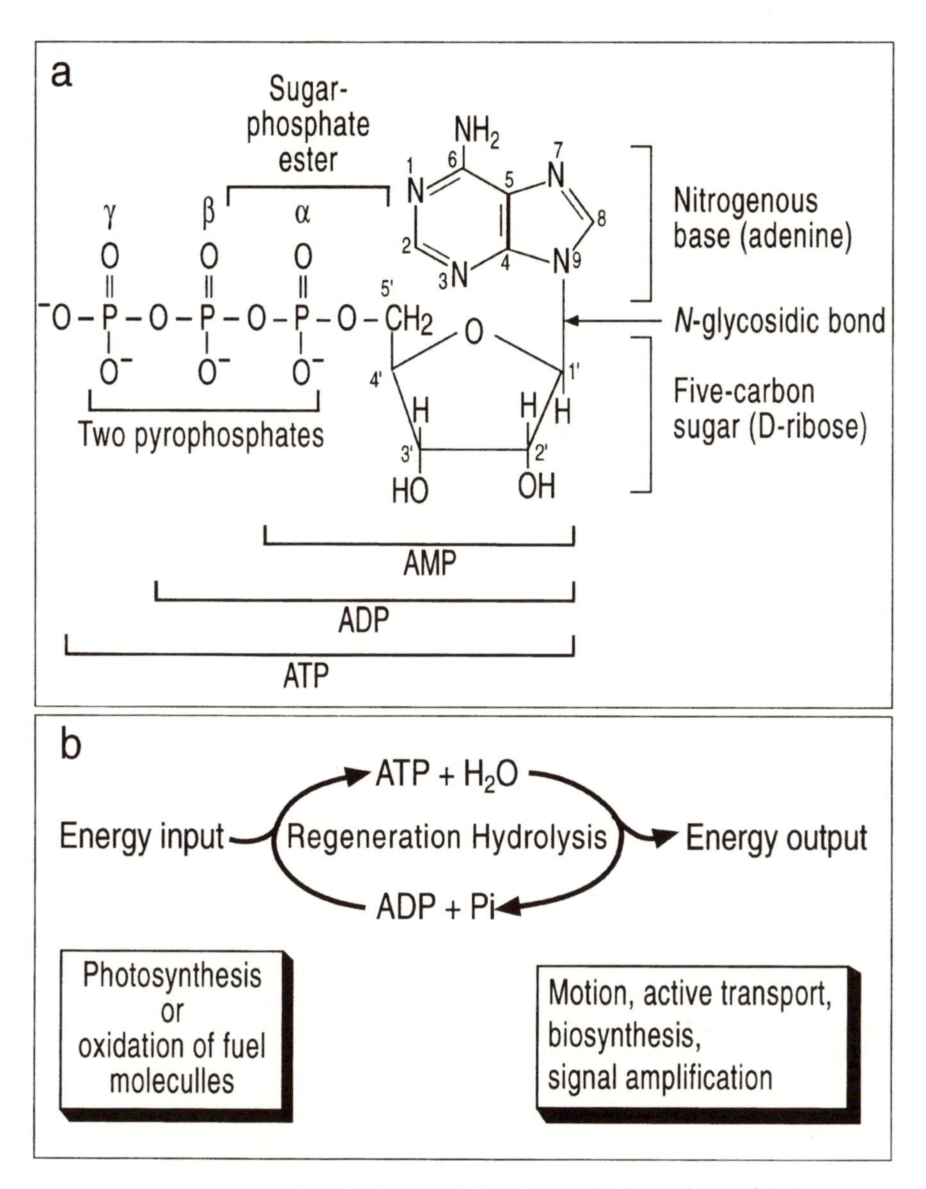

Fig. 9.1. a. ATP, ADP, and AMP. b. The ATP-ADP cycle: hydrolysis of ATP to ADP and Pi. (Following de Duve, 1991)

Other energy-rich molecules

A variety of energy-rich bonds other than the phosphoanhydride bond are always found in the cell. These include glucose 6-phosphate, glucose 1-phosphate, glycerol 3-phosphate, phosphoenolpyruvate, and thioesters. Some of these energy-rich molecules have a higher energy content and some a lower energy content than ATP.

However, it is the latter that is used in most cell reactions. Presumably, ATP was selected for at a very early stage of chemical evolution.

Energy source for high-energy bond formation

Plants obtain the energy needed to synthesize high-energy bonds by trapping solar radiation through photosynthesis. The molecule that carries out the actual trapping is chlorophyll (from the Greek *chloros,* "pale green," and *phyllum,* "leaf"). This pigment is found in the cell's chloroplasts; it absorbs part of the light energy and uses it to synthesize ATP from ADP and Pi. Photosynthetic bacteria obtain energy from light in a similar manner. Much of the ATP thus produced is used in the transformation of CO_2 to glucose ($C_6H_{12}O_6$) in this reaction:

$$6CO_2 + 6H_2O \xrightarrow{ATP \rightarrow ADP + Pi} C_6H_{12}O_6 + 6O_2 \tag{9.4}$$

Animals and nonphotosynthetic microorganisms obtain the energy they need by oxidizing "nutrient" (sometimes called "food") molecules in a process known as respiration, where the predominant source of energy is glucose. In this enzymatic process, the metabolism of one molecule of glucose is coupled with the synthesis of 36 molecules of ATP from 36 molecules of ADP and 36 molecules of Pi, as follows:

$$C_6H_{12}O_6 + 6O_2 + 36Pi + 36ADP \rightarrow 6CO_2 + 6H_2O + 36ATP \tag{9.5}$$

Dehydration condensation of building blocks

As a general rule, building blocks in the biosynthetic process are joined by the removal of water:

$$X\text{—}OH + Y\text{—}H \rightarrow X\text{—}Y + H_2O \tag{9.6}$$

X and *Y* are the building blocks in this equation.

This is a general reaction describing the dehydration-condensation of, for instance, proteins from amino acids, polysaccharides from sugars, and nucleic acids from mononucleotides (whose constituents are pentoses, bases, and phosphates). These biosynthetic reactions result in a decrease in entropy. For example, when amino acids are linked to produce a peptide, they lose much of their freedom of movement in the solution. The formation of a peptide, a rather rigid and ordered molecule, imposes restrictions on the free movements of its building blocks. These restrictions are associated with an increase in the order of the system or a decrease in entropy.

Thus, the condensation of building blocks into biopolymers is not a spontaneous process. Dehydration-condensation is endergonic—that is, it goes against the spontaneous direction that in the aqueous environment of the cell, is hydrolysis of the biopolymers. To perform dehydration-condensation in the cell, energy input is required. As a rule, the energy needed to carry out the dehydration-condensation reaction is provided, directly or indirectly, by the hydrolysis of ATP to either ADP or AMP.

It should be noted that prebiotic dehydration-condensation could not have taken place by sophisticated mechanisms such as those in extant cells. One obvious solution to this problem is condensation reactions by means of environmental dehydration, such as during diurnal hydration- dehydration cycles. These are discussed in part IV.

Oxidation-reduction (redox) reactions

The original definition of reduction was the addition of hydrogen to an ion or molecule. According to the more recent definition, the loss of electrons from an atom or molecule is called oxidation, whereas the addition of electrons to an atom or molecule is called reduction. These reactions may be presented in various ways (see de Duve, 1991). Thus, the oxidation of ferrous ion, Fe^{2+}, into ferric ion, Fe^{3+}, may be presented either as

$$2Fe^{2+} + \rightarrow 2Fe^{3+} + 2e^- \tag{9.7}$$

or as

$$2Fe^{2+} + 1/2O_2 \rightarrow 2Fe^{3+} + O^{2-} \tag{9.8}$$

Obviously, equation 9.8 tells us that the ferrous ion lost an electron and that the lost electron was taken up by an oxygen atom. In biological reduction, protons are generally added together with electrons, which make up the equivalent of the hydrogen gas that would be used in a reduction reaction in the original sense. This is exemplified in the following reactions:

1. The oxidation of hydrogen gas, H_2, as the result of a release of electrons and hydrogen ion (proton) according to this equation:

$$H_2 \rightarrow 2e^- + 2H^+ \tag{9.9}$$

Here the electrons, which cannot exist as separate entities in the solution, must be part of atoms or molecules. Thus, this equation should be supplemented by a second equation, to which it is coupled:

$$1/2\ O_2 + 2e^- + 2H^+ \rightarrow H_2O \tag{9.10}$$

Since these two equations describe two halves of one reaction they are called half reactions. When the first oxidation reaction is coupled with the second reduction reaction, the following overall balanced reaction is:

$$H_2 + 1/2\ O_2 \rightarrow H_2O \tag{9.11}$$

2. Conversion of aldehyde (R—CHO) into alcohol (R—CH_2OH) (see de Duve, 1991):

$$R{-}CHO + 2e^- + 2H^+ \rightarrow R{-}CH_2OH \tag{9.12}$$

3. The conversion of succinate $(COO{-}CH_2{-}CH_2{-}COO)^{2-}$ to fumarate $(COO{-}CH{=}CH{-}COO)^{2-}$:

$$(COO{-}CH_2{-}CH_2{-}COO)^{2-} \xrightarrow[\text{two H atoms are lost}]{-2e^-\ -2H^+} (COO{-}CH{=}CH{-}COO)^{2-} \tag{9.13}$$

Electrons are neither created nor destroyed in a chemical reaction, nor do they occur freely in nature. This means that electrons and protons are always provided by one molecule (a donor molecule) and taken up by a second molecule (an acceptor molecule). The reaction is called a *redox* (reduction-oxidation) reaction or, more commonly, an electron transfer reaction. Obviously, then, whenever an atom or a molecule becomes oxidized, another one must become reduced.

More generally, consider a donor molecule DH_2 in a reduction-oxidation reaction. In the presence of an acceptor molecule A, DH_2 is oxidized in the process to form D, while the acceptor A is reduced to form AH_2:

$$DH_2 + A \rightleftharpoons D + AH_2 \qquad (9.14)$$

Autotrophic biosynthesis

The autotrophic biosynthesis of organic molecules from inorganic molecules requires considerable amounts of reduction (for details see de Duve, 1991). Thus, in order to make the backbone of a glucose molecule out of CO_2 or to convert NO_3^- to the amine group ($—NH_2$) of an amino acid ($RCNH_2HCOOH$), a large number of electrons and protons are needed. This is exemplified in the following reactions, which show the overall chemical equations of the biosynthesis of the sugar glucose (equation 9.15) and the amino acid glycine (equation 9.16):

$$6CO_2 + 24e^- + 24H^+ \rightarrow \underset{\text{Glucose}}{C_6H_{12}O_6} + 6H_2O \qquad (9.15)$$

$$2CO_2 + NO_3^- + 14e^- + 15H^+ \rightarrow \underset{\text{Glycine}}{CH_2NH_2{-}COOH} + 5H_2O \qquad (9.16)$$

Similar equations describe the reduction of other molecules (de Duve, 1991).

Heterotrophic biosynthesis

Since heterotrophic organisms utilize preformed molecules previously reduced by autotrophic creatures, they use reduction reactions to a much lesser extent. One such biosynthetic pathway is the reduction of carbohydrates to form fats.

Energy manipulation: Coupling unfavorable with favorable reactions

The ultimate source of energy for all life is solar radiation. A small part of this radiation is captured by the process called photosynthesis and converted into chemical energy, to be stored in specific molecules, usually carbohydrates. Using a series of reactions that release the stored energy gradually, the cell is able to manipulate the captured energy and use it in all the processes that need energy input.

Coupling reactions

The underlying principle in performing unfavorable reactions is the coupling of a reaction with a positive ΔG to a reaction with a negative ΔG of larger magnitude. If the two reactions take place in a close enough proximity, a large portion of the energy released by the favorable reaction can be useful, while the rest is dissipated to the environment as heat. The unfavorable reaction with the positive ΔG will utilize the extra energy it needs from the energy released by the favorable reaction. Thus, rather than calculating the ΔG of the unfavorable reaction, one must take into account the sum of the ΔGs of the coupled reactions. If this sum is negative, then the unfavorable reaction can proceed, utilizing most of the energy of the favorable reaction.

Universal electron carriers: NAD^+, FAD^+, and $NADP^+$

Special electron carriers are mediators in oxidation-reduction reactions in the cell; they can assume oxidized and reduced forms. The three most important electron carriers are nicotinamide adenine dinucleotide (NAD^+), nicotinamide adenine dinucleotide phosphate ($NADP^+$), and flavine adenine dinucleotide (FAD).

Both NAD^+ and $NADP^+$ can accept two electrons at a time to yield their respective reduced forms, NADH and NADPH, according to the half reactions given in equations 9.17 and 9.18:

$$\underset{\text{oxidized}}{NAD^+} + 2H^+ + 2e^- \rightarrow \underset{\text{reduced}}{NADH + H^+} \tag{9.17}$$

$$\underset{\text{oxidized}}{NADP^+} + 2H^+ + 2e^- \rightarrow \underset{\text{reduced}}{NADPH + H^+} \tag{9.18}$$

FAD can accept either one or two electrons, together with one proton in each such reaction, as follows:

$$\underset{\text{oxidized}}{FAD^+} + 2H^+ + 2e^- \rightarrow \underset{\text{reduced}}{FADH_2} \tag{9.19}$$

The free energy of oxidation of the reduced forms of these electron carriers is very high. In the presence of a complex series of enzymes, they form most of the ATP in the cell. ATP is not synthesized in a single reaction; rather, the free energy is released in small increments by a step-by-step transfer of electrons via a series of proteins. These proteins constitute the *electron transport chain.*

Cofactors

Some enzymes require additional components called cofactors. These may be either one or more inorganic ions, such as Fe^{2+}, Mg^{2+}, or Zn^{2+}, or a complex organic or metalo-organic molecule. The latter is called a coenzyme. Some enzymes require both a coenzyme and one or more metal ions for their activity.

Coenzyme A (CoA): A universal carrier of acyl groups

Coenzyme A (fig. 9.2), a heat-stable cofactor, is required in many enzyme-catalyzed acetylations. As such it is a central molecule in metabolism. The active group in CoA is the terminal sulfhydryl (SH) group, to which acyl groups are linked by a thioester bond to form an acyl CoA. An acyl group often linked to CoA is the acetyl unit CH_3CO—, forming acetyl-CoA. The hydrolysis of this molecule is an exergonic re-

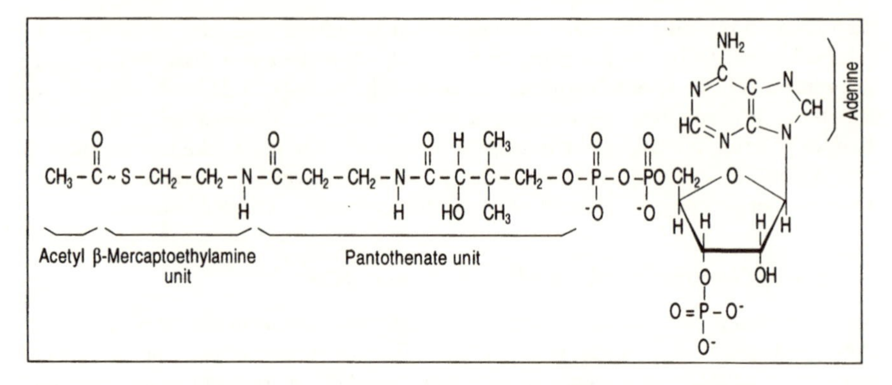

Fig. 9.2. Structure of acetyl coenzyme A (acetyl-CoA).

action, releasing 7.5 kcalories per mole (similar to the amount released upon hydrolysis of ATP):

$$\text{Acetyl CoA} + H_2O \rightleftharpoons \text{acetate} + \text{CoA} + H^+ \qquad (9.20)$$

$$\Delta G^{0\prime} = -7.5 \text{ kcal/mol}$$

Thus, acetyl CoA is a carrier of an activated acetyl group, just as ATP carries an activated phosphoryl group.

Metabolic cycles

The citric acid cycle

The citric acid cycle (also known as the Krebs cycle or the tricarboxylic acid cycle) is considered one of the most ancient metabolic cycles. It is both the final stage in the catabolism of food and, at the same time, the starting point of the biosynthesis of new cellular monomers. These attributes make it of special importance to us, and we shall come to it in part IV.

In the overall pattern of the citric acid cycle (fig. 9.3), which was first worked out by Hans Krebs (1900–1981) in the late 1930s, a four-carbon compound (oxaloacetate) combines with a two-carbon acetyl moiety to yield a six-carbon tricarboxylic acid (citrate). In the nine steps that make up the cycle, the citric acid loses one carbon atom (decarboxylation) to yield a five-carbon α-ketoglutarate; additional decarboxylation yields succinyl-CoA, which is transformed into succinate. In the next three steps, the succinate is transformed into other four-carbon compounds, namely, fumarate, malate, and, finally, oxaloacetate, which is ready to start the next cycle. Many enzymes as well as energy-rich molecules are involved in the operation of the citric acid cycle, and oxidation-reduction and hydrolysis reactions take place at various steps. The net reaction of the citric acid cycle is as follows:

$$\begin{aligned} &\text{Acetyl-CoA} + 3\,NAD^+ + FAD + GDP + Pi + 2\,H_2O \rightarrow 2\,CO_2 \\ &\quad + 3\,NADH + FADH_2 + GTP + 2\,H^+ + \text{CoA} \end{aligned} \qquad (9.21)$$

In the four oxidation-reduction reactions of the cycle, three pairs of electrons are

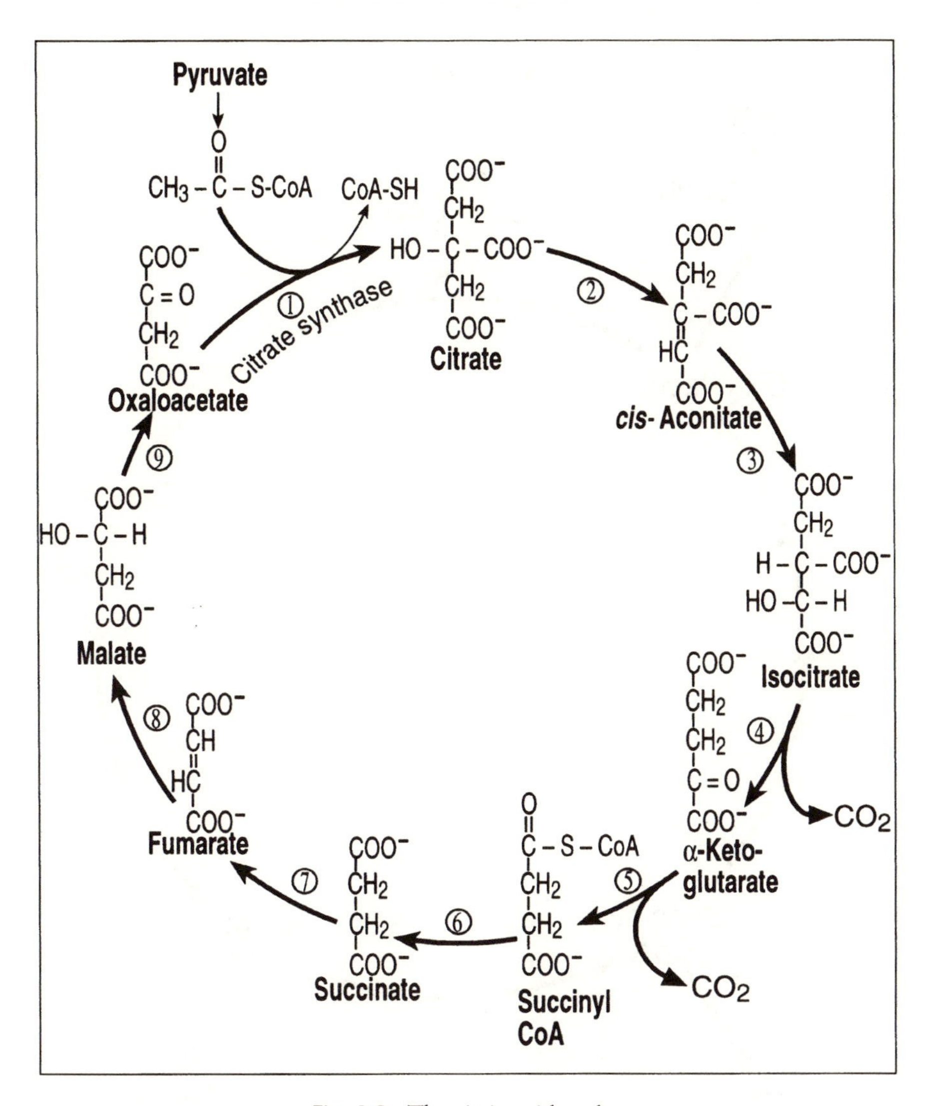

Fig. 9.3. The citric acid cycle

transferred to NAD+ and one pair to FAD. These reduced electron carriers are subsequently oxidized by the electron-transport chain to generate 11 molecules of ATP. One additional high-energy phosphate bond is directly formed in the cycle, and a total of 12 high-energy phosphate bonds are generated for each two-carbon fragment that is completely oxidized to CO_2 and H_2O. The cycle operates only under aerobic conditions, where O_2, the final electron acceptor, serves as the last constituent of the electron-transfer chain. Thus, the degradation of acetyl residues by this cycle yields CO_2 and electrons, and the transfer of these electrons to molecular oxygen is coupled to the phosphorylation of ADP to ATP.

The role of the citric acid cycle in early stages of evolution was recently dealt with by Zubay (1996) and is discussed in chapter 17.

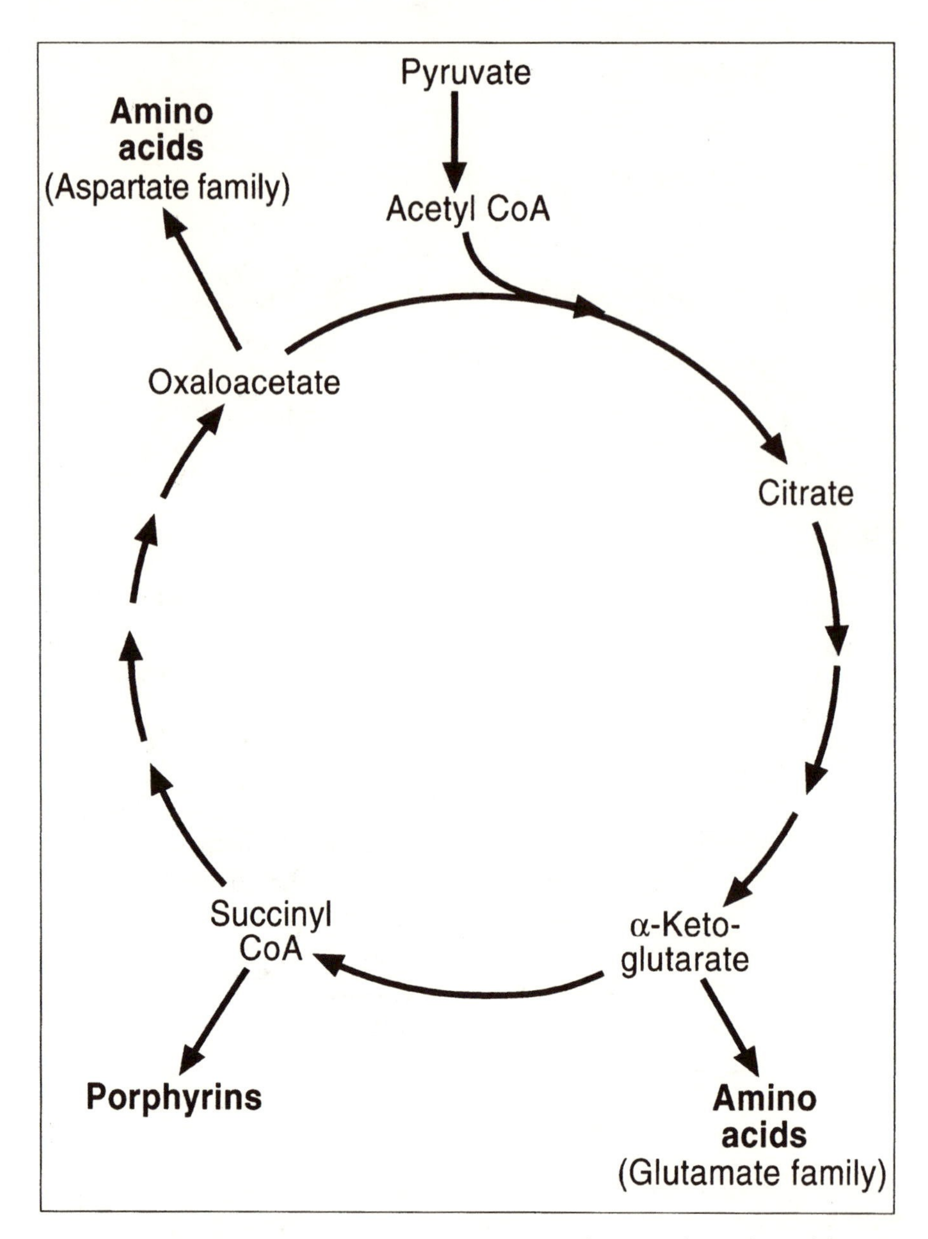

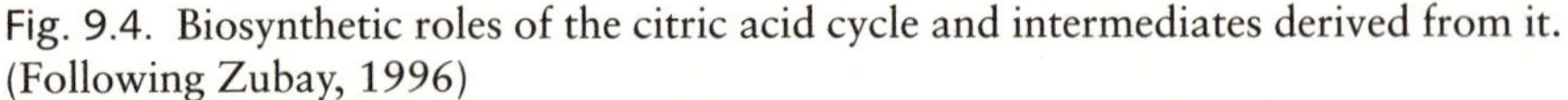

Fig. 9.4. Biosynthetic roles of the citric acid cycle and intermediates derived from it. (Following Zubay, 1996)

Amino acid synthesis from intermediates of the citric acid cycle

An important feature of the citric acid cycle is its role as an amphibolic pathway (see Lehninger et al., 1993): that is, it is involved in both catabolic and anabolic processes. Thus, in addition to functioning in oxidative catabolism, it provides the intermediate compounds that serve as precursors of many biosynthetic pathways. Indeed,

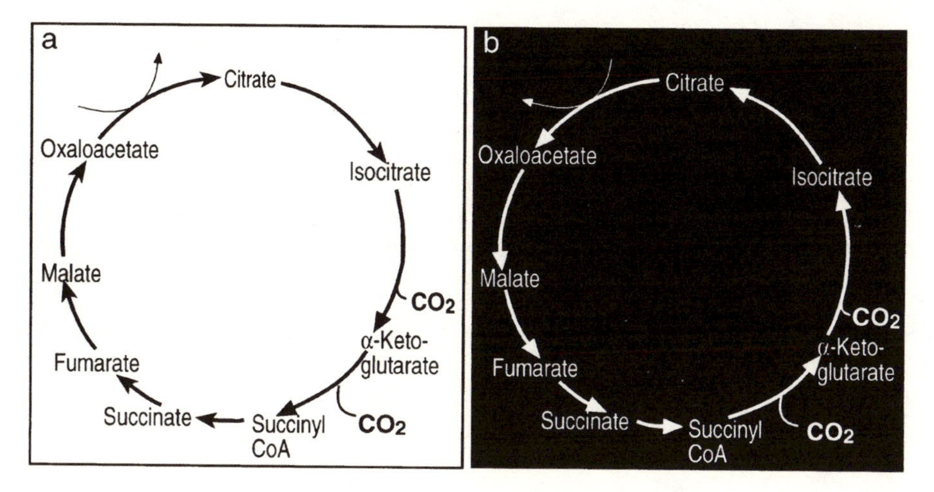

Fig. 9.5. Schematic of (a) the citric acid cycle and (b) the reductive citric acid cycle. (Following Zubay, 1996)

many of the amino acids are derived from α-ketoglutarate and oxaloacetate (fig. 9.4; see also Taylor and Coates, 1989; Stryer, 1995; Zubay, 1996). The latter feature makes the citric acid cycle of special interest to origin-of-life researchers, who have suggested that the biosynthesis of amino acids may be related to early stages of the evolution of the genetic code. This is further discussed in chapter 17.

The reductive tricarboxylic acid cycle

The citric acid cycle liberates CO_2 and generates NADH, the aerobic oxidation of which yields ATP. It does not operate in reverse as a biosynthesis pathway for CO_2 assimilation. Such reverse cycles have been discovered, however: they are called "Calvin's reductive pentose phosphate cycle" and the "reductive citric acid cycle" (fig. 9.5; see also Buchanan, 1992; Zubay, 1996). It has been suggested that the latter cycle is more ancient than the former one (Buchanan, 1992). The reductive carboxylic acid cycle was suggested as a very ancient metabolic cycle, as discussed in chapter 21. The evolutionary implications of the reductive carboxylic acid cycle are related both to the antiquity of this cycle and to its importance in providing building blocks, directly or indirectly, for all other cellular constituents. Thus, "acetyl-CoA may be used directly for lipid synthesis or be converted to pyruvate which in turn may give rise . . . to amino acids (e.g., alanine, aspartate, glutamate) or to sugar phosphates" (Buchanan, 1992). This will be further discussed in chapter 21.

Clues to the study of life's origin

Theories of the origin of life should take into account the predecessors of both enzymes and ribozymes as candidates for the role of primordial catalysts and the evolutionary processes through which biocatalysts could have emerged. If minerals are assumed to have been the primordial catalysts, as suggested by Bernal (1951; chapter 5), then the theory of the evolution of biocatalysis should bridge the gap between the minerals and these biomolecules. This is further discussed in chapters 20, 21, and 23. Similarly, if pyrophosphate could have functioned as energy shuttle in prebiotic reactions, as suggested by Lipmann (1965), then theories of the origin of life should show the evolution of the role of an energy shuttle and its transition from the inanimate to the animate (chapter 19). And by the same token, if prebiotic dehydration-condensation is assumed, as suggested by Lahav et al. (1980), then theories of the origin of life should show the evolution of biological dehydration-condensation and its transition from the inorganic to the biological domain (chapters 20 and 23).

Theories of the origin of life should also address the initiation of extant cellular reaction cycles. Thus, in addition to the problems of both catalysts and free energy for the various steps of such cycles, these theories should address the plausibility of their establishment under prebiotic conditions.

10

Biological Conservatism and Continuity and the Phylogenetic Tree

Since there exist characteristic archaeal, bacterial, and eukaryal versions for just about every universal macromolecular function so far characterized in the cell, there can be no doubt that all life on this planet is organized into three, very distinctive groupings.

Carl R. Woese and Norman R. Pace, "Probing RNA structure, function, and history by comparative analysis"

On the irreversibility of evolution

It was Louis Dollo who concluded, based on paleontological observations, that evolution is unidirectional. The irreversibility of evolution, also known as Dollo's Law (see also Huxley, 1953, pp. 31–32), was recently treated statistically at the molecular evolution level by Marshall et al. (1994). As noted by Dawkins (1986, p. 94),

> Dollo's Law is really just a statement about the statistical improbability of following exactly the same evolutionary trajectory twice (or indeed any *particular* trajectory), in either direction. A single mutational step can easily be reversed. But for larger numbers of mutational steps . . . the chance of two trajectories ever arriving at the same point becomes vanishingly small.

For a discussion on the irreversibility of evolution and its relatedness to thermodynamic considerations, see Berry, 1995, and Dennett, 1996, p. 122.

Dollo's Law is inherently related to the conservative nature of biology: The structure, reaction networks, and functions of the biomolecules of the cell are very complex. Therefore, only relatively small changes can be introduced by mutations at one time. Those mutations that are introduced into (and "accepted" by) the genome are either neutral or beneficial to the survivability of the living creature and its descendants. Big and abrupt changes are much more likely to be lethal; they may take place, however, under certain circumstances, as discussed later.

Directionality of the evolution process is another aspect of the complexity of living entities (see also Gatlin, 1972): The direction of a process leading to the evolution of a complex structure or reaction pathway is unlikely to be reversed. More likely, an alternative system may evolve, either in conjunction with the old system or in conjunction with another system that is beneficial to the living entity during its evolution process. Such a system may then replace, partly or completely, the first and less-competitive system. Indeed, biological evolution is more likely to come up with a new structure or chemical pathway added to an existing one than to dispose of an old structure or chemical pathway without leaving traces. Thus, biological systems are very conservative—that is, they not only change very slowly with time but also retain their basic features. In the words of Szent-Gyorgyi (1972; cited by Hartman, 1975),

> Life has developed its processes gradually, never rejecting what it has built, but building over what has already taken place. As a result the cell resembles the site of an archaeological excavation with the successive strata on top of one another, the oldest one the deepest. The older a process, the more basic a role it plays and the stronger it will be anchored, the newest processes being dispensed with most easily.

The conservatism of biology is reflected also in the biogenetic law of Haeckel: Ontogeny recapitulates phylogeny. Thus, embryonic development of an individual is a rapid recapitulation of evolutionary development of the group of animals to which the individual belongs. This topic is further discussed in chapter 16.

The principle of biological continuity and the biological record

> Understanding the functions and structures of the array of proteins expressed in living organisms is a fundamental goal of molecular biology. Our hope of attaining this goal stems largely from the unifying theme of shared evolutionary ancestry: related organisms have similar proteins and, within an organism, different proteins of related function are often wholly or partly similar in sequence.
>
> Philip Green et al., "Ancient conserved regions in new gene sequences and the protein databases"

The steps via which biological systems change gradually in evolutionary processes are related to the *biological continuity* in the transition from one evolutionary stage to the other. It should be noted, however, that unlike mathematical continuity, biological continuity represents a series of finite and discrete changes, which constitute the evolutionary process. It is also worth noting that not all biological evolution processes proceed by small steps; in some cases major changes can take place rather rapidly, as discussed in the following paragraphs.

At the molecular level each step of the biological evolutionary process is based on the introduction of a change into the genome (genetic system) of a living creature, followed by a "test" of the survivability of the altered system, under the prevailing conditions. This "test" includes competition with other living creatures, where the Darwinian "survival of the fittest" postulate prevails (for comments on the tautological nature of this expression see, for instance, Eigen, 1971, 1992; Gatlin, 1972). The change thus introduced, a *mutation,* is normally assumed to be random, under natural conditions. However, whether mutations are random or not (see Lenski and Sniegowski, 1995; Lenski and Mittler, 1993), some mutations are deleterious or lethal.

De Vries coined the term "mutations" for "jumplike" changes of organisms. Focusing on the discontinuity formed by mutations, Schrödinger (1944, p. 36) noted the analogy between the mutation theory and the quantum theory, suggesting that the first should be considered, figuratively, "the quantum theory of biology."

The apparent discrepancy between the discontinuity of the microsystem of the mutation site and the broader picture of the population of living organisms was phrased by Mayr (1994, p. 42). According to him, population evolution involves the accumulation of small changes in the total gene pool and, as such, is a gradual process.

Size of discrete evolutionary steps

> Evolution cannot see way down the road, so anything it builds must have an immediate payoff to counterbalance the cost.
>
> David C. Dennett, *Darwin's dangerous idea*

Following McLachlan (1987), the genetic mechanisms that bring about changes in the coded proteins can be divided into two main classes, as follows:

Minor steps

The smallest discrete change in a genome may be considered the result of a change introduced into one building block of a cellular polynucleotide; this, in turn, would cause the substitution of one amino acid in a protein—a single point mutation. If subjected to Darwinian selection, such a change may be considered an "elementary evolutionary step." It is convenient to also include in this class small changes such as those that cause deletions and insertions of just a few amino acids at loops in the structure: "The overall domain structure is left almost unaltered and is still recognizable after many such steps have occurred" (McLachlan, 1987).

In view of the extraordinary fidelity of DNA replication and the scarcity of nucleotide substitutions in the genome, point mutation alone is too slow a mechanism to explain actual evolutionary rates.

Major steps

A much more efficient mechanism is genetic recombination, by which genomes undergo expansion or contraction by duplication, insertion, or deletion. By this modular mechanism various domains of genes may be shuffled, and duplicated copies of genes may be formed, ready to diverge and undergo further evolutionary processes (Alberts et al., 1994, p. 386). Major steps are "evolutionary discontinuities" without which "many proteins would reach an evolutionary stagnation point in which every point mutation was slightly harmful" (McLachlan, 1987).

The importance of these processes is twofold: First, they explain the evolutionary rate better than the mechanism of point mutations alone. Second, they can be used in the study of the very beginning of our phylogenetic tree (Gogarten et al., 1989; Iwabe et al., 1989). Moreover, it is tempting to suggest that the principle of biological continuity may be extended to the very first molecular evolution processes. Thus, it may be assumed that gene duplication and primordial "genetic recombination" are as old as the origin of template-directed synthesis itself.

It should be noted that horizontal gene transfer is another mechanism that introduces dramatic evolutionary changes into the genome of a cell. This is discussed later in this chapter.

Biochemical continuity is biological continuity at the molecular level

The modern understanding of the principle of biological continuity can be traced to Darwin (at the morphological level; see Eigen, 1992); with the advent of molecular biology, it has become an integral part of biology at the molecular level. Orgel (1968) suggested that the process may be guided by a "principle of continuity which requires that each stage in evolution develops continuously from the previous one." More recently, the principle of continuity was phrased by Morowitz (1992, p. 27) as follows: "For any postulated stage in biogenesis there must be a continuous path backward to the prebiotic state of the Earth and forward to modern organisms. To introduce molecular structures or processes that are not subject to continuity is once again to violate Ockham's razor."

The "'continuity thesis' as a philosophical assumption that unites researchers of the origin of life" was discussed by Iris Fry (1995). Her conclusion was that "scientists quite often tend to deny any relevance of philosophical considerations to their specific work. Based on the analysis in this paper, the unique importance of philosophical assumptions and arguments in the origin-of-life field cannot be doubted. Philosophy, in this case, goes to the core—to the very 'right of existence' of this scientific endeavor."

One implication of this postulate is that contemporary cells should bear a signature of their evolutionary history because of preservation of the basic chemical features of the cell. Weiner and Maizels (1991) phrased an important aspect of this principle as follows: "Molecules can be frozen in time because the very complexity of the living process itself prevents their further evolution." Preserved molecules or segments of molecules are sometimes called molecular fossils; their function or structure "reflects an ancient origin, and gives us clues about the origin and history of life" (Weiner and Maizels, 1991). Such signatures, which may be considered a biological record, are related to biochemical composition, functions, and reaction pathways, as well as to cellular structures. Moreover, they can be used in the establishment of the order of appearance of various molecular and cellular features in the history of biology, namely, the phylogenetic tree.

In the spirit of Szent Gyorgyi, some people call the molecular evolutionists who discover molecular fossils "molecular archaeologists" (Soll, 1993).

Carl Woese (1991) phrased the evolutionary aspect of the biological record as follows: "Organisms, in a very important sense, are historical documents: their structure at all levels reflects (is determined by and records) their evolutionary history. Therefore, knowledge of evolution (and evolutionary relationships) is an integral part of explanation and understanding at virtually all levels of biology."

The evolutionary history of a certain organism cannot be obtained from the properties of that specific organism alone. It is the comparison of the biological records of an assemblage of many different organisms that enables one to estimate their evolutionary history. The latter feature of such an assemblage of organisms is their evolutionary relationships, as expressed in their phylogenetic tree (see Woese, 1980b).

Where should one search for the biological record?

> The methodology of cladistics [from *clade,* all of the species derived from a common ancestor. Cladistics is the recognition of phylogenetic relationships by means of shared derived properties] provides a powerful

> way of making explicit statements of phylogenetic relationship. . . . One of the central tenets of the method is that the evolutionary process produces a hierarchical, or nested, distribution of evolutionary innovations.
>
> Charles R. Marshall et al., "Dollo's law and the death and resurrection of genes"

The biological record has to be looked for at the molecular level of the genetic material of the cell, namely, the sequence of the mononucleotides in DNA or RNA molecules, or its translated form, the sequence of amino acids in protein; the two are related by the genetic code. The advent of the molecular biology era and the comparative sequencing of amino acids in homologous proteins (proteins of the same origin) have brought about a deeper understanding of the chemical basis of evolution. As sequences of more and more homologous proteins became available, their evolutionary implications were revealed. The technology of nucleic acid sequencing was developed more than a decade after the beginning of protein sequencing. Today, both kinds of sequences serve as the basis for the most recent phylogenetic systematics of all known creatures.

A determination of the phylogenetic relationships between different creatures is based on the comparative sequencing of defined segments of homologous biopolymers of either proteins or nucleic acids, or both, of the creatures under study. Due to the conservatism of biological systems, the specific arrangement of the building blocks of their blueprint reflects their phylogenetic relationships. The basic tenet of these relationships is that if homologous biopolymers of different creatures are found to have a similar sequence, that sequence is most likely to have been inherited from a common ancestor. The extent of similitude can be evaluated statistically, and various algorithms have been developed for this purpose.

An interesting comparison can be made between biology's back-extrapolation to very early evolutionary states and psychology's Freudian treatment. The latter "is a personal voyage in time, to our own past, as well as to the past of all humanity. He [Freud] believed that in Psychology, like in Physics and Biology, the back-exploration in time reveals the original source . . . from which both the personality and character of a human being have emerged" (Elitzur, 1994b, p. 140).

Back-extrapolation to the beginning of the phylogenetic tree of all extant creatures

Based on morphological features, the phylogenetic relationships of multicellular creatures were established by Haeckel in 1866 (chapter 4). Moreover, Darwin and Haeckel had already hypothesized that the seemingly simple unicellular organisms should be located at the beginning of the phylogenetic tree of all living forms. However, the smaller organisms—the bacteria—defied morphological systematics because they lacked the definitive features of whole organisms. And even though Haeckel's three-kingdom scheme of classical phylogeny was further refined and divided into five kingdoms by later generations of scientists, the scheme remained based essentially on the characteristics of whole organisms.

The use of cellular attributes in the classification of living entities was suggested in 1937 by Chatton (see Woese, 1987), who divided the living world into two main groups, namely, eukaryotes (organisms that possess a nucleus, i.e., plants, animals, fungi, and amoeba) and prokaryotes (organisms that lack a nucleus, including eubacteria, halobacteria, methanogens, and eocytes). The criterion that Chatton used—the true cellular nucleus, which has a membrane—reflects the technological progress of his time: electron microscopy, as well as various molecular characterizations, which made the distinction between the two groups possible.

The logical disadvantage of this classification has to do with the negative definition of the prokaryotes. Thus, whereas the presence of a nucleus in the eukaryotes is a *positive* feature of the entire monophyletic group (a group containing all descendants from a common ancestor, including that ancestor), the lack of a nucleus is a *negative* feature that cannot be used for the definition of a monophyletic group. Recognition of this logical disadvantage brought about attempts to improve the definition of prokaryotes by the use of positive cellular features. These were molecular properties, the analytical determination of which became possible as new technologies became available.

The idea of nucleic acids as information carriers of the cells became accepted soon after the discovery of the DNA structure by Watson and Crick in 1953. The recognition of species specificity of proteins dates back to the beginning of the twentieth century. And the understanding that proteins can be used for taxonomy (protein taxonomy) was expressed by Crick in 1958 (cited by Ambler and Daniel, 1991). The first phylogenetic tree based on a comparison of protein sequences, by Doolittle and Blomback, appeared in 1964 (cited by Ambler and Daniel, 1991).

Thus, the basic ideas and technologies to be used in phylogenetic investigations of the two central biopolymers—proteins and nucleic acids—were laid down during the 1950s and 1960s. The technology of protein sequencing was developed first, and that of nucleic acid sequencing became available some time later. One of the problems faced by phylogeneticists has thus become to select the appropriate molecule for sequence analysis. The second problem would be, of course, how to interpret the sequencing data.

Molecular clocks

One exciting theoretical possibility for interpreting the sequencing data has to do with the time scale of the evolutionary process. In 1965, in their classic paper entitled "Molecules as documents of evolutionary history," Zuckerkandl and Pauling suggested that if sequences evolved at a constant rate, they can be used as molecular clocks. It soon became obvious, however, that despite all the beauty of this suggestion, its application depends on the validity of the assumption of a constant evolutionary rate. Nevertheless, the idea was sown, and numerous attempts to use this approach in evolutionary studies have sprouted since. We shall come back to this problem in chapter 16.

Jukes (1994) noted that Pauling and Zuckerkandl's idea of measuring evolutionary time was not entirely new: "Actually, the idea was old and well established. The first molecular comparisons of hemoglobins were made in 1907 and 1909 by E. T. Reichert and A. P. Brown. They found, for example, that cats, dogs, a bear and a seal could be placed in four divergent groups on the basis of axial ratio in crystals of their reduced hemoglobins. Their findings were widely quoted in textbooks."

Juke's comment suggests the existence of a kind of "continuity" in the evolution of the idea of "molecular clocks," tempting one to make some generalizations in this regard, but does not belittle the ingenuity and beauty of Pauling and Zuckerkandl's idea.

The phylogenesis of extant living forms: the primary kingdoms of life

Ribosomal-RNA-based phylogeny

Pioneered by Woese (see Woese, 1987; Wheelis et al., 1992), ribosomal RNA (rRNA) has become the basis for sequence analysis data. The 16S (and 23S) rRNA molecule is found in all extant organisms, and its function through evolution seems to have remained constant. Moreover, it is encoded for by the genomes of all prokaryotes, eukaryotes, and their organelles. Thus, these molecules were already functioning before the three kingdoms of life diverged. These molecules are also not assumed to be likely candidates for horizontal gene transfer (see Doolittle et al., 1991; Lazcano et al., 1992).

In addition to the antiquity of the ribosome, certain domains of the 16S rRNA molecule, which contain about 1,500 nucleotides, are very conserved—that is, they have changed very little during billions of years of evolution. Other domains, interspersed between the conserved ones in the same molecule, are more variable. The conserved domains are useful in looking for distant phylogenetic relationships, whereas the less-conserved are valuable for close relationships. Indeed, the comparison of homologous sequences of organisms from different groups enabled Woese and his colleagues to establish the three primary kingdoms of life, which they called eubacteria, archaebacteria, and urkaryotes (from the German *ur,* "ancient, primitive").

Additional work by other researchers (see Doolittle et al., 1991; Garcia-Meza et al., 1994; Gupta and Golding, 1996) focused on other molecules with preserved domains. The comparative sequence analyses have supported Woese's main conclusion, also suggesting that archaebacteria and eukaryotes are closer phylogenetically than either of these kingdoms is to eubacteria. In 1990 Woese renamed the primary kingdoms ("domains") Eucarya, Archaea, and Bacteria (chapter 6). It should be noted that both the term "archaebacteria" and the more recent term "archaea" are misleading to some extent because they seem to imply that this domain is older than the other domains. This is probably one of the reasons that until now no agreed-upon nomenclature for these domains has been universally adopted by evolutionists (see also Doolittle, 1995). It should be noted that this issue is still controversial among scientists (see, for instance, Baldauf et al., 1996; Gupta and Golding, 1996; Saccone et al., 1995).

The theory pertaining to the phylogenetic relatedness between Archaea and Eucarya has gained much support recently by the sequencing of a very ancient protein (the so-called TATA-binding protein, TBP) by Rowlands and his colleagues (1994; see also Barinaga, 1994). Their conclusion was phrased as follows:

> Taken together with previous studies, our findings suggest that the transcriptional machineries of archaebacteria and eukaryotes are fundamentally homologous; they are also consistent with phylogenetic comparisons that place eukaryotes closer to archaebacteria than to eubacteria. . . .The eukaryotic-type transcriptional apparatus must therefore have already been established in the last common ancestor of eukaryotes and archaebacteria, before the emergence of nucleated cells.

The conservatism of biology, as reflected by the sequence analyses of the archaebacterial TBP, is demonstrated in its remarkable resemblance to the corresponding human protein. For instance, in certain sequences the similarities between eukaryotes and archaebacteria are in the order of 40% (Rowlands et al., 1994). Such similarity implies a common function for the corresponding proteins in eukaryotes and archaebacteria. Thus, it seems that these proteins in both archaebacteria and eukaryotes function in a similar way in the promotion of transcription. It should be noted that the two kingdoms are assumed to have diverged more than 3 billion years ago!

Symbiosis-based phylogeny

An approach differing from that of Woese was taken by Margulis (1996), who suggested symbiosis-based phylogeny. According to Margulis's school, "cells contain at least five types of nucleic acid (DNA, mRNA, tRNA, small and large subunit rRNA) and at least five hundred different proteins . . . no single one of which is usable alone as an adequate phylogenetic marker for history of the lineage." The phylogenetic tree controversy seems to be mainly irrelevant to our interest in the origin of life, since the beginning of the phylogenetic tree deals with a relatively advanced stage of life, not its beginning. However, the symbiosis process may be relevant to the search for the origin of life, as discussed later.

Beyond the most ancient extant organisms

The hypothetical last common ancestor

A number of methods have been used to construct phylogenetic trees from ribosomal sequences. The universality of the ribosome, including the genetic code and its machinery, suggests that all known living forms on Earth are derived from a more ancient creature called the last common ancestor.

The last common ancestor of any group of species is the most recent species from which all of them descend. The last universal ancestor is the hypothetical last common ancestor of all extant species (Castresana and Saraste, 1995).

The term "cenancestor" has been recently suggested, where the Greek prefix may be translated as both "last" and "common" (see Fitch and Upper, 1987). This creature is characterized by a protoribosome and a genetic code basically similar to the contemporary one. The rooted phylogenetic tree stems from the last common ancestor. Thus, the key to the understanding of the relationships between the three kingdoms is their last common ancestor (Woese, 1987). This hypothetical creature may be considered a single cell or a population of similar cells. Presumably it disappeared eons ago and was survived by its descendants, which were more adapted to the changing planet. The estimated age of the universal ancestor is discussed in part IV.

Note that the possibility that primitive creatures unknown to us, such as the hypothetical common ancestor, can prosper *now* in certain environmental niches of our planet has never been published. Darwin's "vision of the warm little pond" (chapter 5) is probably one of the reasons for our belief that primitiveness is always equated with low competitiveness. According to I. Friedman (personal communication), living entities at evolutionary stages similar to that of the legendary common ancestor may still be active on Earth. Being unable to compete with other, more evolutionaric-advanced living creatures in their environments, they may still occupy some of the harshest ecological niches, places in which no other living creature can exist. An interesting clue for this speculation may be found in Wächtershäuser's "pyrite-world" theory (chapters 17 and 21), with its primitive environment.

Until very recently, organisms belonging to the Archae domain (archaebacteria) were thought to be a significant part of the microbial population of only a few extreme ecological niches, characterized by features such as hypersaline, very hot, or strictly anoxic conditions. As a result of the introduction of novel methods in which molecules rather than cells are counted, DeLong et al. (1994) found that major components of the coastal Antarctic oceanic surface waters, picoplankton (cells smaller than 1 micrometer), are members of the Archae group. Thus, these pelagic Archaea represent a significant portion of the earth's biota.

The hypothetical progenote

The progenote was suggested by Woese and Fox (1977) to account for early living entities' having rudimentary relationships between their phenotype and genotype. According to Woese (1987), extant organisms characterized by precise links between phenotype and genotype may be called genotes. Prokaryotes, in which the information is stored in long, juxtaposed strings of genes, are called genomic organisms. Assuming continuity in evolution, organisms simpler than the genomic ones, in which the information was presumably arranged in linear, contiguous arrays, are considered and called genotes. The "progenote" was suggested by Woese (1987) as a theoretical, primitive entity that preceded the genote. Discussing this hypothetical creature, he suggested that the last common ancestor of the three kingdoms of life may still be a progenote (see also Doolittle, 1995).

The exact characterization of the hypothetical progenote was discussed by a number of scientists (Lazcano et al., 1992; Zillig et al., 1992), some of whom concluded that the progenote was the last common ancestor. According to de Duve (1991), the progenote was the first primordial protocell, characterized by features from the three kingdoms of life. Gordon (1995) made a distinction between the progenote and the last common ancestor, where the latter is presumably characterized by DNA replication and a genetic code.

Woese (1987) also mentioned higher, more complex organisms, for which he designated the name "supragenomic entities."

The Yellowstone hot springs are a source of thermophilic microorganisms that are considered among the most ancient of known creatures. In one of Yellowstone's hot pools, a new microorganism was recently discovered by Susan B. Barns and Norman Pace that is probably closer to the common ancestor than any other organism identified so far (Milstein, 1995).

Relevance of horizontal gene transfer, gene duplication, and intron-exon relationships

Horizontal gene transfer, gene duplication, and intron-exon relationships are central issues in the study of early stages of life. Their relevance to the origin of life, however, has hardly been addressed so far. It seems necessary to include a brief discussion of these features at the present stage in order to explore in part IV the possibility that their predecessors may be relevant to the search for the origin of life.

Horizontal gene transfer

The most famous example of the endosymbiosis theory (from the Greek *endon,* meaning "within"; *syn,* meaning "together"; and *bios,* meaning "life") deals with the origin of mitochondria and chloroplasts. The basic idea is rather old, dating back to 1883, in what seemed at the time wild speculation. It was Lynn Margulis, in 1970 (published as Lynn Sagan; see Margulis, 1981), who developed this speculation into a viable theory. According to this theory, the eukaryotes arose from the fusion of eubacteria with archaebacteria at a very early stage of the phylogenetic tree. Margulis later expanded this theory to include flagella as well (see de Duve, 1991).

Hartman (1984; see Hartman 1992a) suggested that the nucleus of the eukaryotic cell comes from a prokaryote endosymbiosis. More recently, Lake (1991), based on sequence analysis of rRNA, suggested a "sister relationship between the eukaryotic nucleus and eocytes" (eocytes are extreme prokaryotic thermophiles that live at temperatures higher than 85°C and metabolize sulfur). A number of possible instances of horizontal gene transfer between the kingdoms, beyond those involved in the acquisition of organelles, have been also suggested. For a recent review of this controversial area see Smith et al., 1992.

Gene duplication

Most evolutionists today agree that the genomes of modern eukaryotes descend from those of prokaryotes known as Archaea. The number of genes in all extant prokaryotes is on the order of 3,000 per organism, where the range is from about 500 to 5,000. The number of genes in complex multicellular eukaryotes range from 30,000 to 100,000. Moreover, it is also assumed that ancient prokaryotes had no more genes than their ancient predecessors. The question is, then, what is the mechanism governing the formation of this large number of genes (Doolittle, 1994)?

Rather than assuming that genes arise de novo, most researchers assume that this process, which took place more than 3.5 Ga ago, has not occurred since then (Doolittle, 1994). The primordial, restricted "prokaryotic gene set" gave rise, over the eons, to more and more complex organisms, including extant animals and plants, by duplication and divergence. According to this proposition, extra genes appeared occasionally by a chance mistake (Doolittle, 1994). These extra copies accumulated in cells, where they could have brought about, by chance, the formation of new proteins with new reactions. The subsequent evolutionary processes brought about new genes, which diverged to form new creatures.

An ingenious use of duplicated genes in the study of the phylogenetic tree was suggested by Schwartz and Dayhoff (1978) and later used simultaneously by Gogarten et al. (1989) and Iwabe et al. (1989). These researchers used gene duplication as a means of determining the root of the phylogenetic tree, and they came to the same conclusion, namely, that this root was located between Archaea and Eucarya on the one side, and Bacteria on the other side (Hilario and Gogarten, 1993). The duplication of the genes under consideration took place before the appearance of the last common ancestor.

Antiquity of the introns

There are two schools of thought regarding the antiquity of introns; "intron early" and "introns late." According to the first school, introns are ancient remnants from a very early stage of the molecular evolution era (Doolittle, 1994). According to the "intron late" school, the introns are invaders that were introduced into the eukaryotic genome more than 2 Ga ago (Doolittle, 1994).

This controversy is still going on, as can be seen, for instance, from the debate in

Science (Senapathy, 1995; Bertolaet et al., 1995; Stoltzfus et al., 1995). For further information see Shih et al., 1988; Hurst, 1994; Stoltzfus et al., 1994; Long et al., 1995; Gilbert et al., 1997).

Gene duplication, exon shuffling, and horizontal transfer are "evolutionary-modular mechanisms"

The mechanisms of gene duplication and exon shuffling increase the likelihood of formation of new functional genes. However, the repertoire of new sequences formed by duplication of entire genes or shuffling of preexisting exons seems to be limited (Golding et al., 1994). A recent example of exon shuffling—the combination of exons from different genes—as a molecular evolutionary mechanism for the formation of new genes with new functions—was reported by Long et al. (1995). Thus, these mechanisms introduce modularity—the use of prefabricated sequences—into evolutionary processes. Together with horizontal gene transfer, these mechanisms may be called modularity mechanisms in evolution.

The acquisition of "prefabricated" strands of informational molecules can be considered an ancient evolutionary feature applicable to the very early stages of the molecular evolution era, as discussed in chapter 23.

Speculative "biological worlds"

It can be assumed that besides the biology that happens to be ours, there are more, unexplored, possible kinds of life. Presumably, at each stage of their evolution, a different set of biopolymers (adapted to their environments) would be selected, giving rise to different "biological-like worlds." Based on our own biological world, the number of constituents needed for other biological-like worlds does not have to be large. Moreover, it is extremely unlikely that during the evolutionary process of extant living organisms, Nature explored all possible biopolymers and their suitability to function in evolving biological-like systems. Thus, it is conceivable that different kinds of biological-like worlds will be discovered in the future, both on other planets and in our laboratories. However, the principle of biological continuity is expected to characterize all evolutionary biological-like systems.

Clues to the study of life's origin

The biological record is rich in clues to early evolutionary stages of the phylogenetic tree. These provide a powerful tool for the establishment of phylogenetic relationships among all known organisms in their three kingdoms, as well as the hypothetical last common ancestor. However, even the latter living entity was far more advanced than those primordial entities that were involved in the transition from inanimate to animate. Thus, the search for the origin of life should be based on additional research methodologies and assumptions.

11

Biological Life

A Multitude of Points of View

> Instead of asking "What are the necessary properties of the components that make a living system possible?" we ask "What is the necessary and sufficient organization for a given system to be a living unity"?
>
> Francisco Varela et al., "Autopoiesis"

Problems and approaches

With so many features needed to characterize living creatures, it is not surprising that "life" is very hard to define. Indeed, it is practically impossible to agree upon the definition of life even among scientists of the same generation, As a matter of fact, the great majority of scientists prefer to use an operational description of what living entities can perform rather than an intrinsic definition that deals with what life really is (Wicken, 1987).

Two aspects of "life"

The scientific approaches to the characterization of life may be divided according to two aspects of life: One view focuses on the molecular level, whereas the second is a cell-centered view. Most of the characterizations and definitions in appendix A can be ascribed to the first point of view. Certain aspects of these are also discussed in relation to the hypotheses of the origin of life (part IV). The second view is represented by the notion of *autopoiesis* (a Greek word that means "self-producing"; see Fleischaker, 1994; Luisi, 1994; Mingers, 1995; Varela et al., 1974; Varela, 1994), according to which life cannot be characterized by means of just one component or attribute of its complex pattern.

Real (absolute) life criteria and potential life criteria

This approach was discussed by Ganti (1987), who differentiated between attributes that a living system must inherently possess (inherent unity, metabolism, inherent stability, information-carrying subsystem, and program control) and "potential life" criteria (growth-reproduction, capability of hereditary change and evolution, and mortality). The "potential life" seems to be similar or even identical with "the capacity to evolve." The problems involved in the use of the latter were discussed by Fleischaker (1990a) and Chyba and McDonald (1995).

Scientific definitions and characterizations of biological life

Is it possible to define "life"?

Obviously, one needs an operational definition to describe what life is and how can it be characterized and, at the same time, to have a historic description relating life as we know it to its history (Wicken, 1987, p. 31). Our inability to give a complete definition of life was discussed by Küppers (1990; see also Santoli, 1997). However, in spite of this conclusion, as well as similar conclusions reached earlier by other researchers, the scientific literature is abundant with attempts to define life (appendix A; see Chyba and McDonald, 1995, for comments on this feature of life). It is thus reasonable to resort to disciplinary points of view, such as physiological, metabolic, biochemical, genetic, and thermodynamic perspectives, in characterizing life (Emmeche, 1994).

Darwinian definitions imply origin of life

Interestingly, the great majority of the definitions and characterizations do not deal with the origin of life. The origin problem may be inferred by introducing evolution (Gatlin, 1972). The attribute of evolution has also been incorporated into the definition of life by Frank (cited by Ganti, 1987); its importance as a central feature of the origin of life was also adopted by the general working definition that was used by NASA (for a critical discussion see Fleischaker, 1994). Several important implications of the Darwinian definition were discussed by Chyba and McDonald (1995), who noted that this definition does not differentiate between biological life and computer life. Additional aspects of the definition of life are discussed by Chyba and McDonald (1995), Dyson (1985), Emmeche (1994), Kamminga (1988a, 1988b), and Küppers (1990).

Rhythm as a fundamental feature of life

Living organisms are also characterized by a rhythm. This feature may be considered an implied part of, for instance, metabolism or interaction with the environment. It should be distinguished as an independent feature, however, because it signifies the presence of a specific regulating system. Moreover, as will be discussed later (chapter 20), it is argued that this feature is probably as ancient as life itself.

Historical outlook

It is instructive to arrange scientific attempts to characterize and define life and its hypothetical origin in chronological order and thus to follow the development of the understanding of what life is over recent generations (appendix A).

The earliest characterization in appendix A was phrased just before Pasteur's famous experiment (1862) and the advent of Darwinism. It not only reflects the debate on the spontaneous generation theory but also shows the level of understanding of the graduality that characterizes the emergence of life as it is understood now, as well as the importance of the environment of primordial Earth in these processes. Additional definitions of the ninteenth century have been reviewed by Spencer (1884).

The understanding of the gradual transition from the inorganic to the organic and the application of the known laws of physics and chemistry to the problem of the origin of life had already started long before Darwin. In retrospect, these initiations reached their peak in the Oparin-Haldane paradigm, at the beginning of the twentieth century. It is interesting to note that our ability to explain the origin of life in scientific terms has always been challenged by some scientists. In appendix A they are represented by Virchow, Liebig, Woodger, and Bohr. Two prominent scientists of our time who denied the possibility of explaining the origin of life scientifically are Monod and Polanyi (for more information see de Duve, 1991; Küppers, 1990; Morowitz, 1992).

Many authors begin their discussion of life with the description ascribed to Oparin, namely, life is characterized by self-replication, metabolism and mutability. The term "self-replication" was recently discussed by Fleischaker (1994) and Luisi (1994). An exchange of matter and energy with the environment (metabolism) implies insulation, which is universally obtained in contemporary organisms (with the exception of viruses) by a lipid membrane. Catalysis and autocatalysis were already visualized as essential criteria of life by Alexander (1948, p. 88), whereas Perrett (1952) stressed the stepwise characteristics of biochemical catalyzed reactions. The inherent importance of the environment was stressed by Kahane (see Ganti, 1987, p. 32).

The understanding of the importance of complexity was stressed, for instance, by Bertalanffy (1933, p. 49) and Perrett (1952), as well as by Eigen and Winkler-Oswatitsch (1981a, 1981b) and Morowitz (1992). Schuster (1984) focused on the simultaneous existence of many features that characterize life, whereas Haukioja (1982), extending the original idea of von Neumann (discussed later), worked out the concept of the automaton. The aspect of cybernetics was stressed by Frank (cited by Ganti, 1987).

Information as a fundamental tenet of living systems was stressed by Gatlin (1972), Fong (1973), Yockey (1973, 1992), and Elitzur (1994a). It should be noted that when information processing is introduced as a criterion of life, crystals are excluded from the domain of living entities.

Wächtershäuser (1994a) denies the need for the concept of "information" (chapter 21). Template-directed synthesis as the mechanism of biological information transfer was treated by Katz (1986, p. 84).

Teleonomy as a most general feature of animate matter was stressed by Lifson (1987) and Mayr (1988).

Communication as a basic attribute of the living state at all its levels was suggested by De Loof (1993). According to him, life is defined as the integral of population, communication, and cell organelles.

Processing information about the environment by living entities, feedback, and autonomy were discussed by Elitzur (1994a). Information processing (signal transduction) was also stressed by Hucho and Buchner (1997), whereas Baltscheffsky (1997) focuses on the dynamics of energy, matter, and information. Root-Bernstein and Dillon (1997) focus on complementarity, which is related to information transfer as well as to other interactions. Matsuno (1984) discusses the graduality of the transition from inanimate to animate, thus focusing on a central aspect of the origin of life.

In addition to all of these characteristics, it should be recalled that the attribute of molecular homochirality (enantiomeric homogeneity) of biological amino acids (with the exception of glycine) and sugars has been known since Pasteur's classical experiments. The scientific debate regarding the possible mechanisms of the emergence of this feature of biology is still continuing and will be discussed further later.

Finally, it should be recognized that there are two approaches to defining life: One focuses on the properties of the individual organisms, whereas the other is much more global and ecological in character (Morowitz, 1992). The first approach, already applied by Aristotle, encompasses all of the characterizations listed in appendix A. The second one was discussed by Morowitz (1992); these ideas are the essence of the Gaia hypothesis suggested by James Lovelock in 1988 (chapter 13).

The main attributes of living entities

We may proceed now to attempt to answer the question "What is life?" remembering that unlike in the cases of succinct physical laws, no equivalent laconic definition of life or living entities has ever been proposed. In the absence of a concise definition of that kind, it is necessary to use a list of features that cover the most important attributes of life (see de Duve, 1991; Emmeche, 1994).

We are asking, "What are the minimum number of elements that we have to add to the non-living physical systems to have the minimum living system?" (Goddard, quoted in Gerard, 1958, p. 133).

Using the first generalization, as phrased by Morowitz (1992), we can summarize the main attributes of living entities as follows:

Living entities are complex, far-from-equilibrium structures maintained by the flow of energy from sources to sinks. They are compartmentalized, organic, homochiral entities, closely associated and communicating with their environment (including other living forms) and at the same time separated from it by a boundary (in extant organisms, a lipid bilayer), and dependent in their activities on a continual flux of energy and matter through this membrane, from their environment. They can replicate, mutate, exchange matter and energy with their environment, and evolve, in processes that are catalyzed by a large arsenal of organic catalysts. The characteristics of most or all of these processes and molecules, as reflected by their chemical cycles, regulation, communication, complementarity, and rhythms, as well as potential life criteria of each organism, corroborate with the principle of continuity. Having evolved from inanimate matter, they constitute autocatalytic, evolvable, teleonomic organic systems that can transfer, store, and process information, based on template- and sequence-directed reactions, all of which characterize autopoietic entities.

The scientific picture is not the only point of view

There exist fundamental differences among various disciplines of human mind activity, such as science, philosophy, medicine, art, sociology, politics, and theology, regarding what life is and how it is related to other biological and cosmological phenomena. Each of the viewpoints of these disciplines not only may be used as a starting point for a discussion on the nature of life but also has an impact on the other disciplines. In particular, a change in the scientific view regarding the origin and nature of life is a challenge to all other disciplines, which tend to adapt to it. This has not always been the case, as evidenced from the history of the search into the origin of life.

Examples of philosophical points of view

> Since a man must needs live before he can be a philosopher, no problem of philosophy is more fundamental than the nature of life.
>
> Lancelot Hogben, *The nature of living matter*

Organisms

An example of a more philosophically oriented approach to the basic attributes of life is given in the following quotation from a paper entitled "The myth of the putative 'organism'" (Fleischaker, 1991, pp. 23–24): "The 'organism' is a fundamental and essential myth of Darwinian biology. In contemporary biology, the 'organism' is conventionally taken as a single discrete and autonomous individual—an observable, genetically-determined entity enclosed within a continuous structural boundary." According to this point of view, an organism is defined in terms of its topological features. The inadequacy of such a definition becomes obvious when the microbiota that is essential to many animals are considered. Obviously, they are not included in the definition. Thus, "this traditional organismic notion is totally inadequate to understand the very real complexity of living systems. In the non-traditional view . . . 'life' appears and persists not as a sum of multiple discrete entities but as a single *ecology*" (Fleischauer, 1991).

Teleology and teleonomy

The term "teleology" has a very long history in philosophy. Teleology is the study and explanation of things, processes, or events in terms of their purposes (for a discussion of this term see Sattler, 1986). A common example of the use of teleology is Aristotle's teachings. It should be noted that Mayr (1988, p. 47) does not accept this interpretation of Aristotle's teleology but instead stresses that Aristotle's end-directed processes are very similar to the modern approach suggested by Pittendrigh (discussed later) and thus may be called "teleonomic."

Implied in the term "teleology" is the existence of a designer, or God, that directs processes and changes. With the advent of modern scientific thinking, teleology had to be discarded, to remain only in the realm of religion. Indeed, it is a very significant merit of Darwin's evolution theory that the teleological aspect of pre-Darwinian biology could be replaced so nicely by a purposeful-like explanation. The term "teleonomy" was coined by Pittendrigh in 1958 (see Sattler, 1986). It is a directed and functional behavioral characteristic of living organisms and life processes (for the equivalent term "teleomatic" and more detailed discussion, see Sattler, 1986, and Mayr, 1988, p. 47). An example of the use of teleonomy in the characterization of life is Lifson's (1987) definition (appendix A). As Julian Huxley pointed out,

> It was one of the great merits of Darwin himself to show that the purposiveness of organic structure and function was apparent only. The teleology of adaptation is a pseudo-teleology, capable of being accounted for on good mechanistic principles, without the intervention of purpose, conscious or sub-

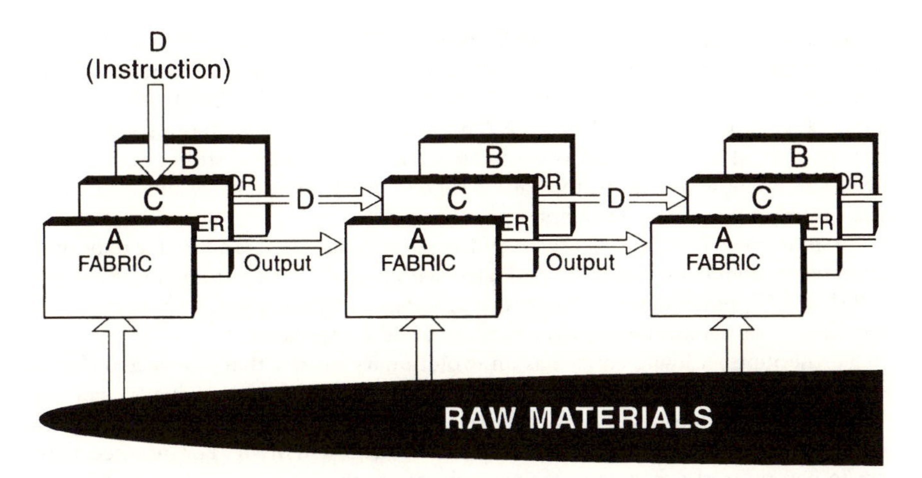

Fig. 11.1. A scheme of von Neumann's (1948) automaton and its self-reproduction (from left to right). (Adapted from Emmeche, 1994)

> conscious, either on the part of the organism or of any outside power. (quoted by Girvetz et al., 1966, p. 324)

For more about the philosophical point of view, see Ruse, 1997.

Example of a mathematician's point of view: The automaton

The first and most popular mathematical description of life is that of John von Neumann, which was presented in 1948 (Dyson, 1985; Emmeche, 1991; fig. 11.1). The analogy between living organisms and mechanical automata (for more information see Rosen, 1989) was an outgrowth of von Neumann's insight into electronic computers. The automaton he suggested has two essential components, which have their analogues in living cells, namely, hardware and software for processing information. In living cells, proteins are the hardware and nucleic acids are the software. Proteins are the essential components of metabolism, whereas nucleic acids embody information. The model describes a self-reproducing automaton moving around in an environment containing raw materials and expresses the logical relationships between the four components, designated *A, B, C,* and *D*.

A is a kind of "factory" that is given a description βx of the desired output x. Based on this description, it selects the appropriate available raw materials in its environment to manufacture the specified product x.

B is a duplicator, a copying automaton, which is given the description β as input and produces a copy of it, β', as output.

C is a controller. When given an instruction $\beta(x)$, it passes the information to *B* for duplication. The first copy it then passes to A, which carries out the construction. The remaining copy is attached to the output x of *A* and is then released as $(x + \beta(x))$ from the machine $A + B + C$.

D is a description that enables A to manufacture the combined system $A + B + C$. *D* is thus the machine's self-description.

The biological functions of the four components of von Neumann's automaton could be fully identified as a result of Watson and Crick's discovery of the DNA structure and the establishment of the discipline of molecular biology. The identification is as follows (Hartman et al., 1985): "D is the genetic material, DNA; A is the ribosomes; B is the enzymes RNA and DNA polymerase; and C is the repressor and derepressor control molecules and other items whose functioning is still imperfectly understood. So far as we know, the basic design of every micro-organism larger than a virus is precisely as Von Neumann said it should be."

In spite of von Neumann's elegant and attractive analysis, it is advisable to be more attentive to arguments raised by biologists with regard to the automaton. Emmeche (1991, p. 253) noted that von Neumann's analogy is not complete, since in a real system there is no separation between functions and components.

Furthermore, a living entity has an evolutionary history that is reflected, for instance, in the principle of continuity. Thus, it is not clear to what extent von Neumann's automaton, which was conceived without any evolutionary or historical considerations, can be of help in the search into the origin of life. For instance, in order to pertain to the problems of the origin of life the "instruction" should be "invented" in the first place. Without expanding on this issue, suffice it to say at present that attempts to mathematize life and its evolution will probably continue to be controversial in the foreseeable future.

Are viruses living entities?

At this point it is necessary to mention the viruses, which are biomolecular complexes capable of replicating in appropriate host cells. If one adopts Joyce's definition (1994; appendix A), then viruses are excluded from the category of life since they are not self-sustained. Some scientists consider them as entities standing at the borderline between inanimate and animate matter. Others consider viruses as latecomers to biology, with possible originations many times during evolution (Eigen, 1992; Margulis, 1993b, p. 36), or as mere remnants of cells (de Duve, 1995b, p. 81).

This issue is still controversial, and the problem of their origin cannot be answered at the moment. For instance, comparative sequencing of certain DNA domains of prokaryotes, eukaryotes, and viruses has been interpreted recently by Forterre (1992) as an indication that the origin of viruses may have preceded the last common ancestor. Moreover, using a different approach that combines the RNA world and the hypercycle (chapter 17), Nemoto and Husimi (1995) conclude that the early appearance of viruses in evolution cannot be discarded. The possible role of viruses in evolution was suggested also by Weiner and Maizels (1991) and by Wächtershäuser (1992a, 1994b; chapter 21).

Clues to the study of life's origin

Living entities are characterized by clues to their evolutionary history. In contrast, the great majority of the attributes of biological life discussed in the present chapter are directly related neither to early stages of life nor to the origin of life. However, once an origin-of-life scenario is suggested and tested, such a discussion may be useful in order to analyze the relationships between the entities under consideration (synthetic, biological, or otherwise) and "life." Moreover, such a discussion is necessary in any attempt to study forms of life based on another chemistry and unexpected—that is, extraterrestrial—environmental conditions.

Appendix A: Definitions and characterizations of life

1855

L. Buchner: "Spontaneous generation exists, and higher forms have gradually and slowly become developed from previously existing lower forms, always determined by the state of the earth, but without immediate influence of a higher power."

1855

R. Virchow: "Life will always remain something apart, even if we should find out that it is mechanically aroused and propagated down to the minutest detail."

1866

E. Haeckel: "Any detailed hypothesis whatever concerning the origin of life must, as yet, be considered worthless, because, up till now, we have not satisfactory information concerning the extremely peculiar conditions which prevailed on the surface of the earth at the time when the first organisms developed."

1868

T. H. Huxley: "The vital forces are molecular forces."

1868

J. von Liebig: "We may only assume that life is just as old and just as eternal as matter itself. . . . Why should not organic life be thought of as present from the very beginning just as much as carbon and its compounds, or as the whole of uncreatable and indestructible matter in general."

1869

J. Browning: "There is no boundary line between organic and inorganic substances. . . . Reasoning by analogy, I believe that we shall before long find it an equally difficult task to draw a distinction between the lowest forms of living matter and dead matter."

1871

L. S. Beale: "Life is a power, force, or property of a special and peculiar kind, temporarily influencing matter and its ordinary force, but entirely different from, and in no way correlated with, any of these."

1872

H. C. Bastian: "Living things are peculiar aggregates of ordinary matter and of ordinary force which in their separate states do not possess the aggregates of qualities known as life."

1878a

C. Bernard: "Life is neither a principle nor a resultant. It is not a principle because this principle, in some way dormant or expectant, would be incapable of acting by itself. Life is not a resultant either, because the physicochemical conditions that govern its manifestation can not give it any direction or any definite form. . . .

None of these two factors, neither the directing principle of the phenomena nor the ensemble of the material conditions for its manifestation, can alone explain life. Their union is necessary. In consequence, life is to us a conflict."

1878b

C. Bernard: "If I had to define life in a single phrase . . . I should say: Life is creation."

ca. 1880
F. Engles (following Hegel, 1842): "No physiology is held to be scientific if it does not consider death an essential factor of life. . . . Life means dying."

1884
H. Spencer: "The broadest and most complete definition of life will be 'the continuous adjustment of internal relations to external relations.'"

1897
W. Pfeffer: "Even the best chemical knowledge of the bodies occurring in the protoplasm no more suffices for the explanation and understanding of the vital processes, than the most complete chemical knowledge of coal and iron suffices for the understanding of a steam engine."

1908
A. B. Macallum: "When we seek to explain the origin of life, we do not require to postulate a highly complex organism . . . as being the primal parent of all, but rather one which consists of a few molecules only and of such a size that it is beyond the limit of vision with the highest powers of the microscope."

1923
A. Putter: "It is the particular manner of composition of the materials and processes, their spatial and temporal organization which constitute what we call life."

1924
A. I. Oparin (quoted in Bernal, 1967): "Life may be recognized only in bodies which have particular special characteristics. These characteristics are peculiar to living things and are not seen in the world of the dead. What are these characteristics? In the first place there is a definite structure or organization. Then there is the ability of the organisms to metabolize, to reproduce other like themselves and also their response to stimulation."

1929
J. H. Woodger: "It does not seem necessary to stop at the word 'life' because this term can be eliminated from the scientific vocabulary since it is an indefinable abstraction and we can get along perfectly well with 'living organisms' which is an entity which can be speculatively demonstrated."

1933
L. Bertalanffy: "A living organism is a system organized in hierarchial order . . . of a great number of different parts, in which a great number of processes are so disposed that by means of their mutual relations within wide limits with constant change of the materials and energies constituting the system and also in spite of disturbances conditioned by external influences, the system is generated or remains in the state characteristic of it, or these processes lead to the production of similar systems."

1933
N. Bohr: "The existence of life must be considered as an elementary fact that cannot be explained, but must be taken as a starting point in biology, in a similar way as the quantum of action, which appears as an irrational element from the point of view of classical mechanical physics, taken together with the existence of elementary particles, forms the foundation of atomic physics."

1944

E. Schrödinger: "Life seems to be orderly and lawful behavior of matter, not based exclusively on its tendency to go over from order to disorder, but based partly on existing order that is kept up."

1948

J. Alexander: "The essential criteria of life are twofold: (1) the ability to direct chemical change by catalysis; (2) the ability to reproduce by autocatalysis. The ability to undergo heritable catalysis changes is general, and is essential where there is competition between different types of living things, as has been the case in the evolution of plants and animals."

1952

J. Perrett: "Life is potentially self-perpetuating open system of linked organic reactions, catalyzed stepwise and almost isothermally by complex and specific organic catalysts which are themselves produced by the system."

1956

R. D. Hotchkiss: "Life is the repetitive production of ordered heterogeneity."

1959

N. H. Horowitz: "I suggest that these three properties—mutability, self-duplication and heterocatalysis—comprise a necessary and sufficient definition of living matter."

1965

J. D. Bernal: "All biochemical and biophysical studies lead straight back to the general question of origins. Origin, structure, and function can no longer be separated."

1967

J. D. Bernal: "Life is a partial, continuous, progressive, multiform and conditionally interactive, self-realization of the potentialities of atomic electron state."

1972

L. Gatlin: "Structural hierarchy of functioning units that has acquired through evolution the ability to store and process the information necessary for its own reproduction."

1973

P. Fong: "Life is made of three basic elements: matter, energy and information . . . . Any element in life that is not matter and energy can be reduced to information."

1973

J. P. Yockey: "Life . . . seems to flout the second law of thermodynamics. Biological organisms seem to be something more than chemical systems yet at the same time *measurements* on these systems reveal that natural laws are obeyed. The fact that information is at the same time a quantity which can be defined mathematically and operationally and yet exists in living matter but not non-living matter may perhaps contribute to a resolution of this paradox."

1975

J. Maynard Smith: "We regard as alive any population of entities which has the properties of multiplication, heredity and variation."

1977
E. Argyle: "Life on earth today is a highly degenerate process in that there are millions of different gene strings (species) that spell the one word 'life.'"

1979
C. E. Folsome (Onsager-Morowitz): "Life is that property of matter that results in the coupled cycling of bioelements in aqueous solution, ultimately driven by radiant energy to attain maximum complexity."

1981
E. H. Mercer: "The sole distinguishing feature, and therefore the defining characteristic, of a living organism is that it is the transient material support of an organization with the property of survival."

1981a
M. Eigen and R. Winkler-Oswatisch: "The most conspicuous attribute of biological organization is its complexity. . . . The physical problem of the origin of life can be reduced to the question: 'Is there a mechanism of which complexity can be generated in a regular, reproducible way?'"

1982
E. Haukioja: "A living organism is defined as an open system which is able to fulfill the condition: it is able to maintain itself as an automaton. . . . the long-term functioning of automata is possible only if there exists an organization building new automata. An automaton may serve as such an organization."

1984
P. Schuster: "The uniqueness of life seemingly cannot be traced down to a single feature which is missing in the non-living world. It is the simultaneous presence of all the characteristic properties . . . and eventually many more, what makes the essence of a biological system."

1985
V. Csanyi and G. Kampis: "It is suggested that replication—a copying process achieved by a special network of inter-relatedness of components and component-producing processes that produces the same network as that which produces them—characterizes the living organism."

1986
N. H. Horowitz: "Life is synonymous with the possession of genetic properties. Any system with the capacity to mutate freely and to reproduce its mutation must almost inevitably evolve in directions that will ensure its preservation. Given sufficient time, the system will acquire the complexity, variety and purposefulness that we recognize as 'alive.'"

1986
M. J. Katz: "Life is characterized by maximally-complex determinate patterns, patterns requiring maximal determinate for their assembly. . . . Biological templets are determinant templets, and the uniquely biological templets have stability, coherence, and permanence. . . . Stable templets-reproducibility was the great leap, for life is matter that learned to recreate faithfully what are in all other respects random patterns."

1986
R. Sattler: "Living system = an open system that is self-replicating, self-regulating, and feeds on energy from environment."

1987

S. Lifson: "Just as wave-particle duality signifies microscopic systems, irreversibility and trend toward equilibrium is characteristic of thermodynamic systems, space-symmetry groups are typical for crystals, so do organization and telemony signify animate matter. Animate, and only animate matter can be said to be organized, meaning that it is a system made of elements, each one having a function to fulfill as a necessary contribution to the functioning of the system as a whole."

1993

S. A. Kauffman: "Life is an expected, collectively self-organized property of catalytic polymers."

1993

A. de Loof: "Life as the ability to communicate."

1994

NASA's definition: "Life is a self-sustained chemical system capable of undergoing Darwinian Evolution" (Joyce, 1994, p. xi).

1994

Varela et al.: "An autopoietic system is organized (defined as unity) as a network of processes of production (synthesis and destruction) of components such that: i) continuously regenerate and realize the network that produces them, and ii) constitute the system as a distinguishable unity in the domain in which they exist."

1997

F. Hucho and K. Buchner: "*Signal transduction* is as fundamental a feature of life as metabolism and self-replication."

1997

H. Baltscheffsky: "Life may well be described as 'a flow of energy, matter and information.'"

1997

R. S. Root-Bernstein and P. F. Dillon: "We propose that living organisms are systems characterized by being highly integrated through the process of organization driven by molecular (and higher levels of) complementarity."

Part III

THE ARENA

There are messages being deciphered that tell of the existence of living material in places as inhospitable as the arctic and antarctic as well as in geothermal vents. Were the prebiotic conditions for life to have evolved much more exotic than we had pictured? Planetary scientists, geochemists, astronomers and astrophysicists are searching for new answers.

Mayo J.J.M. Greenberg, preface to
The chemistry of life's origins

12

Our Universe, Galaxy, and Solar System

> Copernicus, in his heliocentric theory, transferred the frame of reference of motion from the Earth to the background of the so-called "fixed stars." As the Copernican revolution worked its way through European science, the conviction arose that the Sun was but one among myriads of similar luminous objects, the stars. If the Sun has planets, must not the other stars also have planets?
>
> Michael W. Ovenden, "Of stars, planets, and life"

Physical and chemical processes relevant to the formation of the biogenic elements in the universe have been taking place ever since the big bang, some 15 Ga ago. Thus, the conditions needed for life to originate, either in the form we know on Earth or in an as yet unknown form, may have been appropriate even before the formation of the solar system. As a matter of fact, the biogenic elements hydrogen, carbon, oxygen, nitrogen, and sulfur are among the most abundant in the universe. Phosphorus is an exception among the bioelements in that it is considerably less abundant than the former elements (Irvine, 1992). As noted by Irvine (1992), it may be relevant to the origin of life that

> in the gas of solar or "cosmic" composition, these "biogenic" elements will exist predominantly in the gas phase or as volatile ices such as H_2O, simple organics, N_2, or NH_3, etc. In contrast, elements such as silicon and iron are apparently locked up in the stellar medium in refractory grains and in a cooling solar nebula in non-volatile rocky material.

Our galaxy

The Milky Way is just one out of about 400 billion galaxies that populate our universe. It is a disc-shaped body, in which we are located about 30,000 light years from the center. The visible part of our galaxy, both in the optical and radio portion of the electromagnetic spectrum, is about 100,000 light years across. If one includes the dark matter, which is probably a dominant component of the galaxy halo, the diameter is more than 300,000 light years. Our galaxy is populated by about 150 billion stars; the sun, which was initiated by the collapse of an interstellar medium, is just one of them.

The interstellar medium, in its gas and solid phases, contains a large variety of molecules. The gaseous compounds include hydrides, oxides, sulfides, organic molecules, ions, and radicals. The interstellar dust grains are quite heterogeneous in structure. They contain a refractory core that consists of amorphous silica and organic material, as well as ice.

Formation of the solar system

At a speed of about 1 million kilometers per hour, our sun completes one circuit every 250 million years as it moves in its orbit. This orbit, determined by centrifugal and gravitational forces, has hardly changed since its formation, some 4.6 billion years ago.

The solar system was formed from a huge, rotating nebula made of gas and dust. In this state, the nebula started to contract under the influence of its own gravitational force. Most of the matter of the nebula concentrated in its center, evolving slowly into a glowing proto-sun. Burning hydrogen ignited a process that has been destined to proceed for billions of years; a tiny fraction of the electromagnetic radiation emitted in this process is destined to serve as an essential source of energy for the biological evolution yet to be invented on planet Earth.

Some of the remaining matter in the nebula underwent a slow evolutionary process as a result of collisions, where the matter coalesced into planetary bodies of various sizes (fig. 12.1). Coalescence of nebular gas in those regions far from the sun (at distances of Jupiter's orbit and farther away) resulted in the formation of the Jovian planets. The remaining uncondensed gas and dust, hovering in the evolving nebula, was driven off into interstellar space by the strong wind emanating from the young sun as it evolved through its young ("T-Tauri") stage (Kutter, 1987).

The evolution of the solar system continued; collisions among planetesimals resulted in accretion into bigger and bigger bodies. The presently known planetary system is thus the result of the evolution of the solar system. The planets were formed from a multitude of planetesimals and dust during a process that took between about 30 and 90 million years, a short prelude to a long history of 4.6 billion years, plus billions of years to come. The collision rate, which was initially rather intense, slowly subsided, allowing a new era—the era of life—to begin.

The solar planetary system

The heavenly bodies of the planetary system orbiting the sun range from dust and pebble-sized objects to planets with radii of tens of thousands of kilometers (Gaffey, 1997). No less importantly, at least some of these heavenly bodies contain organic compounds; this has important implications on the origin of life, as we shall see later. But first, let us briefly discuss these heavenly bodies, which may be divided, somewhat arbitrarily, into three groups (see Kutter, 1987), as follows:

Small-sized planetary bodies

These include asteroids, meteorites, comets, and many of the satellites of the planets, as well as particles composing the rings of Jupiter, Saturn, Uranus, and Neptune (fig.

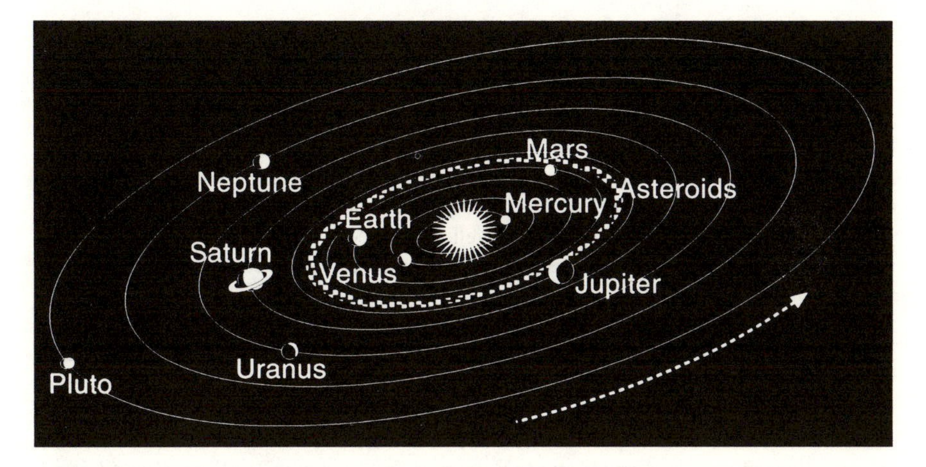

Fig. 12.1. The solar system.

12.1). The objects included in this group are called planetesimals; their radii are up to about 1,000 kilometers, and their masses are up to approximately 1/1,000 of the earth's mass. The composition of these objects varies rather systematically with their distance from the sun, and, according to recent theories, they share a common history: They are hypothesized to be the leftover debris from the pristine material out of which the planets and their large satellites were accreted.

Asteroids

Despite their names, asteroids (from the Greek *aster,* meaning "star," and *eidos,* meaning "form") are not stars at all. The great majority of them are rocky objects, and except for the largest, they are irregular heavenly bodies scarred by collisions they have suffered during their history. Some of them are thought to be the broken-up pieces of larger bodies. According to current theories, asteroids are the debris left-over from the time of the solar system's formation. Most asteroids are located in the Asteroid Belt (fig. 12.1), at an average distance from the sun that is between that of Mars and Jupiter. Some asteroids have different elliptical orbits, extending inward past the orbits of Mars and Earth.

Scientists are still discussing the need to prepare the technological means for the deflection of asteroids that endanger our planet. It is estimated that there are about 1,000 asteroids, each with a diameter of >1.5 km, which may come dangerously close to Earth's orbit during the next decades. A collision of one of these asteroids with our planet could be disastrous to all humankind. According to Sagan and Ostro (1994), "development of this asteroid-deflection technology would be premature. Given twentieth-century history and present global politics, it is hard to imagine guarantees against eventual misuse of an asteroid-deflection system commensurate with the dangers such a system poses."

Still, the idea that mankind on planet Earth is at risk from an asteroid has prompted a number of scientists to suggest the technology needed for either the diversion of such an asteroid from its trajectory or its demolition (Hill, 1995; Morrison, 1997).

Meteorites

Meteorites (from the Greek *meteoros,* meaning "lofty") are asteroids, fragments of asteroids, or fragments of comets, which collide with Earth as it crosses their orbits. It is estimated that every year the earth sweeps up about 10,000 tons of meteorites. Most of these are small enough to be harmless. Some meteors arrive in groups of thousands, forming spectacular showers. Occasionally, meteorites of several kilograms enter the earth's atmosphere, traveling at speeds of up to 10 km per second, and form a bright path before they either hit the earth's surface or break up to fragments.

The earth's geological record shows that during its history, it has been hit by large meteorites weighing thousands of tons or even much more; these meteroites leave huge craters at the impact site.

On the morning of June 30, 1908, an enormous impact hit a forested region of Tunguska, Siberia. The explosion was felt at a distance of 750 kilometers, and the energy released scorched and leveled off most of the trees in a 30-kilometer-wide area. Fortunately, the impact area was not inhabited and no casualties were recorded. Recent calculations by Lyne and Tauber (1995) suggest that the explosion was caused by a carbonaceous chondrite 50–100 m in diameter found to have airburst in the 6–10 km range above the ground.

A much larger heavenly object hit the Yucatan peninsula some 65 million years ago and has been hypothesized to have triggered the mass extinction that wiped out the dinosaurs, among many other species. As a matter of fact, "hordes of asteroids pass close to our planet, some closer than the Moon itself. But, fortunately, most asteroids reside farther from the Sun, in a belt between Mars and Jupiter" (Croswell, 1994).

Meteorites that have survived the passage from interplanetary space through the earth's atmosphere and have been collected are named after the place where they hit the ground. Chemical analyses of such meteorites were already being carried out in the nineteenth century by famous chemists such as Berzelius and Wohler. Based on more recent data, it is now known that the ages of meteorites fall in the range of 4.5 Ga. Obviously, they provide clues to the time of formation of the solar system. In fact, they are the only direct physical evidence of events that can otherwise be studied and modeled only theoretically.

Meteorites are divided into groups according to their composition and history. Being samples of asteroids and comets, they are considered representatives of the pristine materials from which the planets were made. This is especially true for the primitive group, which has changed very little since the formation of the solar system. Even more directly relevant to the origin of life is the carbonaceous chondrite (from the Greek *chondrion,* meaning "small grain") group, which is typified by up to 5% organic compounds. This will be discussed later.

The organic matter in carbonaceous meteorites occurs in various forms. Recent isotopic studies suggest that this organic matter has more than one source and/or more than one production mechanism. Of special importance in this regard is the Murchison meteorite, which was found in Murchison, Australia, in 1969. The detailed study of this meteorite has been the main source of our understanding of the properties of the organic matter in meteorites (Cronin and Chang, 1993).

Several amino acids were found in the Murchison meteorites, the most important of which were glycine, alanine, valine, proline, and glutamic acid. The chirality of these compounds (except glycine, which is a symmetric molecule) was also determined. The D and L forms were found in early analyses in nearly equal proportions. This finding, however, cannot be used simply as an indication that the amino acids in the meteorite are of nonbiogenic origin. It results from the fact that the optical isomers of amino acids racemize at a rate that is a function of temperature. Since the detailed thermal history of the meteorites is not known, the possibility that those amino acids were once in the pure isomeric form cannot be ruled out at present. More recently, evidence for nonracemic meteroritic amino acids was observed by Cronin and Pizzarello (1997) and Engel and Macko (1997).

In summary, as Cronin and Chang point out (1993, p. 246), "It is now clear that the primitive earth accumulated meteoritic matter ranging in size from dust to giant impactors."

Meteorites from Mars

Several rocks found in the Antarctic have been recently identified as meteorites from Mars (McKay et al., 1996). Apparently, they were ejected from Mars as a result of a collision of a heavenly body. Microscopic and chemical analyses of these rocks have been interpreted as an indication that they contain forms morphologically similar to extant blue-algae cells. Some of the implications of these findings are discussed in chapter 15.

Comets

Comets (from the Greek, *kometes,* meaning "long-haired") are primordial bodies consisting of ice, dust particles, and rocks; they populate a remote zone in the outskirts of the solar system. They are hypothesized to have been condensed and accreted in the outer regions of the primitive solar nebula, though the relationships between their components and interstellar matter have not been resolved so far. The name of the zone where the comets come from is the Oort Cloud, after the astronomer Jan Oort, who in 1950 suggested its existence to account for the frequency of appearance of comets in the inner part of the solar system. Astronomers have hypothesized that the deflection of Oort Cloud comets toward the inner part of the solar system may be the result of a very faint companion star of the sun, characterized by an orbital period of 26 million years. When the comets are close enough to the sun, "their ices start vaporising, and gas and dust particles stream outward to form an envelope-the coma and hydrogen cloud-around the central body, the nucleus. . . . Pressure exerted by the solar radiation and solar wind drives the gases and dust many millions of kilometers through the interplanetary space to form the cometary tail" (Kutter, 1987).

This hypothesized companion star of the sun was invoked to explain the seemingly regular geological record of mass extinctions of life during the past 250 million years. Accumulating evidence regarding past mass extinctions was hypothesized to be the result of impacts with heavenly bodies such as comets with our planet. These bolides, which collided with Earth, were speculated to have been deflected from their orbits because of the putative companion star. Though this subject is still controversial among scientists, the mysterious star has been named Nemesis (from the Greek *nemein,* meaning "to distribute, deal out"; the name of the Greek goddess of retribution). (See Steel, 1997.)

The chemical composition of comets is of special interest in the context of the origin of life, since impacts with Earth entail import from space of various compounds, including various organic molecules and water. This will be further discussed later.

The best-known comet is Halley's Comet, named after the astronomer Edmond Halley (1656–1742), which completes one orbit every 75 to 80 years. During its approach to the sun it is seen with the unaided eye and has therefore been known for many generations. According to the present hypothesis, comets consist of relatively unaltered interstellar grains, the collapse of which, together with the gas associated with them, brought about the formation of the solar system. Comets are thus aggregates of these interstellar grains, which are hypothesized (see Greenberg and Mendoza-Gómez, 1993, for a review) to be in the size range of 0.1 micrometer and to consist of silicate cores surrounded by icy mantles. This mantle is rich in refractory compounds and rich in organics.

Unlike Miller, according to whom the chemical precursors of life were formed on Earth, Arnmand Delseme (following Oró's suggestion of 1961) proposed that virtually all the building blocks of life were transported to Earth by comets (see also de Duve, 1995a; Oró and Lazcano, 1997; Chyba and Sagan, 1997).

It is recalled that for a week beginning July 16, 1994, fragments of the newly discovered Shoemaker-Levy 9 comet collided with planet Jupiter after having disintegrated into around 20 fragments as a result of the influence of Jupiter's gravitational field.

Intermediate-size planetary bodies

This group includes the terrestrial planets—Mercury, Venus, Earth, Mars, and the moon—as well as satellites of the big planets Saturn and Neptune. The radii of the bodies in this group are roughly between 1,500 and 6,500 kilometers. They show some similarities in size, mass, and other characteristics. The most heavily cratered planetary bodies of this group are the moon, Mercury, two satellites of Jupiter (Callisto and Ganymede), several satellites of Saturn and Neptune, and the southern hemisphere of Mars. Craters of various sizes, some overlapping, can be observed on these planetary bodies.

Terrestrial planets

This group includes Mercury, Venus, Earth, and Mars. All the terrestrial planets contain substantial amounts of iron and other high-density elements, in addition to rocks. The proportion of iron and other high-density elements is highest on Mercury and decreases progressively with the distance from the sun. This is explained by the existence of a protoplanetary disk, during the formation of the planetary system, that became heated by the sun. The extent of heating became lower as the distance from the sun increased. Near the sun, only the metals and the most refractory minerals survived and were incorporated into the planets. With increasing distance from the sun, where temperatures stayed progressively lower, greater proportions of volatiles survived.

The terrestrial planets are thus nonhomogeneous bodies of metal, rocks, and volatiles. They are characterized by high-density cores made mainly of iron and nickel, surrounded by rather thick rocky mantles, and covered by a thin surface crust of low-density rocks and volatiles. This differentiation took place early on during planet formation, where the energy released from gravitational coalescence on the one hand, and the decay of radioisotopes on the other, brought about the melting of both the rocks and the metals. The concomitant decrease in viscosity enabled the migration of dense metals inward to form the core.

The moon and its record

Hypothetical scenarios of lunar formation varied widely before the Apollo mission, which was followed by intensive study of both return samples and geophysical data collected by instruments placed on its surface. The now-dominant hypothesis suggests that the moon was formed as a result of a collision of a Mars-sized planetoid with Earth some 4.5 billion years ago. Some of the debris from this collision was flung into orbit and became the moon. The impact energy was large enough to melt a substantial part of both the moon and Earth. After this cataclysmic effect, these terrestrial bodies, like many others in the solar system, experienced the next stage in their evolution, namely, impact cratering (see Taylor, 1994).

The giant planets

This family includes Jupiter, Saturn, Uranus and Neptune. Their radii range from 25,000 to 70,000 km, and their masses are 15 to 300 times that of Earth. However, they are much too small to generate energy by nuclear fusion. Still, they do radiate energy at a rate higher than the solar energy they receive.

The Jovian planets are enveloped by thick atmospheres of gases, which are characterized by storms and fast winds. The bottom layers of the atmospheres border with liquid mantles; their centers contain mainly rocks and iron. Thus, unlike the intermediate-size planetary bodies, they carry no external signs of impact craters.

The largest satellite of Saturn, Titan (with a radius of 2,580 km), is the only satellite in the solar system that possesses a dense atmosphere. This atmosphere consists mainly of methane and nitrogen and is rich in organics. Its surface is covered with oceans of liquid methane and ethane. Titan is considered "a natural laboratory for studying on-going prebiotic chemistry" (Raulin, 1992). A joint NASA and ESA mission to explore Saturn and Titan is planned to arrive at Saturn by the next century.

The "prebiotic chemistry" of Titan, as far as we presently know, is not directly connected to the study of the origin of life on Earth. It may be important for the understanding of different life forms, which are at present totally unknown. This topic will not be further discussed here because more data and scientific models are lacking.

Clues to the study of life's origin

The biogenic elements are widespread in the solar system, in our galaxy, and in the universe. Life based on these or other elements may have been formed in our galaxy as well as in other galaxies. Moreover, in addition to terrestrial life as we know it on our planet, additional forms of life in our solar system and elsewhere in the universe may exist. However, at the present stage of our understanding, life and clues to its origin have been found only on our planet.

13

Planet Earth

> Except for major short-term perturbations in surface environments caused by a declining flux of impactors, equable conditions for prebiotic evolution could have existed as early as 4.4 GA.
>
> Sherwood Chang, "The planetary setting of prebiotic evolution"

The primordial earth

The earth is about 4.6 Ga old. At that remote time, known as the Hadean era (fig. 13.1), its surface was very hot as a result of the accretion process, which, according to recent hypotheses, took about a hundred million years: Its temperature, according to recent models, was about 1,500°K. Thus, the surface was molten. The iron-group elements (Fe, Ni, and Co) melted and passed through the lighter silicate molten rocks down beneath the crust in a process known as the iron catastrophe (R. F. Fox, 1988). Gradually the surface, rich in silicates, cooled down as the accretion energy input decreased. Solid rocks started to emerge, forming a thin scum, and a steam atmosphere began to condense and rain down to form the primordial oceans. Surface temperatures at or below 100°C could have developed as many as 4.4 Ga ago (Chang, 1993, 1994).

Fig. 13.1 also includes the earliest indications of living organisms found in ancient rocks. This topic will be discussed in chapter 15.

The oldest known terrestrial sediments on the earth's surface are 3.8–3.9 billion years old, and there is no geological evidence of prebiotic organic chemical processes taking place on the earth's surface prior to this time. Most of the very old rocks on Earth were transformed geologically by plate tectonics. Moreover, the early craters formed by impactors disappeared through erosion processes. Fortunately, despite the destruction of much of the geological record of primordial Earth, some of this evidence has been preserved and shown to be relevant to the study of the origin of life.

Impact craters on Earth and its moon

Intense bombardments of the inner planets of the solar system was widespread between 3.9 and 4.1 Ga ago (Ash et al., 1996). Indeed, most of the craters on the moon date back about 4 billion years. In contrast to Earth, erosion processes on the moon,

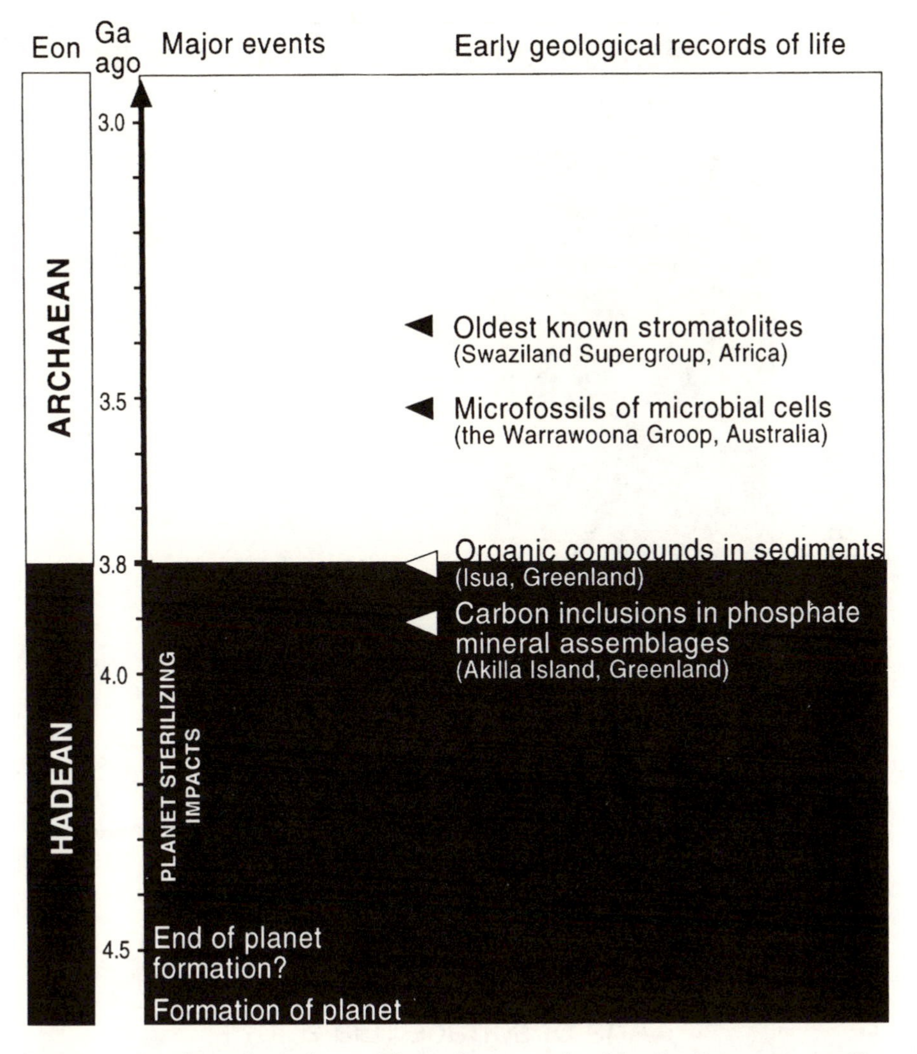

Fig. 13.1. Geochronological eons. The earliest records of fossil cells was reported by Schopf (1992a, 1992b). The earliest chemical records of life were reported by Schidlowski (1988) and Mojzsis et al. (1996), as discussed in chapter 15.

which is devoid of atmosphere, have not taken place. The cratering process can thus be related to the history of the moon, together with the help of radioactive dating of lunar samples and analysis of the cratering record. Scant weathering and the almost complete lack of geological activity have helped in the preservation of these craters since the time of their formation. On the other hand, intensive geological activity, as well as considerable weathering, have wiped out much of the evidence of cratering on Earth. However, the earth must have undergone a similar bombardment process.

Careful analyses of the record of impact craters on the moon have enabled researchers to characterize the changes in both the rate and intensity of the cratering on both the moon and Earth as a function of time during the first several hundred

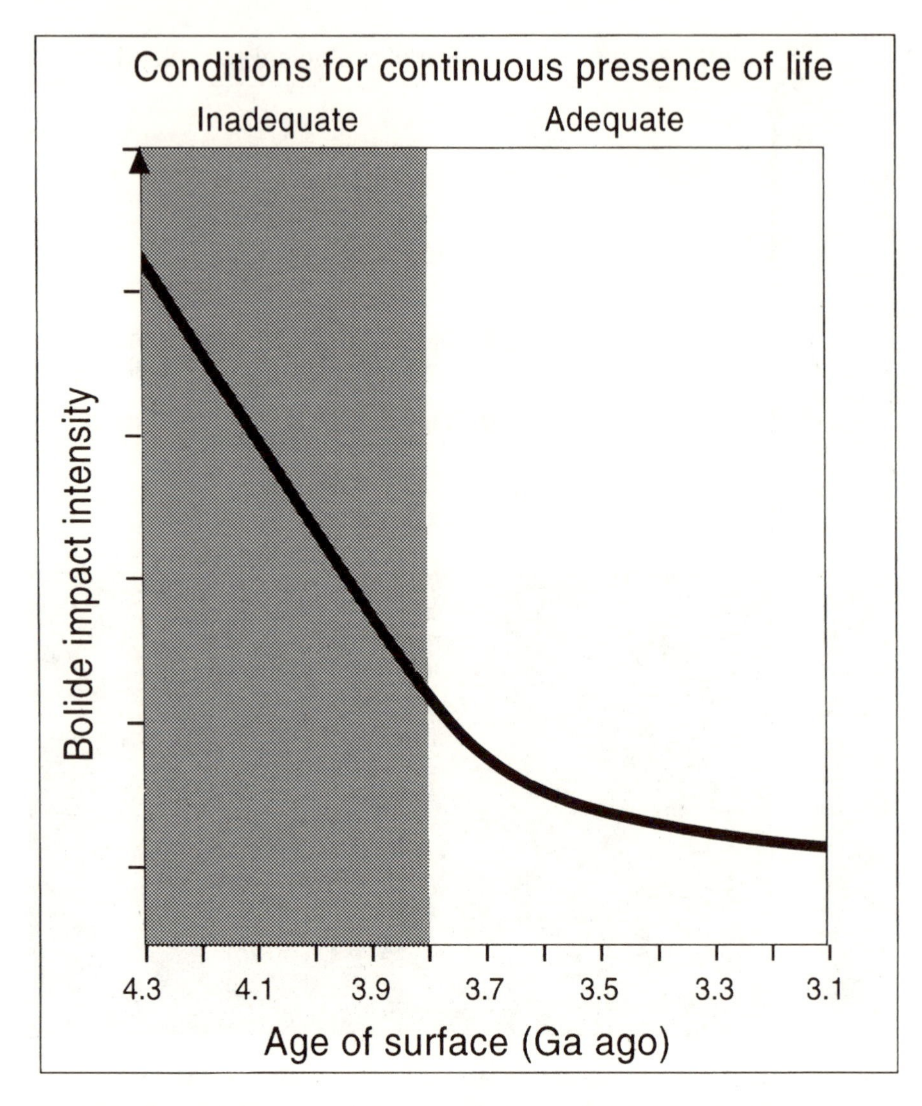

Fig. 13.2. Terrestrial cratering history as a function of surface age for the total number of craters >20 km per km^2. Geological conditions for continuity of life ("geological continuity") are assumed to have started approximately 3.8 Ga ago. (Adapted from Maher and Stevenson, 1988)

million years of the formation of the solar system. A plot of the total number of craters > 20 km per km^2 as a function of age (fig. 13.2) revealed a most important relationships—namely, that during the first several hundred million years of its existence, the earth was bombarded intensively by large and small planetesimals.

The rate of these bombardments declined with time but has not yet reached the zero point. The transition from bombardments by massive planetesimals capable of bringing about total annihilation of Earth's biota to bombardments capable of bringing about only local annihilation occurred about 3.8 Ga ago (Chang, 1993; Tiedemann, 1997).

Hill (1995) has observed, "As humans continue to monitor a sky teeming with hazards, they may one day detect an object headed our way."

The prebiotic sea

Seawater composition

Seawater composition during the Hadean and early Archean eras are not known (Chang, 1993). A most important feature of the early ocean was its reduction-oxidation (redox) reactions. The existence of hydrothermal systems, with circulating seawater, were suggested by several researchers to go back to the early days of the molecular evolution era. In this circulating seawater, soluble Fe^{2+} was extracted from hot igneous rocks to serve as the major reductant of this water. Model calculations suggest that the redox capacity of the prebiotic ocean water far surpassed that of the prebiotic atmosphere, thus serving as a major source of reducing power needed for prebiotic synthesis reactions (ibid.).

Banded iron formations

The banded iron formations have to do with the composition of the prebiotic atmosphere, which was degassed at a very early stage by the so-called T-Tauri wind emanating from the early sun. In various parts of the world, precambrian strata of ~1 mm thick iron-rich bands have been observed and studied, mainly while searching for iron ores. The age of these banded iron formations (BIF) ranges between 1.5 Ga to the earliest geological record of 3.8 Ga. It has been suggested (Braterman et al., 1983) that the formation of these bands was the result of photo-oxidation of soluble ferrous (Fe^{2+}) species in the oceans, with the generation of ferric ions (Fe^{3+}) and molecular hydrogen (equation 13.1).

$$Fe^{2+} + H^+ + h\upsilon \rightarrow Fe^{3+} + 1/2H_2 \qquad (13.1)$$

The importance of this photo-oxidation reaction is that it could have supplied reduced raw materials such as H_2, CH_4, and HCN to the primitive atmosphere. The Fe^{3+} precipitated in the form of sparingly soluble ferric or ferro-ferric oxides and other complexes; the BIF thus served as a sink for the oxygen produced by the photolysis of water vapor in the atmosphere. This process was presumably already going on 3.8 Ga ago, and perhaps even earlier (Mason, 1991). The presence of these gases in the prebiotic atmosphere could have been of great importance in synthesis reactions of various organic molecules.

The iron cycle

The BIFs suggest the occurence of an iron cycle in which Fe^{3+} replaced Fe^{2+} in the lithosphere (de Duve, 1991). The ferric ions could have served as the main electron acceptors in the first bio-oxidation reactions and the establishment of a protometabolism of a kind (Hartman, 1975, 1992a, 1992b; Braterman et al., 1983). The

Fe^{2+}/Fe^{3+} cycle could have served as a transducer of light energy into a biologically useful form, similar to what the H_2O/O_2 cycle does today (de Duve, 1991). So far, no such reactions were reported (Chyba and McDonald, 1995).

Sulfides, carbonates, and silicates of ferrous iron

FeS (pyrrholite), FeS_2 (pyrite), $FeCO_3$ (siderite), and Fe_2SiO_4 (fayalite) are also found in BIF. The older the BIF, the higher their quantities. As will be discussed in chapter 21, the presence of pyrite at certain sites during the prebiotic era has been the basis for an imaginative scenario for the origin of life (Wächtershäuser, 1988).

Hydrothermal vents (submarine vents)

These vents are formed when seawater is forced down several kilometers into the sediments and basaltic rocks of the sea bottom; heated by the magma, thus dissolving a variety of compounds; and pushed through the vents to the sea floor at temperatures of about 350°C. These vents were suggested (Corliss et al., 1981) as possible sites for the origin of life (see chapters 16, 21, and 22 for further discussions). The problem of organic synthesis in these vents is still controversial (see Henley, 1996; Lazcano and Miller, 1996; Simoneit, 1995; Shock et al., 1995).

Prebiotic uranium compounds

Without going into the geochemical processes of uranium in the earth's crust, it is noted that its possible involvement in the origin of life has recently been related to two suggestions. The first is the discovery that natural reactors in the crust of primeval Earth may have been established (Mason, 1991; Maynard Smith and Szathmáry, 1995). The second has to do with the effect of radiation of radioisotopes on organic molecules and their prebiotic evolution, as discussed briefly in chapter 16.

The atmosphere of the primordial earth

The properties of the early atmosphere are crucial to our understanding of the origin-of-life process, whether life originated in the near-surface environments; or in hydrothermal vents; or somewhere in space, then being brought to Earth by extraterrestrial bodies such as asteroids or comets, as the Panspermia hypothesis holds. Typically, this study is based mainly on computer modeling of atmospheric composition, pressure, and climate (see, for instance, Kasting, 1993a, 1993b).

Formation of the atmosphere and the oceans

Our planet was formed by the accretion of solid material from the solar nebula, where any primary, captured atmosphere must have been lost very early in its history. A secondary atmosphere started to build up soon afterward, generated from volatile compounds contained in the solid phases of the planetesimals from which the earth was accreted (Kasting, 1993a, 1993b). This secondary atmosphere has undergone changes during the eons, slowly evolving into the present one. The composition of the primordial secondary atmosphere is considered an essential key to the understanding of the central reactions of prebiotic chemistry that took place on Earth; dealing with such remote processes, it is no wonder that these have been the subject of intensive study and much debate for more than four decades.

Recent models of planetary accretion suggest a relatively short formation time of 10 to 100 million years. The interior of the early earth was initially hot due to large impact events, including the one that has been suggested to have formed the moon. The earth's core is hypothesized to have been formed simultaneously with accretion, implying that metallic iron may have been removed from the upper mantle. Thus, the atmosphere that developed in contact with a planet surface poor in metallic iron was not very reducing; the strongly reducing atmosphere assumed by Urey and simulated by Miller in his classical experiment (1953) is today considered unlikely by most researchers.

According to Kasting (1993b) the early atmosphere was dominated by H_2O, CO_2, CO and N_2. Water was the most abundant volatile, followed by carbon compounds. Based on model calculations, Kasting suggests the possibility of a primitive atmosphere containing 10 bars of CO + CO_2, as well as approximately 1 bar of N_2, during the early history of Earth.

The early earth's climate

The climate of the prebiotic earth can be described by the following interdependent parameters:

1. Mean surface temperature.
2. Radiation provided by the young sun.
3. Seasonal and diurnal fluctuations brought about by the rotation of Earth during its voyage around the sun.
4. Pulsed, irregular, and relatively short-lived perturbations due to impact events.

According to recent climate models, the mean surface temperature of Earth was around358°K. It is reasonable to assume climatological variability in different zones, just as variable zones are found today on Earth. Day-night variations are also expected to have characterized all kinds of climatic regimes. Moreover, the surface of early Earth is expected to have been characterized by nonhomogeneities induced by topographical variations, volcanos, and hot springs, thus forming an almost endless number of niches with characteristic microclimates. These considerations are important to the search into the origin of life, since they offer a large variety of climatic conditions into which prebiotic scenarios can be applied.

The mean temperature of early Earth was higher than the present one (about 288°K) in spite of the smaller radiation rate of the early sun. According to recent models, the early sun was about 30% less luminous than it is today (Kasting, 1993a). The reason for the higher mean temperature of the early earth is that most of its CO_2 (which today is found predominantly in sedimentary rocks) was deposited directly into the primordial atmosphere. Consequently, the CO_2 pressure of the ancient atmosphere could have been as high as 10 Atmospheres. The concomitant greenhouse effect increased the mean surface temperature and more than compensated for the effect of lower solar luminosity.

Freeze-thaw cycles on primordial Earth?

As I have already noted, the luminosity of the young sun, some 3.8 billion years ago, was about 70% of its present value. Theoretical calculations using energy balance models show that, for the extant atmospheric composition, such a reduction in luminosity would have resulted in the permanent freezing of our planet's surface. How can these energy-balance calculations be compatible with the scenario of a warm planet, presumably adequate for molecular evolution processes?

One of the solutions suggested for this paradox was an atmosphere rich in CO_2. Another solution, suggested by Bada et al. (1994), deals with the bolide impacts that presumably characterized at least part of the molecular evolution era. According to these researchers, bolides with diameters of approximately 100 km, impacting Earth more than 3.6 Ga ago, would have caused the ice-covered oceans to melt. The primordial Earth may thus have been characterized by freeze-thaw cycles.

Source of organics and water on early Earth

An extraterrestrial source of organics on Earth was first postulated by Chamberlin and Chamberlin, as early as 1908, and was repeated, quite independently, by Oró in 1961 (see also Oró et al., 1992). It has been established since then that in addition to terrestrial organics, such as those emitted during volcanic exhalation, a sizeable portion of the organic carbon may have been imported to planet Earth from space, either by impact delivery or by interplanetary dust. An estimate of the sources of prebiotic organics depends on an understanding of the composition of the early terrestrial atmosphere. For more information, see Chyba and Sagan, 1992; Kasting, 1993a; and Travis, 1994. At the moment it seems best to conclude this paragraph by citing Chyba and Sagan (1992) as follows: "Endogenous, exogenous and impact-shock sources of organics could have made a significant contribution to the origin of life." The list of organic molecules that can be derived from interstellar and cometary sources includes formaldehyde, acetaldehydes, hydrogen cyanide, and cyanamide (Oró, 1994).

Urey-Miller's reducing atmosphere revisited

The atmosphere assumed by Miller and Urey in their spark-discharge experiment (Miller, 1953) was a mixture of CH_4, NH_3, H_2, and H_2O. Current hypotheses regarding the atmospheric composition of the post-heavy bombardments era exclude CH_4 and NH_3, indicating a "weakly reduced" atmosphere; this is dominated by CO_2, N_2, and H_2O, with small amounts of H_2, CO, SO_2, and H_2S (see Kasting, 1993a). The apparent advantage of the first model (a strongly reducing atmosphere) was the richness and versatility of organic molecules obtained by various energy sources, especially electrical discharges. In contrast, only a few organic molecules can be synthesized in a weakly reduced atmosphere.

One important biological precursor that is formed readily in a weakly reduced atmosphere is formaldehyde (H_2CO). The reaction involves photolysis of H_2O and CO_2 and proceeds via the formation of the formyl radical (HCO) (Kasting, 1993a). However, synthesizing hydrogen cyanide (HCN), which is considered an important precursor to amino acids and nucleotides, under weakly reducing conditions is rather difficult because it requires breaking the strong bonds of both N_2 and CO. Thus, the prebiotic syntheses scenarios based on a weakly reduced prebiotic atmosphere suffer from the implausibility of HCN synthesis.

The prebiotic conditions assumed by Urey and Miller were essentially those of a highly reducing atmosphere. Under slightly reducing conditions, the Miller-Urey reaction does not produce amino acids, nor does it produce the chemicals that may serve as the predecessors of other important biopolymer building blocks. Thus, by challenging the assumption of a reducing atmosphere, we challenge the very existence of the "prebiotic soup," with its richness of biologically important organic compounds. Moreover, so far, no geochemical evidence for the existence of a prebiotic soup has been published. Indeed, a number of scientists have challenged the prebiotic soup concept, noting that even if it existed, the concentration of organic building blocks in it would have been too small to be meaningful for prebiotic evolution.

Accumulation of oxygen

Most researchers agree that molecular oxygen (O_2) was absent from the primordial atmosphere. Some oxygen is assumed to have been formed by photodissociation of water vapor in the upper atmosphere. This oxygen, however, reacted with Fe^{2+} , as well as with other unoxidized reactants available in the prebiotic environment. Only after the evolution of photosynthesizing organisms could the oxygen thus formed have accumulated in the atmosphere. The buildup of oxygen concentrations to its present level of 20% is hypothesized to have taken place more than 2 Ga ago. Three stages in the history of oxygen in the atmosphere, surface ocean, and deep ocean are described schematically in the "three-box" model (Kasting, 1993b), as follows:

1. The atmosphere and the entire ocean are anoxic, with the possible existence of localized "oxygen oases" in the surface ocean.
2. The atmosphere and surface ocean are oxidizing. The deep ocean is still anoxic.
3. Both atmosphere and ocean contain free oxygen, as in the modern earth.

The Gaia hypothesis

> Life makes its own environment.
> Lynn Margulis,
> *Symbiosis in cell evolution*

In the mid-1960s, the chemist James E. Lovelock was trying to develop a general strategy for the detection of life on Mars. Using available data on the composition of the atmospheres of several planets, he concluded that biological life would bring about dramatic changes in the chemical composition of the planet's atmosphere. Thus, the atmosphere of our planet, which does host life on its surface, represents a chemical state far from equilibrium. On the other hand, the atmospheres of both Mars and Venus are in a chemical state close to equilibrium between the gases of their atmospheres and their surface rocks. Two of his conclusions from these calculations were as follows:

1. Mars is lifeless (Lovelock's conclusions became known before NASA's mission to Mars. To date, the results of this mission have in no way suggested the presence of life on Mars). However, this topic is today the subject of intensive study (McKay et al., 1996).
2. The mechanism controlling the chemical dynamics of a planet's atmosphere includes the living organisms themselves.

Lovelock thus concluded that the earth itself can be considered a living unit, regulating its environment in response to changes. Collaboration with Lynn Margulis helped in publicizing the novel hypothesis (see Margulis, 1993b). In response to their critics, Lovelock and Margulis stressed that they had not proposed a teleological hypothesis, in which Gaia is a purposeful entity that acts with a kind of foresight and planning.

Critics of the Gaia hypothesis argue that the ability of our planet to "resist" changes in its atmosphere cannot be considered "regulation." For instance, in the case of the increase of atmospheric CO_2 concentration, the concentration continues to increase without reaching a stable state. Regarding the close-to-equilibrium state of Mars's atmosphere, this by itself is not an indication of a lack of life on this planet, since life can exist on Mars in small ecological niches that have only a slight impact on its atmosphere (Levine, 1993).

What is the relevance of the Gaia hypothesis to the study of the origin of life? The main reason for introducing the concept in this book is the same one that triggered the Gaia concept in Lovelock's mind; namely, it may serve as a basis for a research strategy to identify life activities on heavenly bodies.

The second reason has to do with the concept of an adaptive control system of the environment-plus–living entities, which started its biological evolution on our planet around 3.8 Ga ago. In analogy with the back-extrapolation that we have used in our search for the biological record (chapter 10), we can back-extrapolate from the present planet and its intricate Gaia system, to the beginning stages of the prebiotic system. This primordial system of the beginning of life was heterogeneous, that is, a network of environmental niches in which the evolving biological entities were in an intimate "collaboration" with their environment. In the same way, present life cannot be separated from its environment, so the beginning of life cannot be fully understood without considering the environment in which the first collaboration between organic molecules and their immediate environment coevolved. More on the self-organization of the earth's biosphere can be found in Schwartzman et al., 1994.

As Tickell (1993) observes,

> I think all can acknowledge that the concept of Gaia has been a productive intellectual tool, and research arising from it has contributed to our understanding of the relationship between animate and inanimate nature. . . . Gaia has no particular tenderness for the human or any other species. . . . We cannot consider the evolution of species separately from changes in their environment.

Clues to the study of life's origin

The surface of the primordial earth, its rock, sea, atmosphere, and climate, were rather variable and presumably could have initiated and supported a large variety of organic reactions and processes assumed to be related to the origin of life. The organic molecules needed for these processes were formed from inorganic molecules both on Earth and in space, followed by their transfer to adequate sites where molecular evolution processes took place.

Part IV

BEYOND THE PROGENOTE

Rationale, strategies, scenarios, and models in the search for the origin of life

In the beginning there was simplicity.
Richard Dawkins,
The selfish gene

Life's origin as the key to its nature.
Avshalom Elitzur, "Life and mind, past and future"

The present part is the climax of our effort to comprehend the origin of life. The interdisciplinary enquiry described and discussed below in this part has been conducted in conjunction with hypothetical terrestrial scenarios; however, the methodology developed by scientists since the beginning of the modern search into the origin of life is based on general principles and, as such, can serve as a starting point also to the search of evolution of different forms of life elsewhere in the universe. Such an undertaking has nothing to do with science fiction, as the dramatic August 1996 announcement by NASA about possible life on Mars (McKay et al., 1996) has reaffirmed, in spite of being controversial.

It seems adequate to start the present discussion with a citation regarding the scope of the scientific explorations (Sattler, 1986, p. 41), as follows: "Modern philosophy of science has gone far beyond the naive belief that science reveals the truth. Even if it could, we would have no means of proving it. Certainty seems unattainable. All scientific statements remain open to doubt. . . . We cannot reach the absolute at least as far as science is concerned; we have to content ourselves with the relative."

The next thing to do would be to stress two central attributes of all scientific indications pertaining to the origin of life, namely, remoteness in time and lack of tangible traces. The unavoidable outcome of these features is the speculative character of all scientific propositions pertaining to the origin of life.

The presentation of the present discussion is complex and problematic and tends to be subjective in one way or another. The presentation according to Wächtershäuser

(Davis and McKay, 1996) is logical, and the reader is advised to consult it. The alternative presentation used here seems to be more convenient in dealing with the many subdivisions that should be applied to the main topic, each of which overlaps somewhat with other topics. This presentation is thus arbitrary and is probably biased to some extent.

14

Basic Assumptions and Strategies

What led to life on Earth? So great a question can only be tackled by breaking it down into sections.
James P. Ferris, "Life at the margins"

Our departure point in the voyage toward the beginning of the biological world

The study of the origin deals with hypothetical chemical entities and primordial creatures that *presumably* do not exist today. It focuses on central attributes of the interactions among these primordial entities, that is, the physicochemical reactions of their gradual organization into more and more complex forms in the prebiotic evolution process. The latter encompasses catalysis, template-directed synthesis of oligomers and polymers, mutability, selection, establishment of primordial reaction cycles, compartmentation, and the buildup of complexity, all of which are manifested as evolutionary processes at the molecular level.

But first let us turn to the progenote—the hypothetical creature which is on the verge of the realm of extant biology-and use it as a departure point in our back-extrapolated voyage toward the very beginning of biology.

The last common ancestor and the progenote

We shall try now to look beyond the last common ancestor and the progenote. It is recalled that a glimpse from the last common ancestor still backward into the hypothetical history of biology left us with an uncertainty regarding the ill-defined and more obscure progenote (Woese, 1987). With all its antiquity, the progenote was a rather advanced and sophisticated creature. The ribosomal RNAs are rather complex molecules; the ribosome, the organelle where the cell proteins are synthesized ("polypeptide polymerase," according to P. B. Moore, 1993; see chapter 6), is even more complex. Presumably, the ribosome of the progenote and the last common ancestor were simpler versions of the ones in extant cells; but even a simple system capable of processing the information of primitive nucleic acid analogues is considered a rather advanced stage in the evolution of life (for recent reviews see Lazcano, 1994b, 1994b; Garcia-Meza et al., 1994).

Indeed, the "invention" of template-directed synthesis (which is the essence of the information processing taking place in the ribosome) is considered a great evolutionary hallmark. Moreover, it seems much easier to explain the evolution of the phylogenetic tree starting from either the last common ancestor or the progenote than to explain the emergence of those obscure chemical entities from inanimate matter during the prebiotic molecular evolution era. Furthermore, the central attributes of the progenote, vague as they necessarily are, are assumed to be reflected by the noncontiguity of the genetic material and its template-directed processes. Thus, the hypothetical progenote is assumed to have evolved *after* the emergence of template-directed syntheses of replication and protein synthesis.

The principle of biological continuity

The progenote being such a vague hypothetical creature, there is no guaranteed way to further back-extrapolate from it toward more primitive entities. We are left with the principle of biological continuity as the main biological guide to the reconstruction of the origin of life, without being able to point out with certainty the genuine pathway through which life originated.

The only known way to extrapolate farther backward in the search of the origin of life is to apply the principle of continuity to conserved domains of biopolymers, conserved functions of biomolecules, conserved reaction cycles, and maybe also—and unlikely as it may seem—to inorganic moieties of extant cells. It is appropriate to conclude this section with a citation from Woese (1987): "The progenote is today the end of an evolutionary trail that starts with fact, progresses through inference, and fades into fancy."

It is here, in the obscure and controversial realm of fancy, theories, and speculations, that the sparse pathways of our search into the origin of life should be sought.

Basic assumptions

> What can be done with fewer assumptions is done in vain with more.
>
> William of Occam, cited in Allan Goddard Lindh, "Natural selection"

The scientific search into the origin of life is based on two general, fundamental assumptions: the applicability of the known physical laws and the applicability of the evolutionary theory at the molecular level.

Applicability of the known physical laws

The origin of life can be explained by the basic physical laws of matter and energy, as embodied in the periodic table (see, for instance, Morowitz, 1992). Moreover, possible complexities connected to subatomic quantum physics are neglected (Dennett, 1996, p. 101). Can the origin of life be predicted from first principles? The answer of R. F. Fox (1988, p. 94) to this question is probably shared by all origin-of- life researchers—namely, given the today's know-how regarding the origin of life, the basic attributes of the living state and its formation cannot be predicted from first principles.

Applicability of the evolution theory at the molecular level

The origin of life was a Darwinian evolutionary process in which assemblies of inanimate molecules gradually became living entities. The uniqueness of the Darwinian evolution among evolutionary processes in general was phrased by Lifson (1997) as follows:

> Evolution of galaxies and solar systems means the long-time course of a random drift of macroscopic systems toward ever-increasing entropy, or ever-decreasing free energy. Darwinian evolution of autocatalytic systems is no exception in this respect. However, its random drift is directed by natural selection towards all those specific properties that characterize animate matter, namely, fitness, adaptivity, organization and the like.

The superficial similarities between these processes is a source of much confusion in the context of life and its origin.

Natural selection is thus "the only theory that a biologist needs in addition to those of the physical scientist. Both the biologist and the physical scientist need to reckon with historical legacies to explain any real-world phenomenon" (George Williams, 1985, cited in Dennett, 1996, p. 123).

The term "anastrophe" was suggested by Baltscheffsky (1993; see also Baltscheffsy, 1997), as follows: "An anastrophe is the opposite to a catastrophe, just as, in metabolism, anabolism is building up and catabolism is breaking down, the word anastrophe having been suggested to fill a language gap by describing momentous changes in a constructive, rather than destructive, direction. . . . Biological evolution may be considered to consist of series of anastrophes."

Reconstruction strategies

It is convenient, although somewhat arbitrary, to divide the strategies of the reconstruction of the origin-of-life processes into three groups, namely, (1) cosmogeological, (2) biological, and (3) biogeochemical. The cosmogeological strategies attempt to characterize the physicochemical conditions of the primordial earth during the time of the origin-of-life. The biological strategies attempt to reconstruct the early stages of the origin-of-life processes by using clues from biology. The biogeochemical studies deal with attempts to reconstruct the earliest stages of the origin of life processes by using clues from both biology and geology, focusing on the processes through which the transition from inanimate to animate matter took place.

The cosmogeological strategy

In its broad sense, the aim of this strategy is to look for life in the universe. In a more limited sense, this strategy focuses on the emergence of life as we know it in the solar system. This includes both life on Earth and life on other planets or on planets' moons. In a still more limited, terrestrial sense, which is the main goal of the present book, this strategy focuses on our planet. Thus, this goal may be divided into:

1. The characterization of the processes in which the solar system, and specifically planet Earth, were formed.
2. The identification of traces of possible living entities, and the consequences of their activities.

3. Reconstruction of the prebiotic earth's surface and atmosphere at the presumed time period of the prebiotic evolution process. This is an attempt to characterize the environments in which the first living entities could have evolved.

The biological and biogeochemical strategies

The reconstruction of the early stages of the origin of life can be studied by the top-down and bottom-up approaches (fig. 14.1). These routes of enquiry have been used rather loosely by different researchers. The definition by Maynard Smith and Szathmáry (1995, p. 84) is general and is exemplified in a discussion on the origin of the genetic code: "The bottom-up strategy tries to get from elements with known properties to systems, whereas the top-down approach attempts to figure out the details by considering the pattern of behavior at the system level."

The top-down approach is defined here as back-extrapolation from biology to the earliest identifiable biochemical entities that presumably were involved in the origin-of-life processes. It is based on composition, sequence, structure, interactions, and functions of extant biomolecules and is aimed at the characterization of possible remote predecessors of these molecules and their functions.

The bottom-up approach is defined here as an attempt to suggest possible primordial organic building blocks of the central structures, biopolymers, and metabolic cycles, as well as their inorganic environment, and the study of their interactions and evolution, which led to the transition from the inanimate to the animate. These studies include laboratory experiments and computer modeling designed to simulate prebiotic reaction cycles, processes, and scenarios. Simulation of simplified biochemical and molecular-biological functions and evolutionary pathways are then contrived and manipulated in an attempt to reconstruct plausible early stages of the origin-of-life processes. The constrains imposed by the experimenter and the theoretician in these studies have to do with the presumed prebiotic environment in which the processes under consideration are assumed to have taken place.

The bottom-up approach encompasses models and scenarios with different starting points: Some of them start with oligomers and biopolymers without dealing with the question of their formation; others start with either the building blocks of central molecular-biological entities or inorganic compounds capable of forming these building blocks. Because of how these models' starting points overlap, it is convenient to treat them as one group.

The biological strategy is a top-down strategy

Assuming evolution as the fundamental attribute of the origin of life implies the extension of the biological continuity principle to the earliest possible clues of the beginning of this process. The application of this principle means a search for plausible ancient components, functions, reactions, and processes at the molecular level, in the biological record of extant organisms, in an attempt to characterize the primordial chemical entities and processes that preceded the progenote.

The biogeochemical strategy is a bottom-up strategy

The scope of this approach encompasses the evolution of the Milky Way galaxy and the solar system but focuses on the geological record of our planet, as well as on other

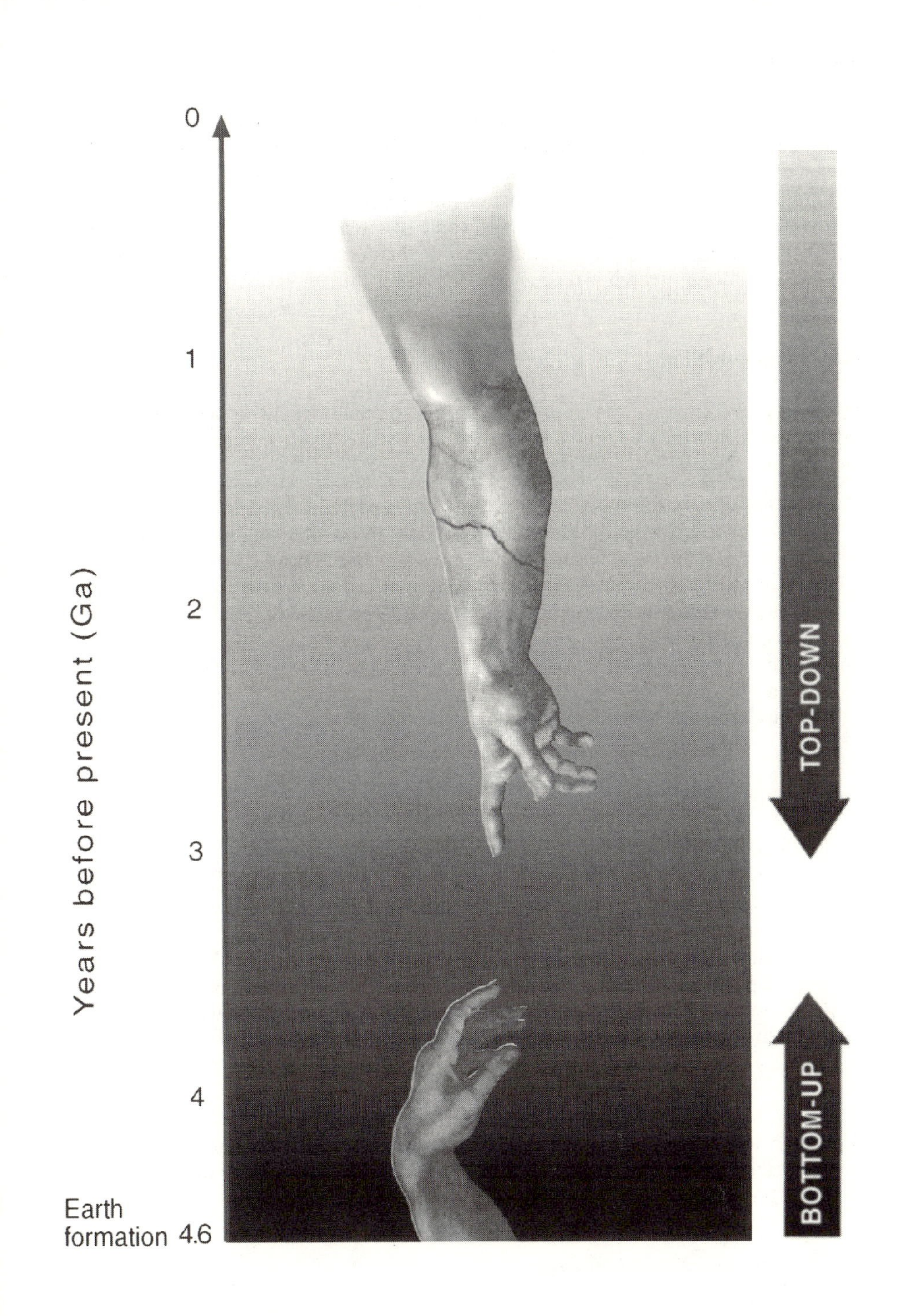

Fig. 14.1. A paraphrase on Michelangelo's *Creation of Man*.

planets; organic syntheses both on the primordial earth and other planets and in space; and the evolution of chemical organic entities and processes, starting with the syntheses of the predecessors of the building blocks of the most important biochemical compounds. The actual pathways of the origin of life will probably never be known to us in all their details; however, some of the most likely kinds of reactions and scenarios, as well as the methodologies of their study, are gradually unfolding, as more and more research is done.

Plausibility guidelines

Scientific propositions in the study of the origin of life must rely on plausibility considerations, which may be presented in the form of three guidelines or principles (see Deamer and Fleischaker, 1994, p. 9), as follows:

1. Biological continuity A scenario that demonstrates a biologically continuous evolutionary pathway between an early and a later stage is considered more plausible than an alternative scenario characterized by a discontinuity in the pathway between these stages.
2. Ubiquity Ubiquitous conditions are considered more plausible than isolated special cases.
3. Robustness A robust reaction, pathway, or scenario is "relatively independent of precisely defined environmental parameters" (Deamer and Fleischaker, 1994).

Experimental methodologies

> Chemistry becomes a still more powerful tool by allowing scientists to ask "what if?" and "Why not?"
>
> Steven A. Benner et al., "Reading the palimpsest"

> When we explain, we reduce known facts to theoretical entities (Popper, 1963): known reaction products to an unknown reaction mechanism or to unknown intermediates; known extant organisms to unknown missing links; known biochemical pathways to an unknown origin of life.
>
> Günter Wächtershäuser, "Groundworks for an evolutionary biochemistry"

According to conventional wisdom of the search into the origin of life, two approaches are possible, namely, the "whole-system" and "constructionist" methodologies (Lahav, 1985).

The whole-system methodology

This methodology was originally suggested by Pattee in 1965 (Cited by Lahav, 1985) and has never been fully explored experimentally. It is founded on the assumption that in a simulated prebiotic environment containing the necessary elements of biochemistry, energy sources, and adequate terrestrial as well as marine and atmospheric environmental systems, a biomolecular evolution process will sooner or later take

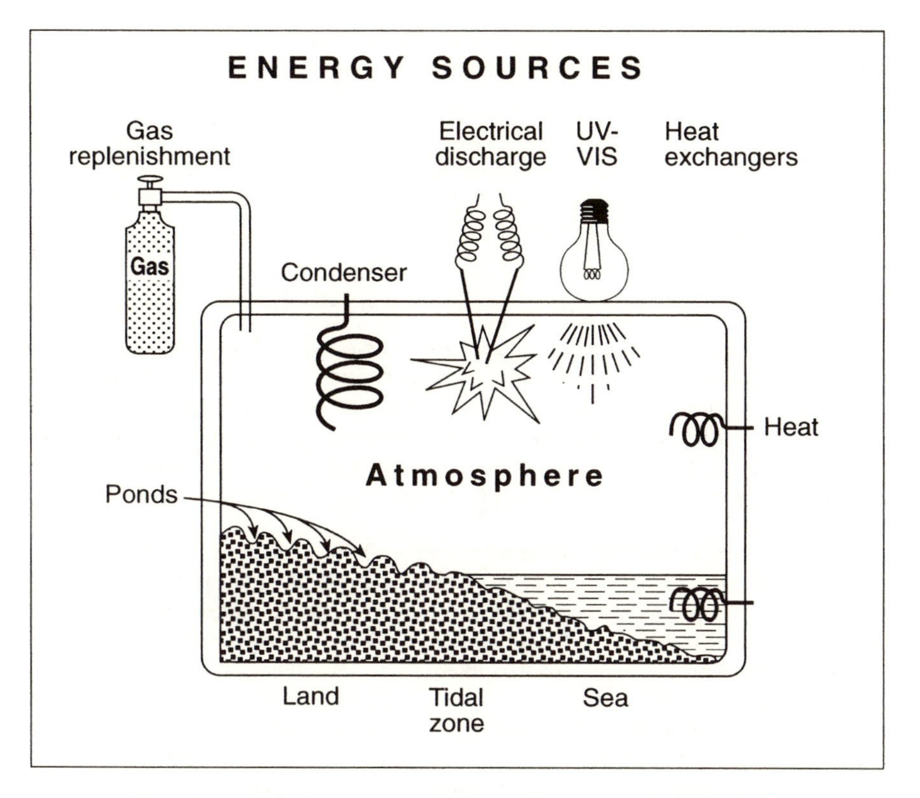

Fig. 14.2. A whole-system reactor. (Adapted from Lahav, 1985)

place. The original evolution process that presumably took place on the primordial earth may have been much longer than any practical experimental system of this kind. However, it is possible to facilitate the evolution rate of the chemical entities, and to some extent to direct the processes taking place in a properly operated whole-system reactor.

A scheme of a reactor to be used according to this strategy is given in fig. 14.2. Obviously, the chemical pathways of the presumed evolutionary process inside the reactor would not be easy to observe and understand. Moreover, the identification of "living" entities inside the reactor poses a rather difficult problem to the experimenter, as discussed by Lahav (1985).

The constructionist methodology

This methodology, which has been adopted by practically all students of the origin of life, can be phrased as follows:

1. Use the already-discussed guidelines from biology and geology.
2. Go step by step from the simple to the more complex, starting from the synthesis of the building blocks of biochemistry or their procecessors and proceeding with their evolutionary organization into more and more complex entities.

3. Combine separate reactions into sequences of reactions.
4. Construct complex systems from simpler subsystems.

Eventually, presumably after a long period of research time, a clear picture of the evolution of living forms will gradually emerge. Finally, the synthesis of simple "living" entities will be understood to a great extent and may become possible.

The two main methodologies employed in the framework of this strategy are laboratory experiments and computer modeling (chapter 23).

Practical considerations suggest humbleness and patience

According to both the whole-system and constructionist methodologies, one has to use the basic laws of matter and energy in order to explore the history of our planet, search for the geological record of the first living entities on Earth, and study the biological record of extant organisms. However, depending on the methodology to be employed, one has to either devise plausible reaction pathways and scenarios for the very first steps of prebiotic evolution or resort to semi-black-box kind of experiments.

The hypothetical reaction mechanisms and pathways resulting from this intellectual endeavor would hopefully lead to a better understanding of the evolution of the earliest living entities. However, unlike in the case of "the light at the end of the tunnel," there is no clearly marked road leading toward the known goal; and not only this road should be invented, but it is likely to remain always a *compatible* rather than the *original* road!

The scientific research of the origin of life is still very far from a stage in which it would be capable of encompassing the emergence of the whole spectrum of phenomena presented by life, in one continuous and coherent scenario. And just as life presumably did not appear suddenly on Earth, so would the scientific research of molecular evolution processes, with its own gradual and "evolutionary" development, lead to a *gradual* understanding of the origin of life. Indeed, the origin of life seems to be too complex to be solvable today by just one set of experiments. However, it does not seem necessary, at the present stage of the research, to invoke "the nature of knowledge itself" (Pattee, 1995; see Preface).

Summary

The search into the origin of life focuses on the chemical entities that preceded the progenote. The rationale of this search may be presented in the form of two basic assumptions, three strategical approaches, three plausibility guidelines, and two experimental methodologies, as follows.

Basic assumptions (1) Applicability of the known physical laws. (2) Applicability of the evolution theory at the molecular level.

Strategic approaches (1) Interpretation of the cosmological and geological record. (2) Biological (top-down) Strategy: Back-extrapolation from biology, including biochemistry and molecular biology, aimed at the characterization of early stages of life. (3) Biogeochemical (bottom-up) strategy: invention of theories of the origin of life, based on (1) and (2).

Plausibility guidelines (1) Biological continuity. (2) Ubiquity. (3) Robustness.

Experimental methodologies (1) Whole-system: an attempt to simulate an entire process, not the individual reactions, in which primordial living entities can be formed. (2) Constructionist: an attempt to focus on separate reactions and combine them gradually into processes and scenarios.

15

Clues and Speculations Based on Cosmology and Geology

> "One good experiment is worth a thousand models (Bunning); but one good model can make a thousand experiments unnecessary."
>
> David Lloyd and Evgenii I. Volkov,
> "The ultradian clock"

The cosmological record and the Panspermia theory

Life in the universe?

The idea of life in planetary systems similar to ours is rather old. If the origin of life on planet Earth is assumed to have been based on organic compounds formed in the atmosphere (the classical Miller-Urey experiment), then life could have formed on other planets as well (Kasting et al., 1997).

The presence of a large variety of organic molecules in space (see reviews by Bada, 1991; Chyba and McDonald, 1995; Chyba and Sagan, 1997; Delsemme, 1997; Kissel et al., 1997; Oró and Lazcano, 1997; Whittet, 1997) suggests that the building blocks of biology are likely to be present there. But whether such organic molecules could interact in a pathway that would result in living creatures elsewhere in the universe is not currently known. Moreover, even if life has originated elsewhere in the universe, its very beginning is still a scientific problem par excellence.

In spite of the impressive list of types of organic molecules in space, there is no direct, unequivocal evidence for the presence of living extraterrestrial entities, either in interplanetary space or on a unique heavenly body. Until compelling evidence is presented to the contrary, we shall adhere in the present book to the assumption-popular among researchers in the field-that life was formed on our planet. But first we have to rediscuss the theory entitled Panspermia (chapter 5), according to which life on our planet was imported from space.

Resurrection of the Panspermia theory?

The Panspermia theory suggested by Arrhenius was mainly the result of an inability to explain the origin of life, not the result of cosmological records of life or theoretical considerations. Arrhenius's courageous attempt seemed to have come to a dead end a few years after it started, by the beginning of the twentieth century (chapter 5).

With the advent of the Oparin-Haldane paradigm (chapter 5), the need for radically different points of view had to be postponed until the accumulating weight of uncertainties and discrepancies implied in that paradigm encouraged scientists to resort to less orthodox alternatives. The renewed interest in the Panspermia hypothesis in recent decades may again have been partially motivated by some scientists' feeling that other existing scientific theories regarding the origin of life had not been satisfactory. The Panspermia speculation has recently been revived by three prominent scientists, namely Francis Crick, Leslie Orgel, and Fred Hoyle and his collaborators.

Evidence for bacteria and viruses in space?

Hoyle based his approach on spectroscopic observations of many kinds of organic compounds in space. His claim of spectroscopic evidence of bacteria and viruslike organisms in space, however, is still controversial among scientists (Henderson et al., 1989). An example of some of his group's work is a paper on comets as habitats for primitive life (Wallis et al., 1992). In this paper it is argued that most of the nucleus of Halley's Comet is covered by an insulating crust presumably made of organic material. The subcrust, however, is warmer, serving as a habitat for primitive replicating organisms. The possibility of life in comets was reviewed recently by C. P. McKay (1997b). For further information the reader is referred to Hoyle and Wickramasinghe, 1986; Marcus and Olsen, 1991; and Hoyle and Wickramasinghe's popular book *Space Travelers* (1981).

Directed Panspermia

An even bolder proposition (or was it a joke at the time?) was suggested in 1973 by Crick and Orgel (see Crick, 1981) and was termed Directed Panspermia. According to this proposition, the history of primordial Earth may not have been long enough for life to develop. Their solution to this problem was the speculation that more advanced creatures developed science and technology to a higher degree than we did. Spaceships of these creatures dispatched germs of life to our planet, thus directing the import of life to Earth. Various relevant considerations involving the chemical composition of living organisms and the primordial sea (Banin and Navrot, 1975) are reviewed by Kobayashi and Ponnamperuma (1985a, 1985b).

In the present book I adopt the approach expressed recently by de Duve (1995b, p. 7): I assume that life was formed on Earth. Moreover, even if one assumes that life was formed elsewhere in the outer space and then came to Earth, the origin of life would still be unknown.

Crick and Orgel's extraterrestrial speculation was suggested at a time when the origin of life was considered by most scientists to have been a very long process, in the order of hundreds of millions of years. According to more recent terrestrial hypotheses and estimates, the "time window" during which life should have evolved on Earth was much shorter, as we will discuss later.

An additional aspect of the Directed Panspermia hypothesis has to do with the difficulties in synthesizing organic compounds by spark discharges using less reducing gases in the reaction chamber than in the classical Miller (1953) experiment, according to more recent models. Thus, according to the panspermia theory, life could have evolved on a planet with more reducing primitive atmosphere (Crick, 1993).

Life on Mars?

The recent evidence for meteorites from Mars, and the possibility that large impacts on planet Earth resulted in ejecta that reached Mars (see Davis and McKay, 1996) and vice versa, may lead one to the idea of living organisms from one planet colonizing another. This is a specific example of "Panspermia" and may be called Solar System Panspermia.

The recent claim (McKay et al., 1996; see Kerr, 1996; Davis, 1996) of the presence of microfossils of bacterialike cells in a meteoritic rock that was ejected from Mars and landed on Earth may give more credibility to the Solar System Panspermia hypothesis. However, more research is needed to examine both the validity of the interpretation and its implications. If life on Mars is a reality, then the Panspermia theory should be reconsidered as a viable possibility for the origin of life on Earth. It should be noted, however, that the presence of life on Mars, either in the past or at present, still tells us very little about *how* life was formed.

McKay (1997a) noted that if indeed life was formed on Mars before Earth, then it preceded and later produced life on Earth. The rationale of this suggestion is that because of the moon-forming impact (chapter 12), emergence of life on Earth was delayed.

Clues from the geological record

Stromatolites

Chyba and Sagan (1997, p. 148) noted that "there is virtually no remaining geological record of terrestrial conditions prior to the existence of a biosphere." Stromatolites are macroscopic, laminated organosedimentary structures found in various sites around the globe, representing almost all of the history of life on Earth, from the very ancient to the contemporary, and serving as relics of microbial mat communities of cyanobacterial and algal microbenthos (from the Greek *benthos,* translating to "bottom-dwelling organisms"). Morphological and isotopic evidence suggests that stromatolite-building phototactic bacteria were already in existence some 3.5 Ga ago (Schopf, 1993), thus forming an unbroken record of life from very early stages of living organisms to present (Schidlowski, 1993).

Doubts

In a recent work, Grotzinger and Rothman (1996; see also Walter, 1996) developed a novel approach to studying certain physical features of the stromatolite. According to their interpretation, the formation of at least some of these structures can be explained by physical processes, without invoking the involvement of living creatures. Because of the generally accepted association of stromatolite with the early stages of life's evolution on Earth, this novel point of view will no doubt become the focus of many future studies.

Microscopic fossils

Ancient rocks of the Early Archaean (>3.0 Ga ago) era contain microscopic forms that are morphologically similar to contemporary cells of blue algae. The great majority of students of these formations interpret them as preserved cellular microorganisms. The geological formations that have been studied rather intensively are the Serializing Supergroup of South Africa and the Bipolar Supergroup of western Australia (Schopf, 1993; Awramik, 1992). Though both of these rocks were subjected to geologic metamorphism, they contain what seems to be decipherable evidence of early life (fig. 15.1). According to Schopf (1992a), the oldest known microbial communities (including photosynthetic prokaryotes) are at least 3.5 Ga old.

Other interpretations

Though the interpretation of Schopf's school with regard to photosynthetic, algae-like microstructures found in ancient rocks has been very popular among scientists, it has not been accepted unanimously in the scientific community. Alternative interpretations have been suggested according to which these forms are inorganic self-organized precipitates, resulting, for instance, "from a crystal aggregation process controlled by a metal silicate membrane" (Garcia-Ruiz, 1994). Thus, this topic is still under debate.

The oldest records of photosynthesis

Based on the fossil evidence, it is generally agreed that photosynthetic organisms existed on Earth 3.5 Ga ago. This evidence includes the stromatolites in western Australia and South Africa. Moreover, the evidence allows the interpretation that those photosynthesizing organisms included prokaryotic, bacteriumlike microorganisms called cyanobacteria (also called blue-green algae, cyanophytes, and myxophytes). This indicates that at that time "nearly all prokaryotic phyla had already evolved and that prokaryotes diversified rapidly on the early Earth" (Awramik, 1992). The evolutionary implications of the early emergence of photosynthesis, as well as the central photosynthetic pathways, were discussed recently by Blankenship (1992), Gogarten and Taiz (1992), Mauzerall (1992), Buchanan (1992), and Hartman (1992b).

The conversion of CO_2 to organic compounds via the biochemical pathways of photosynthesis involves a fractionation that favors ^{12}C over ^{13}C. The ^{12}C enrichment of the ensuing organic matter can be measured quantitatively, and the organic history of ancient sediments can thus be ascertained. Measurements of the $^{13}C/^{12}C$ ratio in samples of kerogen (modified organic matter) from a sedimentary rock at Isua (Greenland), studied by Schidlowski (1988, 1993), suggest that its origin is organic, 3.8 Ga ago. Thus, Schidlowski (1988) advanced the geological record for life processes by about 300 million years by suggesting that carbonaceous material from the Isua rock formation in Greenland is 3.8 Ga old. This finding has been controversial ever since it was published, in 1988. Schidlowski's claim may be supported by recent indications of the presence of biotic carbon in 3.87 Ga old sediments from Akilla Island, Greenland (Mojzsis et al., 1996; see also Holland, 1997). These are the oldest known sediments containing organic carbon, and the age reported by Mojzsis and his associates indicates that during that time period, life was already prolific in certain areas on Earth.

It should be noted that the presently known stormy record of the young Earth raises a few questions regarding the characteristics of the biological evolution during the Hadean era, which encompasses the time period between the Earth's formation (around 4.6 Ga ago) and the first appearance of sedimentary rocks (around 3.8 Ga

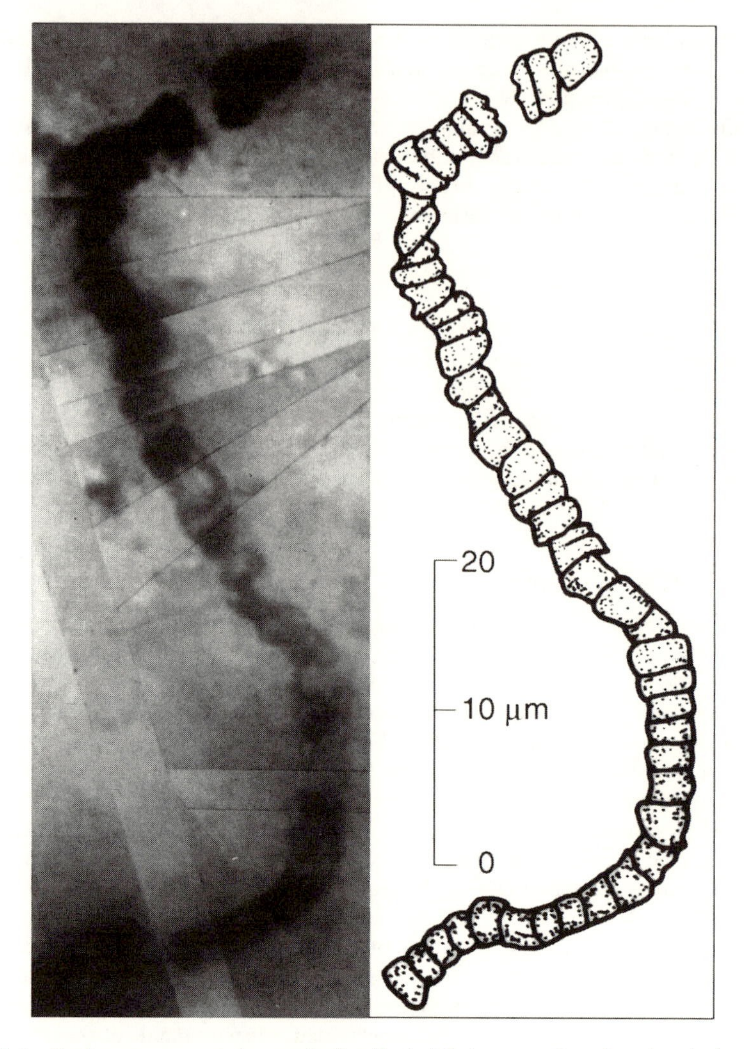

Fig. 15.1. Carbonaceous prokaryotic fossils (with interpretive drawings) shown in petrographic thin sections of the Early Archaean (3,465 million years old) Apex chert of northwestern Western Australia (Schopf, 1993). (Photo courtesy of J. W. Schopf)

ago). For instance, did life emerge after each sterilizing bombardment? Is it possible to estimate the emergence time of life?

Isotopic sulfur evidence for ancient living bacteria and atmospheric oxygen level

Ancient rocks are known in several places around the globe. An analysis of the isotopic sulfur composition ($^{34}S/^{32}S$) of microscopic grains of pyrite (FeS_2) that formed about 3.4 Ga ago in South Africa revealed that the pyrite was formed by bacterial reduction of seawater sulfate. This suggests that sulfate-reducing bacteria were already active 3.4 Ga ago (Ohmoto et al., 1993). Since the geochemistry of sulfur is strongly dependent on the oxygen level in the atmosphere, the results of this study enabled the researchers to estimate the atmospheric concentration of free oxygen. The results imply that the atmosphere at that time contained virtually no free oxygen.

The geological record holds little information regarding the "how"

The geological record pertaining to the oldest known living organisms hold little information regarding the actual evolutionary pathways in which these ancient living entities were formed. It has to do, however, with circumstantial evidence, such as temperature and redox condition, during the molecular evolution era, as discussed later.

Geology and the principle of biological continuity

The following discussion focuses on clues for the origin of life that corroborate with the biological continuity principle. They deal with the prebiotic environment, the time needed for evolution, the availability of the predecessors of the central molecules of biology, and the mechanisms by which they were continuously manipulated at specific sites in an inanimate world, protected from destructive agents.

Is there an Isua-life to biological-life continuity?

It is generally assumed that life on our planet have been continuous for the last 3.8 Ga of its history. This is in large part due to lack of geological evidence for a worldwide sterilizing impact for about 3.8 Ga and due to both the morphological and radioisotopic similarities between the aforementioned ancient microfossils and organic matter on the one hand, and extant blue-algae, on the other. Note, however, that there is no evidence for the proposition that both the ancient and extant types of creatures belong to the same uninterrupted evolutionary line. If various kinds of life did evolve on our primordial planet, then the "fingerprints" of living entities found in the form of $^{13}C/^{12}C$ enrichment (Schidlowski, 1993; Mojzsis et al., 1996) are not necessarily continuous with those microfossils identified by Schopf (1992a, 1992b). Presumably, these life forms could have been initiated during the era of sterilizing bombardments. The possibility of life originating more than once during the earth's history depends on both the length of time needed for the emergence of life, and the conditions needed for this process to take place. Obviously, *if* the emergence of life was a very long process, then its formation is limited to a certain time period, and life is likely to have been continuous only since the decrease in the earth's last heavy bolide bombardments. But if the emergence process is rapid, then it is conceivable that there were many events of initiation and extermination of life during the heavy bombardment era.

Thus, geochemical considerations alone cannot, at our present stage of understanding, resolve the problem of continuity of life. The continuity of life and its monophyletic origin is supported, however, by biochemical and molecular-biological considerations (Lazcano et al., 1992). For a different point of view on the monophyletic origin of life, see Schwabe, 1985. Thus, in spite of strong indications in favor of both continuity and monophyletic origin of life, there are open questions that should wait for more research work.

Width of the time window available for the origin of life: The geological clues are based on circumstantial evidence

Maximal width of the time window

Initially, it was generally assumed that the actual time needed for the emergence of the first living entities was on the order of hundreds of millions of years. The available time interval, the time window width, was estimated by subtracting the age of the first geological records of life from the age of the earth soon after the accretion and cooling-off period. During the 1960s and 1970s, the large time window calculated by this procedure was considered insufficient for the emergence of life by some of the researchers, as reflected by the following quotation (Bernal, 1967, pp. 107–108): "The relatively short time of 1,000 and 2,000 million years for the chemical evolution of life is undoubtedly hard to explain, because it apparently conflicts with the general law of the acceleration of the evolutionary process with time."

Narrower time window

The aforementioned time-window estimates became obsolete when it was realized that decimating bombardments by heavenly bodies plagued our planet until 3.8 Ga ago (Maher and Stevenson, 1988; Sleep et al., 1989; Oberbeck and Fogleman, 1989; Zahnle and Sleep 1997; chapter 13; fig. 15.2). If the last decimating impact happened about 3.8 Ga ago, then the time window available for prebiotic evolution is up to a few million years, according to Schidlowski (1993) and Mojzsis et al. (1996). This figure represents the uncertainties in the time estimates of the methods involved. If the earliest living forms are 3.6 Ga old, as implied by Schopf's (1992a, 1992b) microfossils, then the time window cannot be larger than about 250 million years.

Were bolide impacts needed for the emergence of life?

According to Bada et al. (1994; chapter 13), the primordial earth was characterized by freeze-thaw cycles. These were caused by the lack of sufficient greenhouse gases (and, thus, freezing), on the one hand, and the melting associated with bolide impacts, on the other. Thus, the initial effect of the bolide bombardments was to annihilate terrestrial life; however, additional impacts brought about melting of the ice coverage and the establishment of the conditions needed for the evolution of life.

An upper limit of 10 million years for the time window

Lazcano and Miller (1994, 1996) recently used a geochemical argument in their estimate of the time needed for evolution to proceed from the hypothetical prebiotic soup to the geologically recorded first living prokaryotes. The geochemical considerations in their estimate are based on the destruction of organic compounds during their cycling in the deep-sea vent system, which would reduce the availability of these compounds. The average turnover time of the extant hydrothermal vents is about 10 million years. Assuming the same ternover time for the prebiotic era, Lazcano and Miller (1994, 1996) thus argue that very slow prebiotic reactions are unlikely to have been fruitful for the molecular evolution process, since the organic molecules involved would be destroyed. According to them, the time window is less than 10 mil-

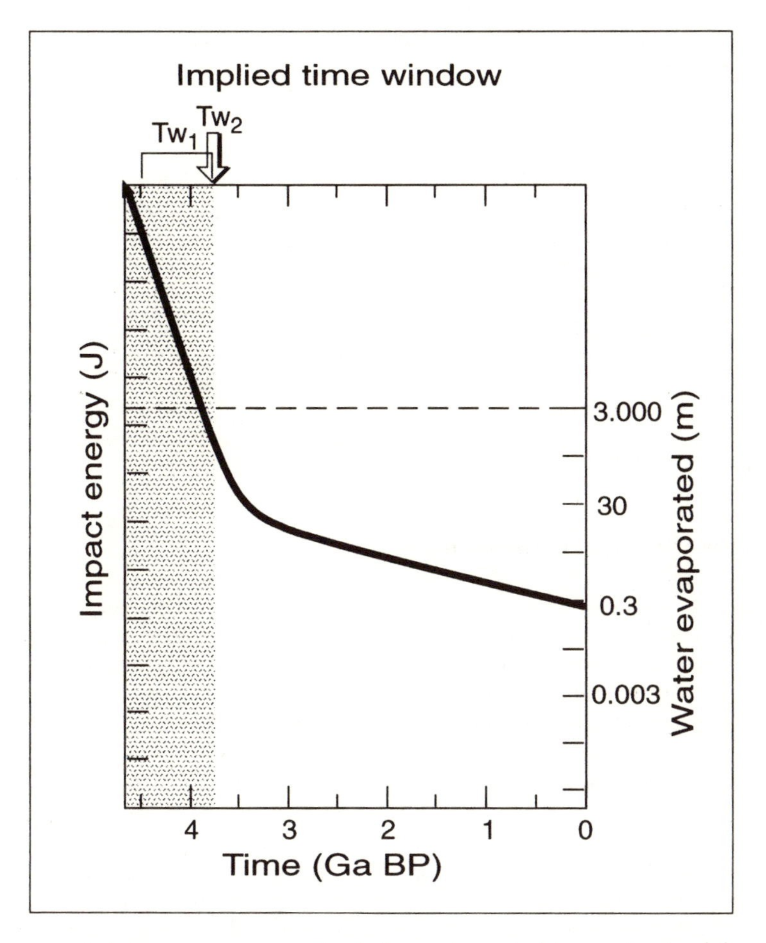

Fig. 15.2. Bolide impact energy on Earth during its early history, depth of the ocean vaporized by the impact, according to Sleep et al. (1989), and the estimated width of the time window for the emergence of life (upper scale). TW_1 is the time window calculated without taking into account the meteorite bombardments. TW_2 represents the narrow time window based on the last sterilizing impact (~3.8 Ga BP) on the one hand, and the geological record of the first living entities according to Schidlowski (1988) and Mojzsis et al. (1996) (i.e., ~3.85 Ga BP), on the other. The uncertainties involved in these estimates result in inaccuracies of the time window. Note that adequate conditions for continuous life characterize the primordial earth since ~3.8 Ga ago. (Adapted from Sleep et al., 1989)

lion years. It is noted that the turnover time is an average value and that its use in such a calculation may be problematic. Furthermore, it does not take into account environmental systems such as lagoons, tide pools and lakes.

Recent measurements of sea-floor vents (Von Damm et al., 1995) show large variations in both temperatures and composition of the vent fluids in a matter of a few weeks. These observations suggest that the present models of hydrothermal fluxes should be revised.

A time window of 6,000 years

According to Greenberg (1995), the collision frequency of comets at the end of the strong bombardment period, some 3.8 Ga ago, decreased to about 1 in 6,000 years. Thus, the time window of his scenario is approximately 6,000 years. Greenberg's scenario is discussed later in this chapter.

Very short emergence time of life

Computer modeling of biogeochemical scenarios for the emergence of the first prebiotic autocatalytic entities suggest emergence times on the order of years or tens of years (White and Raab, 1982). These calculations are controversial (Joyce, 1983; chapter 23).

Relevance of prebiotic environments

The question "In what environment did the first 'living entities' emerge?" cannot be answered at present and is highly controversial, as are the concomitant corresponding scenarios (for a summary of this topic see chapter 22). Adoption of a site for prebiotic evolution scenario should be based on several considerations and includes a variety of geochemical parameters by which the plausibility of each site can be evaluated. Moreover, an evaluation of the adequacy of such sites can be performed only in conjunction with specific scenarios based on a specific set of molecules and reactions. For instance, the effect of radionucleotides on specific molecular-evolution reactions, such as polymerization (Martell, 1992; Negrón-Mendoza and Albarran, 1993; see also Mosqueira et al., 1996), is of interest but should be integrated into a comprehensive scenario to be a viable hypothesis for the emergence of life. On the other hand, the role of iron and sulfur in molecular evolution is connected to rather detailed scenarios and environments (Hartman, 1992b; Wächtershäuser, 1992a, 1992b). Such scenarios are discussed later. Before discussing their details, however, we must draw up the main guidelines for the selection of candidate sites.

The principle of environmental continuity

Analogous to the principle of biological continuity, and in fact, as an implied complementary component of it, the principle of *environmental continuity* reflects the inseparability of living entities from their environment. Biological continuity can be maintained only if at least some of the biological entities are not destroyed or deactivated during evolution. The environmental continuity principle can thus be phrased as follows: For any biological evolutionary process to take place, there must have been concomitant environmental continuity, both chemical-physical and geographical, through which biological continuity could have proceeded without interruption.

Features of plausible prebiotic environments

The following list is general enough to also include some other forms of "life" based on chemical entities different from the biochemical ones.

Availability of C, H, O, N, S, P

Availability of the primary elements of the extant cell—C, O, H, N, S and P—in a suitable chemical form, capable of undergoing molecular-evolution processes, in suitable environments and concentration ranges. A speculative suggestion regarding synchronizing effects of NO in prebiotic chemistry will be dicussed in chapter 20.

Availability of trace elements

These include metal cations such as Zn^{++} or Fe^{++} (see reviews by Kobayashi and Ponnamperuma, 1985a, 1985b).

Availability of catalysts

These include metal elements such as Mg^{++}, minerals, and organic molecules (Böhler et al., 1996).

Availability of activated building blocks of the central biopolymers

Prebiotically plausible activating agents capable of activating amino acids, thus enabling peptide-bond formation, have been suggested by various researchers. These include thioester derivatives (Weber and Orgel, 1979; Keller et al., 1994) and acylated amino acids synthesized by means of thioacids and oxidizing agents (Liu and Orgel, 1997). These reactions take place in aqueous solutions. Attempts to activate mononucleotides under plausible prebiotic conditions, and thus to bring about their condensation, have not been successful so far (see below).

Availability of at least one energy source

Energy sources such as solar energy or chemical energy must be available for molecular-evolution reactions.

Lack of thermodynamic equilibrium

Any hypothesized site must be "contained far from thermodynamic equilibrium by kinetic barriers" (Eschenmoser, 1994), as discussed in chapters 8 and 11.

Appropriate intensity ranges of damaging agents

The intensity of potentially damaging agents such as temperature, radiation, and chemical compounds should never be "lethal" simultaneously for the molecular-evolution processes taking place at the sites under consideration.

Closed microenvironments and open macroenvironment

Whereas the microenvironment should be compartmentalized, the macroenvironment in which the compartmentalized entities and their niches are to be found needs to be open and interconnected. The latter attribute is necessary not only to renew food supplies but also to avoid damaging effects such as coverage by sediments or locking into isolated porous media.

Availability of either organic or inorganic compartmentation means

This consideration is based on the observation that in extant cells the organization pattern of lipid molecules in the membrane is not determined by template-directed processes. This, then, implies that the use of an inorganic compartmentation agent in molecular-evolution scenarios is not necessarily in disagreement with the principle of biological continuity (chapter 21).

Sources of the major elements

According to all existing models (chapter 13), carbon, hydrogen, oxygen, and nitrogen were available in the prebiotic atmosphere. Sulfur may also have been present in small concentrations in the atmosphere and in the sea water, but most sulfur existed in the earth's crust in the form of metal sulfides.

In contrast to the high concentration of phosphorus in living systems, this element seems to be rather scarce in the interstellar medium. Phosphorus seems to be a minor but ubiquitous element in the solar system (Maciá et al., 1997) and was detected in the mineral anionic forms of PO_2, PO_3, and phosphate minerals in interplanetary dust particles. Of special interest from the chemical evolution point of view is the detection of methyl and ethyl phosphonic acids in the Murchison meteorite. Thus, organic phosphorus compounds capable of being involved in the origin-of-life processes could have been delivered to the primordial earth by comets and meteorites. For more details see Arrhenius et al., 1997; Maciá et al., 1997; and Schwartz, 1997a, 1997b.

Prebiotic synthesis of central biochemical molecules

It is convenient to make the distinction between prebiotic synthesis, on the one hand, and the prebiotic reactions undergone by the organic molecules thus synthesized, on the other. Moreover, such a distinction fits quite well into the prebiotic-soup picture and the Oparin-Haldane-Urey-Miller paradigm. It should be noted, however, that there is not always such a clear-cut distinction between different stages in chemical evolution (chapter 21). For simplicity and convenience the present section deals with the synthesis of organic molecules, monomers, oligomers, and polymers, without relating them to primordial metabolic cycles or template-directed reactions.

Energy sources for prebiotic syntheses

Energy is needed for the production of organic compounds from inorganic reactants in what is called prebiotic synthesis. The present sources of energy in the biosphere of our planet include solar radiation, electric discharges, volcanos, shock waves, and chemical energy. The same energy sources presumably also characterized the prebiotic earth, albeit at different intensities. The conditions under which these energy sources, activated molecules, and quenching reactions could have been involved in prebiotic synthesis are discussed by Miller (1992).

Since Miller's classical experiment (1953), with its electrical discharges and reducing atmosphere, a wide variety of energy sources have been tested and various gas mixtures have been employed, according to the model of the prebiotic atmosphere adopted by the researchers. Ultraviolet light was found not to be an efficient energy source for the production of amino acids. Similarly, heating reactions (pyrolysis) of CH_4 and NH_3 gave very low yields (for more details and references see Miller, 1992).

The importance of a given energy source may be calculated as the product of available energy and its efficiency in the organic synthesis reaction. The actual importance of an energy source should be evaluated in relation to a specific environment or reaction. The efficiency factors can only be estimated. Moreover, no single energy source can account for all syntheses on the prebiotic earth.

The Miller-Urey experiment in the light of the principle of biological continuity

The most famous uses of biological continuity, prebiotic plausibility, and testability guidelines were contrived in the classical Miller-Urey experiment (Miller, 1953). In this experiment, as you may remember, electrical discharges were applied to a closed reaction chamber filled with a mixture of gases consisting of water (H_2O), methane (CH_4), ammonia (NH_3), and hydrogen (H_2), thus simulating the then-hypothesized prebiotic atmosphere and one of the presumed prebiotic energy sources (chapter 5). The products of the reaction were collected as a solution (the prebiotic soup) and partially analyzed for various building blocks of biochemistry. These compounds comprised organic compounds of various chemical groups, including hydrogen cyanide, aldehydes, and amino acids (Orgel, 1994). It should be noted, however, that the amino acids were racemic, that is, both D- and L-enantiomers (except for glycine, which is a symmetric molecule).

By looking for amino acids in the solution of his reaction chamber, Miller used the biological continuity principle: amino acids were assumed to have served as prebiotic building blocks of some very primitive biocatalysts, similar in their chemistry, structure, and functions to contemporary proteins. The fact that the amino acids were racemic indicated that mechanisms for the establishment of the amino acid homochirality (L) observed in biology should also be looked for. But most researchers considered this a problem that could be postponed for later research stages.

Following the Miller-Urey experiment, researchers began studying other biochemical building blocks. In the hundreds and thousands of "prebiotic syntheses," and the many speculative scenarios that followed this classical experiment, environments, reactants, and energy sources other than the ones used by Miller and Urey were also employed. The results of these experiments seem to confirm the assumption that most of the important biochemical building blocks could have been synthesized on primordial Earth. The products of such prebiotic syntheses include amino acids, sugars, pyridine and pyrimidine bases, coenzymes, and various carboxylic acids. However, attempts to carry out, under "prebiotic conditions," the synthesis of several important building blocks, most notably D-ribose, have been unsuccessful to date.

Prebiotic synthesis of biochemical building blocks

Under this heading are included all the organic compounds that have been synthesized in the laboratory under "plausible prebiotic conditions." The latter term represents the understanding of the researchers at the time of the experiment. It may change as a result of improved understanding, as has been the case with the models of the primordial atmosphere. Thus, the phrase "plausible prebiotic conditions" is model dependent.

By far the most popular energy source in experiments of prebiotic syntheses has been the Miller's spark discharge. The main products of this experiment are amino acids, hydroxy acids, short aliphatic acids, and urea. A typical feature of these experiment is that a small number of compounds, almost all of them of biological importance, were produced in substantial amounts. Among the amino acids, glycine is always the major product. Another important amino acid in these experiments has been valine.

The central role of HCN in prebiotic synthesis

Most researchers agree that hydrogen cyanide (HCN) is a key compound in the prebiotic synthesis of the central molecules of biology, that is, amino acids, purines, and pyrimidines and their derivatives (for recent works see Miller, 1992; Oró, 1994; Matthews, 1995; Robertson and Miller, 1995a, 1995b; Robertson et al., 1996; Zubay, 1996; Keefe and Miller, 1996b). However, the possible role of HCN in prebiotic scenarios involves some difficulties. One such problem is the following: Cyanide (CN^-) reacts efficiently with formaldehyde (HCHO) to form glyconitrile (NCCOH); due to the efficiency of the scavenging of cyanide by formaldehyde, all cyanide species in the hydrosphere and cryosphere that would not be bound as ferrocyanide would be effectively removed as glyconitrile. This problem exemplifies the difficulties of devising a prebiotic process based on the scant information available from prebiotic experiments. Possible solutions for the problems of cyanide in the prebiotic context were discussed in Arrhenius et al., 1994.

Examples of the syntheses of some important building blocks of biochemistry

Synthesis of amino acids by the Strecker synthesis The mechanism of synthesis of amino acids in the spark discharge experiment was studied by Miller in 1957. It was found that the amino acids were not formed directly in the electric discharge but were the result of solution reaction of aldehydes and hydrogen cyanide, which were synthesized in the discharge (Miller, 1992) according to this equation:

$$\underset{\text{aldehyde}}{RCHO} + HCN + NH_3 \rightleftharpoons \underset{\text{amino nitrile}}{RCH(NH_2)CN} \xrightarrow{H_2O}$$

$$RCH(NH_2)\overset{\overset{\displaystyle O}{\|}}{C}{-}NH_2 \xrightarrow{H_2O} \underset{\text{amino acid}}{RCH(NH_2)COOH} \tag{15.1}$$

where R represents the side chain of the aldehydes or amino acids.

This reaction is called the Strecker synthesis; it requires the presence of the ammonium ion (NH_4^+). NH_4^+ and ammonia (NH_3) are chemically related as follows:

$$NH_3 + H^+ \rightleftharpoons NH_4^+ \tag{15.2}$$

Assuming the ammonium ion concentration and the pH as well as the temperature of the primordial sea, it is possible to estimate the equilibrium concentration of the ammonia in an atmosphere equilibrated with this sea. For instance, when the NH_4^+ concentration of the sea is 0.01M at pH8 and 25°C, the partial pressure of NH_3 (which is needed for the amino acids synthesis) in the atmosphere would be 4×10^{-6} Atmospheres (Miller, 1992).

Amino acids that have been found in relatively large concentrations and yields in Miller-type experiments are considered "prebiotically plausible" by many researchers, meaning that their presence in prebiotic environments (the primordial soup) is considered likely. The list of these amino acids includes glycine, alanine, aspartic acid, and valine.

An interesting feature of the Miller-Urey experiment is the production of more than twenty nonbiological amino acids (Miller, 1987; see also Maynard Smith and Szathmáry, 1995). This observation will be discussed later (chapter 19) in relation to the question, "Why were only 20 amino acids selected by biology?"

Synthesis of purines and pyrimidines The mechanisms of the prebiotic syntheses of purines and pyrimidines were recently reviewed by Miller (1992), Oró (1994), and Zubay (1996).

Synthesis of sugars The synthesis of sugars from formaldehyde under alkaline conditions, the *formose reaction*, is considered a prebiotic reaction because both the reactant, formaldehyde, and the reaction conditions are considered prebiotically plausible. The reaction depends, however, on the presence of an inorganic catalyst, such as $Ca(OH)_2$, $CaCO_3$, or clay minerals (see Joyce, 1989; Miller, 1992). The reaction is autocatalytic and proceeds through a series of stages that include a variety of sugars (Miller, 1992).

Two problems with regard to the formose reaction as a prebiotic source of sugars were addressed by various researchers. The first one has to do with the stability of sugars. The estimated decomposition time of pure sugars under presumed prebiotic conditions is a few hundred years or less at 25°C (Miller, 1992). Pure ribose is short-lived, depending on temperature and pH. But sugars are known to be stabilized by converting to glycosides (attaching the sugar to purine or pyrimidine base). Mineral-induced synthesis of ribose phosphates was reported by Pitsch et al. (1995), and Krishnamurthy et al. (1996) and may be also involved in increasing the ribose stability. Thus, the stability problem may not be a difficult problem in certain cases.

The second problem is more serious; it was raised forcefully by Shapiro (1988) as follows: The formose reaction gives a wide variety of sugars. The number of these sugars, both straight-chain and branched, is more than 40, including ribose, the sugar used in RNA. Therefore, the involvement of ribose in the prebiotic production of nucleic acids is unlikely. The conclusion adopted by most researchers is thus that the presence of ribose-containing nucleotides in the prebiotic environment is unlikely (Miller, 1992). In order to overcome this difficulty, a number of ribose substitutes have been suggested (Miller, 1992; Maynard Smith and Szathmáry, 1995). Another problem connected to the chirality of ribose will be discussed in chapter 17.

Prebiotic amphiphiles A major problem in the prebiotic synthesis of amphiphiles is that so far no abundant source for these compounds has been identified. The Miller-Urey synthesis is probably an unlikely source of long-chain hydrocarbons under prebiotic conditions (Deamer, 1997; Deamer et al., 1994; Deamer and Volkov, 1996). The fatty acids found in these experiments were branched and straight fatty acids C_2 through C_{10}; they do not produce the vesicular structures called liposomes that are considered possible precellular entities. Prebiotic formation of liposomes will be further discussed in chapter 19.

Catalytic effects of micelles, such as on prebiotic polymerization of organic molecules, were suggested already by Oparin. They were recently proposed theoretically for membrane-water interface (Pohorille and Wilson, 1995) and shown experimentally for certain micelles (Böhler et al., 1996).

Prebiotic synthesis of coenzymes Coenzyme M is a simple molecule, 2-mercaptoethanesulfonic acid: $HS\text{-}CH_2CH_2\text{-}SO_3^-$. It was discovered in methane bacteria, and is involved in methane production from CO_2. Since methanogens are very early bacteria coenzyme M was investigated as a possible prebiotic compound (Miller and Schlesinger, 1993a).

The ubiquity and central role of coenzyme A (CoA; see chapter 9) in many enzymatic reactions suggest its early involvement in the origin of life. More specifically, it could have been involved in the peptide synthesis that is related to the thioester world (chapters 17, 19), which, according to several researchers, is assumed to have preceded the RNA world (de Duve, 1991). Miller and Schlesinger (1993a, 1993b) suggested that CoA-type enzymes could have evolved under plausible prebiotic conditions and that the precursor of CoA contained most of the components of the present coenzyme. Keefe, Newton, and Miller (1995) have shown that the panetheine moiety of coenzyme A can be synthesized under prebiotic conditions, which include a temperature of 40°C and the presence of various organic molecules that are considered "prebiotically-plausible." The high solubility of these molecules immediately suggests a fluctuating environment such as a lagoon or tide pool, where evaporation brings about elevated temperatures and increased reactant concentrations.

The assumption adopted by Keefe and his collaborators with regard to the synthesis of coenzyme A is that the structural units consisting of the coenzyme under study were formed independently of RNA molecules. Following their independent synthesis, coenzymes were then incorporated into the metabolic activity of ribozymes (RNA catalysts) (see Ferris, 1995). The synthesis of coenzymes and their incorporation into the primordial metabolism of the emerging life is still under debate among researchers.

In reviewing the scant experimental work that has been done on this topic, Ferris (1995) noted that the reason for the apparent lack of interest in this synthesis may be simply that most origin-of-life researchers focus on the synthesis of peptides and oligonucleotides.

The last note by Ferris should not surprise us, since this is probably a rather common feature of the advancement of science: In the absence of a coherent paradigm regarding the entire field of the origin of life, most researchers are attracted to hypotheses that seem to be consistent within their own small domains and, even more so, to areas that seem to be central to the field in general. The relevance of these small domains to the entire field will be known only in hindsight. Examples of such scientific domains are the Miller-Urey prebiotic experiment, the "clay world," and the RNA world (chapters 17, 19). In spite of their acknowledged contribution to the scientific progress in the field of the origin of life, their place in the scientific understanding of the origin of life is still not clear.

Other prebiotic compounds

Numerous other organic compounds have been synthesized under "plausible prebiotic conditions." These compounds include dicarboxylic acids, tricarboxylic acids, and imidazoles. Other compounds that may have been involved in prebiotic polymerization reactions include derivatives of cyanide and phosphate polymers (Miller, 1992) and cytosine and uracil (Robertson and Miller, 1995a, 1995b).

Compounds that have not been synthesized so far by "plausible prebiotic reactions"

Since the prebiotic plausibility is up to the researcher, these compounds should be reconsidered repeatedly, and their list is expected to change in the future. Some important biological compounds that could not be synthesized under prebiotic conditions include several amino acids (arginine, lysine, and histidine), porphyrins and riboflavin (Miller, 1992).

Non-templated-directed prebiotic polymerization

The following brief discussion on prebiotic polymerization reactions is limited to reactions that take place without the intervention of one-dimensional organic templates, such as the extant nucleic acids. The present reactions will thus be referred to as non-template-directed reactions.

Following the prebiotic synthesis of building blocks, the next stage of prebiotic molecular evolution is the condensation of the building blocks into oligomers and polymers. Since the major biopolymers are proteins and nucleic acids, the two chemical bonds that are of the greatest importance in the present discussion are peptide bonds formed between the amino and carboxyl groups of amino acids, resulting in a peptide, and ester bonds formed between an acid and an alcohol group, linking phosphate and pentose sugars (phosphodiester linkages) in strands of nucleic acids. In extant organisms the template-directed polymerization of amino acids into proteins and nucleotides into the polymers DNA and RNA consume a large part of the energy produced by the cell: The dehydration-condensation reactions, in which a water molecule is removed between the two functional groups, require energy input.

Peptide formation by dry heating

One of the most common sources of energy for condensation reactions in the prebiotic environment is likely to have been heat, followed by dehydration. For instance, the polymerization reaction studied mainly by S. Fox (Fox and Dose, 1977) involves heating of a mixture of amino acids, characterized by an excess of aspartic and glutamic acids, at 150°C to 180°C for a few hours. The polypeptide formation is described (see Miller, 1992) as follows:

$$n \text{ amino acids (with excess asp and glu)} \rightleftharpoons \underset{\text{(proteinoid)}}{\text{polypeptide}} + nH_2O \tag{15.3}$$

The plausibility of the reaction conditions (dry heating) has been questioned by several researchers (see Miller, 1992). The relevance of the resulting "proteinoids" to the origin of life will be discussed in chapter 22.

Peptide formation in solution

Peptide-bond formation can take place also in solution, with the help of chemical condensing agents, such as cyanamide (H_2N—CN) or dicyandiamide (H_2N—CNHNH—CN).

Minerals as catalysts in peptide-bond formation

Condensation of glycine in the presence of kaolinite clay mineral at temperatures below 100°C in a fluctuating system undergoing a hydration-dehydration process was reported first by Lahav et al. (1978) and was followed by other researchers.

A recent study has shown (Ferris et al., 1996) that a combination of glutamic acid, illite mineral, and a condensing agent (carbonyl diimidazole) resulted in a very efficient condensation reaction of the glutamic acid. Aspartic acid plus hydroxylapatite in the presence of the condensing agent carbodiimide gave polymers of very high molecular weight. The possible role of minerals in the condensation of amino acids was recently reviewed by Lahav (1994), Bujdak et al. (1995) and Bujdak and Rode (1995, 1996). More details about prebiotic condensation reactions of amino acids can be found in Collins et al., 1988, and Miller, 1992. The presumed relevance of minerals to the origin of life will be discussed in chapter 20.

Prebiotically plausible nucleic acids formation

Aqueous phase Polymerization of nucleotides in an aqueous phase to form nucleic acids under prebiotic conditions is problematic, yielding oligomers that are predominantly smaller than the pentamer (Ferris et al., 1996; Von Kiedrowski, 1998).

Clay-catalyzed condensation of nucleotides The most successful attempts to synthesize polynucleic acids under "plausible prebiotic conditions" were carried out by Ferris and his coworkers (1996; Ding et al., 1996; Ertem and Ferris, 1996). In one of these studies (Ferris et al., 1996) a montmorillonite clay mineral was used as an adsorbent and catalyst. The nucleotide monomers were activated by imidazole (fig. 15.3).

Polymerization on the surface of iron(III) hydroxide oxide A major problem in the polymerization reactions carried out in the presence of water, such as condensation of polynucleotides, is the destruction of the condensing agents and reactive intermediates by water, thus preventing the production of large polymers. Based on earlier works, Weber (1995) used iron(III) hydroxide oxide (Fe(OH)O) in the oxidative polymerization of 2,3-dimercapto-1-propanol (fig. 15.4). In addition to the prebiotic plausibility of the reaction under study, it was noted by Weber that "redox reactions could have provided the energy for the earliest type of polymer synthesis involved in the origin of life." Weber's experiment plus theoretical considerations suggest that polysulfides could have had important functions in prebiotic reactions. Additional redox reaction on solid surfaces in which iron is involved are discussed in chapter 21.

Doubts and uncertainties

It should be noted that the Oparin-Haldane-Urey-Miller paradigm has been under attack in the last decade or so, mainly because according to new and more reliable models (chapter 13) the prebiotic atmosphere is likely to have been less reducing than the one assumed by Urey and Miller. Under such conditions, many organic molecules that were synthesized in the Miller-type experiment, with its reducing atmosphere, would not be formed, or would be formed in very small quantities. Moreover, as was noted earlier, so far no geological evidence for a prebiotic soup of kind has been was found. In addition to these geochemical problems, the Miller-Urey model was criti-

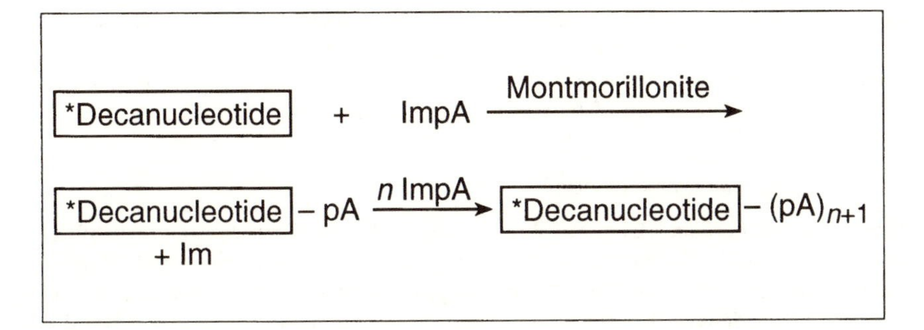

Fig. 15.3. Elongation of ^{32}P-labeled decanucleotide (*Decanucleotide) by reaction with phosphorimidazolide of adenosine (ImpA). The ^{32}P-labeled phosphate group was used to identify the reaction products. Im = imidazole; pA = adenosine-5′-phosphate; pdA = 3′-deoxyadenosine-5′-phosphate. (Adapted from Ferris et al., 1996)

cized by some researchers because it was heterotrophic. According to these critics, autotrophic models are more plausible than heterotrophic models, as discussed in chapter 21.

In view of these recent developments, the Oparin-Haldane paradigm seems to have lost some of its appeal, as well as its leading position in the scientific discipline of prebiotic chemistry. And even though it is still rather popular among researchers, it has become one out of several conjectures that have been adopted by various schools of thought.

Extraterrestrial sources of organic compounds

It is interesting to note that the most glamorous (though pseudoscientific) modern suggestion of extraterrestrial source of large amounts of organic compounds on Earth was made by Velikovsky (1950). Even though his work does not corroborate with scientific criteria, it certainly incited the imagination of many people, judging by the (short-lived) great popularity of his books in the early 1950s. Could it have influenced some scientists, triggering the adoption of novel and bold ideas that were, at that time, close to the realm of pure imagination?

As noted above (chapter 12), import of organic compounds of extraterrestrial origin to the primordial earth and its involvement in the origin-of-life processes were independently suggested by Chamberlin and Chamberlin (1908) and by Juan Oró (1961) at a time when it was considered a wild speculation. At that time analysis of meteorites was problematic, both because of lack of analytical tools and methods and

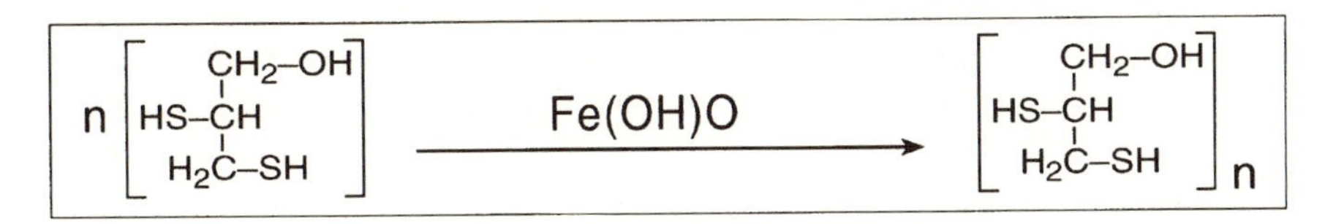

Fig. 15.4. Oxidative polymerization of 2,3-dimercapto-1-propanol on the surface of iron(III) hydroxide oxide. (Adapted from Weber, 1995)

because of contamination by terrestrial organic compounds. Many years had to elapse before this speculation became an established field of scientific study among students of the origin of life. More information about organic matter in meteorites can be found in Cronin and Chang, 1993; Delsemme, 1997; and Kissel et al., 1997.

Extraterrestrial organic carbon

Several researchers (see Chyba, 1997; Chyba and Sagan, 1992; Anders 1989) estimated the infall of extraterrestrial material delivered to the earth's surface. Conservative estimates are 10^6–10^7 kg per year for cometary delivery and 10^8–10^{10} kg per year for delivery of interplanetary dust particles. Over a period of 100 million years following the primary accretion process of our planet, some 4.6 Ga ago, the total amounts of organic carbon added to Earth by these two sources have been estimated as 10^{16}–10^{18} kg. For comparison, the total organic carbon of extant organisms on our planet is estimated to be 6×10^{14} kg. The extraterrestrial sources of organic carbon during the time period in which life is hypothesized to have evolved are thus significant.

Meteoritic amino acids

The first amino acids isolated from meteorites were suspected as terrestrial contaminants. These uncertainties were resolved when a carbonaceous meteorite fell in Murchinson, Australia; it provided the first demonstration for the existence of amino acids in the solar system, outside Earth. Careful analysis revealed that those amino acids that contained a chiral carbon were present in the meteorite as racemic mixtures. Moreover, the dominant amino acids included some that rarely occur in organisms and some that had never before been detected in organisms. No less interesting is the fact that the mixture of meteoritic amino acids was very similar to that found experimentally by the Miller-Urey experiment (Bada, 1991).

Amphiphilic molecules in carbonaceous chondrites

As explained in chapter 13, organic compounds in this group of meteorites are relatively abundant. Moreover, it was demonstrated by Deamer (1994, 1997) that amphiphilic substances could be extracted from the Murchison meteorite (which is a carbonaceous condrite), could be assembled into membranes, and, under certain circumstances, could encapsulate aqueous solutes. However, it was argued by Anders (1989) and by Chyba and Sagan (1992) that carbonaceous condrites would have contributed a negligible amount of amphiphiles to the primordial earth.

Soft impact of a cometary nucleus?

It was speculated by Clark (1988) that life originated on Earth in a small, shallow body of aqueous solution of organic compounds formed by a rare "soft impact" of a cometary nucleus. This interesting possibility has not been explored so far.

Greenberg's scenario

According to Greenberg (1995), the organic molecules that initiated the molecular evolution process were parts of comet nucleus, forming a mantle on interstellar grains. During the collision tail-off period about 3.8 Ga ago, the collision frequency fell off to about 1 in 6,000 years. Since the age of the oldest geological records of life seems to overlap with that of the end of the heavy meteorite bombardments, the transition from inanimate molecules to molecules with self-replicating features occurred during a very short time. Moreover, the very dense CO_2 atmosphere of the prebiotic earth provided a cushioning effect for these molecules.

Greenberg also suggested a mechanism for the formation of an enantiomeric excess in the organic compounds, proposing that the clouds of interstellar dust passed near neutron stars. The circularly polarized UV radiation emitted from neutron stars thus formed an enantiomeric excess in the organic grain mantles, which was preserved for a long period of time. Upon landing on the primordial Earth the organic molecules started an unspecified chemical evolution process which resulted eventually in living entities.

Whether or not this mechanism is plausible, it is recalled that enantiomeric excess of amino acids in meteorites was recently observed by Cronin and Pizzarello (1997) and Engel and Macko (1997).

Summary

The classical alternative to the proposed terrestrial origin of life is the Panspermia theory and its modern version of 'directed panspermia'; together with the 'solar-system-panspermia' they represent theoretical constructs which do not offer any mechanism for the origin of life. As a matter of fact, even the controversial spectroscopic observations of bacteria in space and the "fossils" in rocks from Mars are not likely to provide many clues to the understanding of the origin of life.

The geological record does not carry clues regarding the chemical composition of primordial "living" entities or regarding mechanisms and reaction pathways of the transition from inanimate to animate. It does, however, provide clues regarding the morphology and age of microstructures believed by most researchers to represent fossils of ancient microorganisms, as well as evidence of specific chemical signatures regarding the activities of ancient living organisms. It also provides information regarding the possible geological environments in which life could have emerged, and the time window during which life could have emerged, on the primordial earth.

Clues for the origin of life from both cosmology and geology also include terrestrial and extraterrestrial origin of the molecules from which living organisms are made. Based on theoretical considerations and experimental observations, it is likely that extraterrestrial import has been a major source of organic compounds, water, and phosphorus to the primordial earth, whereas the environment in which these reactants were contrived to bring about the emergence of life was terrestrial. Thus, rather than the Panspermia thesis, and the terrestrial-origin-of-life antithesis, the origin of life is better understood as a synthesis of the two.

Based on laboratory experiments of prebiotic synthesis, it has been esstablished that many, but not all, of the major building blocks and central molecules of biology could have been synthesized under "terrestrially plausible" prebiotic conditions. Non-template-directed condensation reactions of amino acids and nucleotides into their respective oligomers and polymers have also been carried out under "plausible" prebiotic conditions.

16

Clues from Biology

Evolution, Conservatism, Continuity, and Their Implications

There is a constant urge in man to seek beginnings.
Samuel Granick, "Speculations on the origins and evolution of photosynthesis"

Two properties of living systems that are unique from the standpoint of physics, namely, self-replication and homochirality, may serve as Ariadne's thread in the labyrinth of hypotheses concerning this [the origin-of-life] problem.
Vitalii Goldanskii, "Chirality, origin of life, and evolution"

The language of the origin of life is chemistry

Echoing his contemporary scientific know-how which focused on the chemistry of life at the molecular level, Handler (cited in Rosen, 1989) wrote, "Life can be understood only in the language of chemistry." Indeed, the application of the principle of continuity to the chemistry of the origin of life is a most important tenet of the search into the origin of life. This search should be based on the properties of the very building blocks of biochemistry, the organic entities of which contemporary biopolymers are made.

Constituents

Chemical elements

The chemical elements of the hypothetical first living entities are assumed to have been those of biochemistry. These are the major elements C, O, H, N, P, and S, or at least the great majority of them. The secondary and microconstituent elements were also recruited for their functions from the prebiotic environment. The relative abundance of these elements in extant cells probably reflects solubilities of chemical constituents in the prebiotic environment (Wächtershäuser, 1992a), and is related to other physicochemical properties of the chemical elements (Banin and Navrot, 1975).

Metal clusters

The possible relationships between extant metal clusters and prebiotic chemistry will be discussed later in connection to the origin of photosynthesis and the pyrite-world theory (chapter 21).

Universal biomolecules

At least some of the universal biomolecules (table 7.2), and probably most of them, or their chemical precursors, are assumed to have been used by the first living entities. This, then, implies certain limits regarding the environments in which the first stages of life could have taken place.

The origin of biological homochirality

> The bias for a single handness in biopolymers from homochiral L-α-amino acids and D-sugars is still regarded as a most remarkable hallmark in nature.
>
> Isabelle Weissbuch et al., "Lock-and-key processes at crystalline interfaces"

The biological symmetry-breaking has been a controversial problem since its discovery by Pasteur. The chirality of biological amino acids and sugars poses a problem for researchers of the molecular evolution processes, because in all known laboratory prebiotic syntheses studied so far, the products were racemic. It has thus been assumed by most researchers that the prebiotic syntheses of these molecules, some 4 billion years ago, also were characterized by racemic products. Considerations of stereochemistry on the one hand, and the principle of continuity on the other hand, suggest that the first "living" entities were homochiral. How can one explain the emergence of the first homochiral organic oligomers and polymers, which presumably constituted the first "living" entities, from the prebiotic racemic mixtures? The three possible answers to this question have been that biological homochirality originated either before, during, or after the origin of life.

It is convenient to divide the hypotheses on the origin of chirality in the biological world into two categories, namely, biotic and abiotic. In addition, it is convenient to differentiate between enantiomeric amplification induced by terrestrial mechanisms, such as adsorption on mineral surfaces and enantiomeric amplification induced by extraterrestrial mechanisms, such as circularly polarized ultraviolet radiation from neutron stars remnants of supernova (chapter 15).

The biotic school

According to this school, the ability to select the enantiomers of chiral biochemical molecules has been an inherent property of "living" entities. This attribute evolved sometime during the molecular evolution era.

The abiotic school

According to this school, chiral segregation of the building blocks of the biopolymers was carried out prior to the formation of these biopolymers. Moreover, "such asymmetry could have emerged provided one can amplify a small fluctuation from the racemic state to magnitudes useful for biotic evolution. The small enantiomeric excess can be generated either through a determinate or a chance mechanism" (Weissbuch et al., 1994, pp. 173–174).

According to Goldanskii and Kuzmin (1991), "The chiral purity observed in the biosphere was achieved at the stage of prebiotic evolution and was a necessary condition for the subsequent development of self-replication. . . . Chiral purity resulted not from the gradual (evolutionary) accumulation of an enantiomeric excess, but by a process in which mirror symmetry was broken spontaneously." Moreover, "the sign of chiral purity in the bio-organic world (L-amino acids and D-sugars) is random."

Recent analyses obtained from the Murchison meterorite have shown the possibility of abiotic enantiomeric enrichment (Cronin and Pizzarello, 1997; Engel and Macko, 1997; Mason, 1997). Since the Murchison meteorite was formed about 4.5 Ga ago, the formation of this enantiomeric excess may have preceded the origin of life (see also Bada, 1997).

The origin of chirality is still unsolved

The problem of the origin of chirality has not been settled so far. It will receive more attention in chapter 17. For recent reviews see Bonner, 1991, and 1995; Navarro-Gonzalez et al., 1993; Popa, 1997; Ponnamperuma and Chela-Flores, 1993; and Weissbuch et al., 1994.

Quid pro quo: Chemical stability in exchange for replication, exchange, and interactions

Life may be viewed as a succession of generations of complex chemical entities undergoing evolution, where exchange, mutability, and replication processes of the molecules compensate for their low stability. "Life is a general example of the Delphian boat, this boat made of planks that would have to be replaced one after one, as they start to rot away, so that the boat, a permanent structure persists despite the fact that all of its original planks have been changed" (Danchin, 1992). This is a very general statement regarding the "strategy" by which living creatures have maintained their presence on our planet for eons, in spite of their inherent low chemical stability. This strategy may be viewed as a kind of trade-off between longevity of the chemical building blocks of life and their polymers, on the one hand, and the ability of these molecules to replicate, exchange with other molecules, or undergo alterations, on the other. Applying the principle of biological continuity, this feature is likely also to have characterized the very first universal biomolecules and their assemblies, in their molecular-evolution processes. Thus, it is not necessary to look for very stable chemical entities for the primordial entities that eventually evolved into the first forms of life, as long as this strategy is obeyed.

Cells and even biopolymers could not have arisen in one step

Size and simplicity of first living entities

Based on thermodynamic considerations, Elitzur (1994a) concluded that life could have started only as a microscopic system. The same conclusion is reached by applying the guideline of simplification, which seems to stem from the biological continuity principle. A different approach, according to which life started in a relatively complex system was advanced by Kauffman (1993) and will be discussed later (chapter 17).

Prokaryotes and eukaryotes

If simplicity reflects an early evolutionary attribute, then prokaryotes would seem to be more similar to early living creatures than eukaryotes.

Cells, organelles, metabolic cycles, and information-processing systems could not have arisen at once

The celebrated biologist Jacques Monod used the expression "Chance and necessity" as the title of his 1971 book, in which he discussed the probability of the emergence of life. This expression was coined by the Greek philosopher Democritus more than two millennia earlier in order to describe the accidental combination of the various kinds of atoms. In treating the origin of life, Monod concluded that because of the very low probability of constructing a living entity out of its building blocks in one "lucky" move, the emergence of life on our planet was a rare event that happened only once, "which would mean that its a priori probability was virtually zero." It should be noted that the application of Monod's approach to even single biopolymers, such as nucleic acids and proteins, would result in a similar conclusion, namely, that the chance formation of such biopolymers has an extremely low probability. Moreover, according to Monod, "The universe was not pregnant with life," and life is compatible with the laws of physics but does not stem from them. The great majority of students of the origin of life do not agree with the conclusions reached by Monod, since it is beyond the domain of science and invokes a miracle (de Duve, 1991).

To those who like Monod's conclusion, the impossibility of synthesis of a living cell out of its building blocks has been compared (by Fred Hoyle) to the construction of a Boeing 747 in a single step by spontaneously assembling its constituents from a junkyard (see de Duve, 1991). The use of this argument is superfluous. As noted earlier even the formation of a small molecules like chlorophyll cannot be achieved via random chemistry (Mauzerall, 1992). On the other hand, the evolution of chlorophyll can be partially explained, as shown by Granick (1957), Hartman (1992b), Zubay (1996), and others. This is discussed in chapter 17.

The origin of life is an inevitable, step-by-step process

In contrast to Monod and the school of thought that he represents, Eigen (1971), and in fact the great majority of origin-of-life researchers, consider the emergence of life, under appropriate conditions, an inevitable event based on physics. Moreover, such conditions presumably exist on other planets elsewhere in the universe. The statistical argumens of the kind used by Monod do not apply to the origin-of-life process. This process is characterized by an enormous number of very small steps, each of which is likely to happen.

Various features of the environment in which life originated may be inferred from extant life by assuming biological continuity

The very attributes of life, including the principle of continuity, also suggest various characteristics of the environment in which the first stages of the origin of life took place. This statement stresses the inseparability of life from its environment (for thermodynamical considerations see Elitzur, 1994a, and Matsuno, 1984). The following examples are instructive.

Lack of chemical equilibrium, preferred reactions, hydrated environments, damaging radiation, and solar radiation

Anionic nature of central biopolymers

According to several researchers, the anionic nature of central biopolymers reflects certain aspects of the environment in which the first "living" assemblies of molecules evolved, as exemplified in the following suggestions:

Preferential stability

According to Scott (1981), hydrated electrons in the prebiotic soup selectively enhanced the rate of decomposition of cationic organic molecules, thus bringing about selection of molecules characterized by anionic groups, such as nucleic acids.

Preferential adsorption

According to Wächtershäuser (1988), the first reactions of the origin of life took place on pyrite crystals. Since these crystals are positively charged, only anionic molecules, which could be adsorbed onto the surfaces of these crystals, were selected for the evolution process of the first living entities. This topic is further discussed in chapter 21.

Hydrated environments

Life evolved in systems that were aqueous at least part of the time. This conclusion allows for combinations between liquid, solid, and gas phases, as well as their dynamics, their variations with time (Brack, 1993c). The very first reactions that led to the origin of life should have involved polymerization of the building blocks of those hypothetical primordial oligomers and polymers. However, at high-water activities hydrolysis rather than polymerization is favored, particularly at elevated temperatures. This stumbling block may be circumvented in specific environments, such as systems fluctuating between hydrated and dehydrated states, or in hydrophobic sites, such as in amphiphilic vesicles or organic scum (Pace, 1991). The use of chemical condensing agents may be another method to enhance condensation in the presence of water. Carbodiimides (general formula R—N=C=N—R) can be used as condensing agents in aqueous systems, provided the two R groups are adequate (Brack, 1993b). The prebiotic relevance of these condensing agents with regard to stability and concentration has yet to be studied.

Damaging radiation

Life is not likely to evolve in environments with high intensity of damaging radiation. As a first estimate, the upper limit of radiation in the prebiotic environments in which the primordial molecular evolution processes took place was presumably in the vicinity of the upper limit characterizing extant life. Avoiding UV radiation damages can be done, for instance, by screening, such as by minerals that absorb in the UV range.

Another possibility for avoiding UV damages has to do with evolutionary processes where organic molecules that are transparent to UV radiation are selected. This is further discussed in chapter 19.

The effect of ionizing radiation emanated from long-lived radionucleotides within clays, particularly ^{40}K, on nucleic acid bases adsorbed on clays was studied by Mosqueira et al. (1996). They observed that the overall effect was enhanced degradation.

Solar radiation flux and the beginning of photosynthesis

In order to support photosynthesis, solar radiation should have been available in appropriate environments. The exact role of chlorophyll, or its predecessor, in the primordial stages of the origin of life has not yet been established. A rather detailed scheme of the evolution of photosynthesis, dealing mainly with the biochemical aspects of this evolutionary process, was recently discussed by Zubay (1996). The geochemical point of view of this discussion is connected to both the sources of electrons for the photosynthetic process and the solar radiation. Several scenarios for the evolution of photosynthesis deal with central biogeochemical aspects and specific prebiotic environments in which these processes could have taken place. One such environment, the surface of the primordial sea, was suggested by Morowitz (1992). A different environment, along with a general mechanism for the transition from inorganic to organic reactions, was suggested by Granick (1957) and by Hartman (1992b). These will be discussed in chapters 21 and 22.

Temperature

Upper temperature limit

As a first estimate of the upper limit of temperatures of the prebiotic environment that supported the primordial molecular evolution processes, it may be assumed to have been similar to that known in modern organisms. Though the upper limit of this boundary has not yet established, it is certainly above 100°C, perhaps higher than 120°C (Stetter, 1992, 1994).

Did life evolve at relatively high temperatures?

The beginning of the universal phylogenetic tree is characterized by hyperthermophilic organisms, which grow optimally in the temperature range between 75°C and 100°C. Does this imply a "hot origin of life"?

Thus, extant hyperthermophily is the result either of adaptation to high temperatures or the preservation of hyperthermophilic properties since the time of the origin of life. Fig. 16.1 shows schematically two alternative temperature regimes for the origin and early evolution of life on Earth (see Miller and Lazcano, 1995). Like most topics in the study of the origin of life, this one is highly controversial (see also Nisbet and Fowler, 1996; Forterre 1992, 1996; Forterre et al., 1995). It is convenient, though somewhat arbitrary, to divide the arguments regarding the temperatures of the origin of life as follows.

Chemical argument I: Stability One of the arguments involved in this problem comes from organic chemistry considerations: If it is assumed that sugars, nucleotides, and

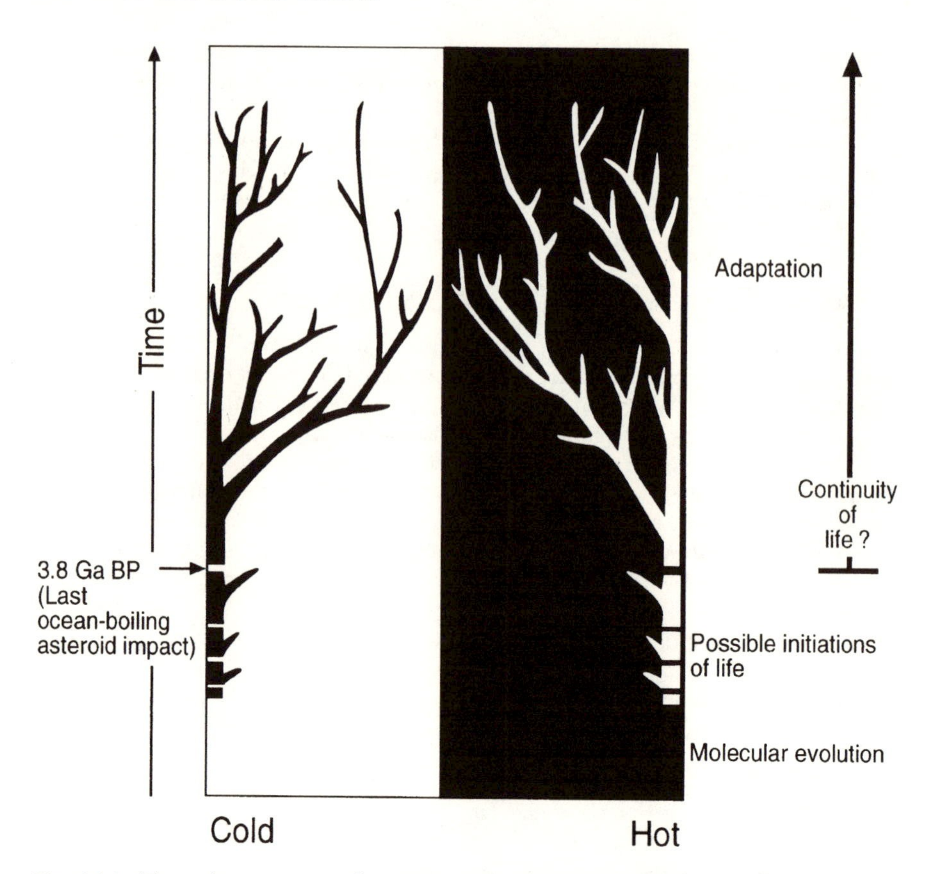

Fig. 16.1. Two alternative evolutionary patterns according to the corresponding temperature regimes for early evolution of life, before and after the last asteroid impact that boiled the ocean. The hot origin of life is followed by adaptation to lower temperatures, and the low-temperature origin of life is followed by adaptation to higher temperatures, but only the secondarily adapted hyperthermophiles would survive an ocean-sterilizing impact of an asteroid. (Following Lazcano and Miller, 1996)

their derivatives were involved in the beginning of life, then high temperatures are not likely to have characterized the sites of the first molecular evolution reactions. The reason is that these compounds are unstable at the upper range of the temperature scale given earlier. Thus, chemical considerations suggest that the accumulation of polymers and complex compounds would be favored at low temperatures. Various aspects of this problem were discussed in Bada et al., 1995; Chang, 1993, 1994; Miller, 1992; and Miller and Bada, 1988.

Chemical argument II: Hydrogen-bond formation Template-directed reactions are the heart of replication and coded-peptide synthesis reactions of molecular biology; they can take place only below the melting temperature of polynucleotides, which are the contemporary universal templates. The temperature range for these reactions is approximately 0°C to 35°C. Assuming that template-directed reactions are as old as life itself, as some researchers do, the origin of life, in the absence of additional interactions, should have occurred at low temperatures (see Miller, 1992).

Biochemical arguments Contemporary hyperthermophilic organisms are characterized by special biochemical adaptations, which by themselves seem to represent a rather advanced evolutionary stage. Indeed, rather than assuming primordial emergence of hyperthermophilic life, some researchers suggest a more complex history of the evolution of hyperthermophility, which started before the first divergence of our phylogenetic tree (García Meza et al., 1994; Forterre et al., 1995).

Phylogenetic arguments An interesting approach to this problem was suggested recently by Gogarten-Boekels et al, (1995). Interpreting the molecular record of the three domains of life, they suggest that

> life emerged early in the Earth's history even before the time of the heavy bombardments was over. Early life forms already had colonized extreme habitat which allowed at least two prokaryotic species to survive a late nearly ocean boiling impact. The distribution of ecotypes on the rooted universal tree of life should not be interpreted as evidence that life originated in extremely hot environments.

More about this problem can be found in Cowan, 1995; Forterre, 1992, 1996; Holm, 1992; Lazcano, 1994a; Lazcano and Miller, 1996; Miller and Lazcano, 1995; Pace, 1991; Stetter, 1992, 1994.

Enzymological, molecular-structural, and metabolic arguments In a recent manuscript entitled "The Thermal Irreversibility of Evolution," Wächtershäuser (personal communication) came up with a bold suggestion according to which "evolution from hyperthermophily to mesophily did obviously occur, but the reverse evolution from mesophily to hyperthermophily is impossible." The reasons for Wächtershäuser's generalization are as follows:

1. Since the temperature optimum of many enzymes is very close to their maximally tolerable temperature, the upward adaptation from mesophily to hyperthermophily would require a simultaneous increase of the optimum of many enzymes, which is unlikely. On the other hand, downward adaptation is much more likely, since enzymes retain their catalytic activity considerably below their optimal temperature.
2. Considerations of optimized folding interactions suggest that hyperthermophiles' homologous sequences would be much more constrained in their folding than mesophiles' sequences. Therefore, the transition from hyperthermophily to mesophily is much more likely than the opposite transition.
3. "Enzymes of hyperthermophiles must consist of small discrete units, separate or in domains . . . but they are often fused to large units in mesophiles. . . . Reversal of these fusions in upward evolution is unthinkable."
4. Because in hyperthermophiles metabolic pathways are kinetically guarded against degradation of unstable intermediates, the likelihood of simultaneous reinstallation of such kinetic contrivances in a transition from mesophily to hyperthermophily is extremely low.

In conclusion, the scant information about the creatures that can be identified as the beginning of the universal phylogenetic tree cannot be used unequivocally, at the present stage of our understanding, as a guide for the evolutionary events that preceded its documented history. The possible role of hyperthermal conditions in the origin of life will be further discussed later in relation to the only scenario suggested so far for a hot origin of life, by Wächtershäuser (1988, 1992a).

Consequences of the thermal-irreversibility-of-evolution postulate

Whether Wächtershäuser's postulate is true or not is still considered an open question by many origin-of-life researchers. However, if it is plausible, its consequences are far-reaching, as explained by Wächtershäuser:

1. If organisms are never more thermophilic than their ancestors, then the common ancestor and the progenote were hyperthermophilic.
2. It is generally accepted that our phylogenetic tree is monophyletic, and in some cases it was assumed that the mutational rate of certain homologous sequences throughout the evolution time was unchanged. The latter assumption, which was used in evolutionary time calculations (i.e., Eigen et al., 1989), is thus unwarranted. In one case (Doolittle et al., 1996), calculations based on the relatively rapid evolutionary rate of mesophiles would result in a too-short evolutionary time.
3. Inasmuch as the genetic code and its origin were related to protein folding, the deciphering of its origin should be based on hyperthermophilic proteins.
4. The first proteins had a very small number of allowed conformations; this fits into Wächtershäuser's scenario, which is based on "surface metabolism" on pyrite (chapter 21).

Wächtershäuser's postulate will probably raise another scientific controversy. One of the topics to be discussed would be the relevance of the recent finding (L'Haridon et al., 1995) of hyperthermophilic archaea in Alaskan oil fields. One possible explanation for their occurrence in a continental petroleum reservoir some 1,700 m below the surface is that "they may have deposited with the original sediment and survived over geological time." The relevance of this issue to the origin of life is unclear at the moment.

Low temperatures

The synthesis of complementary strands of oligonucleotides at a temperature of −18°C was studied by Stribling and Miller (1991). The use of such low temperatures enabled the researchers to lower the reactant concentrations in their nonenzymatic template-directed synthesis. Their conclusion was that such temperature regimes could be used as models for prebiotic environments. Moreover, they suggested that the prebiotic plausibility of these freezing conditions may be higher than that of higher temperatures that are normally used in this kind of experiments.

The issue of prebiotic synthesis at very low temperatures has hardly been investigated. It may be connected to the thaw-freeze cycles proposed by Bada et al. (1994; chapter 13); freezing temperatures, however, may be a part of a thermal regime such as in environments fluctuating between high and low temperatures. Much more work is needed in order to substantiate this suggestion as a plausible prebiotic model, especially with regard to the kinetics of organic reactions and diffusion rates in such thermal regimes.

Clues based on conserved sequences and functions of biochemical entities and domains of early forms of life

Back-extrapolation based on conserved domains of biopolymers have led us to the last common ancestor (chapter 10). The same approach can be used in an attempt to characterize chemical entities and their age, biosynthetical pathways, and functions at the molecular level of the origin of life process.

A common origin for various RNAs

Several research groups found sequence homology in different RNA molecules, which are interpreted as an indication for common origin. For instance, Bloch et al. (1983) suggested a common ancestral origin for tRNA and rRNA molecules. Eigen and Winkler-Oswatitsch (1981a, 1981b) suggested a related origin for tRNA and mRNA (see also chapter 23).

Conserved functions of molecules attest to their antiquity

Conservation has to do with vital functions of certain domains or molecules that are not likely to be replaced by other chemical entities. Thus, the results of a recent study on the aminoacylation of minihelices designed to explore recognition of domains on simiplified tRNA models (Hipps et al., 1995) "suggest that the operational RNA code for glycine has been preserved from bacteria to man, if the positions of the critical nucleotides are considered and not the nucleotides *per se.*"

The antiquity of coenzymes Theoretical biologists suggested as early as the late 1960s that RNA molecules had a central role in the most fundamental processes of the cell—information processing and catalysis—during very early stages of evolution. The case of coenzymes has been one of the first specific suggestion to highlight this point.

Many enzymes cannot function properly without the presence of an additional chemical entity tightly bound to the enzyme molecule, called coenzyme (chapter 9). It was known for a long time that many of the coenzymes are nucleotides or contain cyclic nitrogenous bases derivable from nucleotides. In line with the principle of continuity, it was suggested by H. B. White as early as in 1976 that coenzymes are "vestiges of nucleic acid enzymes which preceded the evolution of the ribosomal protein synthesis." These early catalysts have been little changed during hundreds of millions, probably billions, of years of evolution, and they attest to the antiquity of the biochemical reactions in which they are involved in modern cells. Presumably, during the eons of evolutionary time since the emergence of coenzymes, many catalytic functions and structures gradually emerged and were *added* to the more ancient biocatalytic coenzymes.

The principle of many users According to H. B. White (1976), it is extremely unlikely for a structure that serves as a cofactor in many reactions to change while maintaining efficiency and specificity. Most cofactors are relatively small, with molecular weights less than about 600, and enzymes can bind and orient them for their catalytic functions by recognizing the cofactor surface. Indeed, this requirement is reflected in deep cofactor-binding clefts found in modern protein enzymes. With such specific recognition elements, cofactors are not likely to change easily (see Giver et al., 1994).

Can the dynamics of changes of conserved biomolecular domains be used as "evolutionary clocks"?

Homologous (evolutionally related) biopolymers of various organisms have the same function. For instance, ribosomal RNA molecules of different creatures differ from each other to some extent, but their function remains the same. This observation is the basis for the concept of "evolutionary clock," suggested first by Zuckerkandl and Pauling (1965). According to this hypothesis, the differences in monomer sequences of homologous biopolymers among various creatures can be related to the time in which these differences were acquired, under certain assumptions (chapter 10).

Though the realization of this concept is still problematic and controversial, it is interesting to illustrate it by the age-estimate of the tRNA molecule, carried out by Eigen and his collaborators (1989). tRNA is a relatively small RNA molecule (70–93 monomers), which has a central role in the template-directed synthesis of proteins (chapter 7). By comparing tRNA molecules of many different creatures, it was found that certain domains in these molecules were conserved—that they are molecular fossils. Changes in sequences are the result of mutations; *assuming a constant rate of mutation formation during the evolutionary history of the conserved domains*, and using sophisticated statistical analyses, the researchers estimated the age of the tRNA molecule. Since the tRNA molecule is a central component of the template-directed protein synthesis machinery of the cell, its age is considered to be the age of the "hardware" of this machinery as well as that of its "software," the genetic code. The estimated age of the tRNA is 3.6 ± 0.6 Ga. According to Eigen et al. (1989), "the genetic code is not older than, but almost as old as our planet."

This estimate thus seems to supports the terrestrial-origin-of-life hypothesis but is too rough to be useful in calculating the age of the template-directed synthesis of proteins. Furthermore, as stressed by Wächtershäuser (1996), the evolution rate of the hyperthermophiles, which are the most ancient creatures, was much slower than that of the latecomers, the mesothermophiles. Thus, the assumption of constant rate of mutation cannot be accepted without reservation (see also Gogarten-Boekels et al., 1995).

It should be noted that the emergence of the tRNA molecule is not necessarily identical with the "invention" of template-directed synthesis, since the latter also includes replication (sometimes called "self-replication," or "self-copying").

Conserved biosynthetic pathways are repositories of evolutionary events

The biosynthesis chain represents evolutionary history

The idea that metabolic pathways are another kind of memory that may hold information about the origin of life was pioneered by Horowitz (1945) and later generalized by Granick (1950, 1957). Basically this postulate is related to the stepwise way by which evolution proceeds and to Dollo's Law (chapter 10). It should be noted, however, that the direction of the formation of the biosynthetic chain according to Granick is opposite to the one suggested by Horowitz (chapter 5). Thus, according to Horowitz, the evolution of biosynthetic pathways began with useful products and developed backward by using simpler molecules (Mauzerall, 1992). Horowitz's concept was based on the view that prebiotic synthesis had established large numbers of the molecules necessary for evolution. As this pool gradually dwindled, other paths would be discovered by means of mutations and selection, until some basic, ever-present substrate was found. Granick (1957) adopted Horowitz's first idea but rejected his second one, regarding the direction of evolution.

The forward direction of evolution suggested by Granick (1957) can explain the evolution of complex molecules that could not have evolved by random mutation (Mauzerall, 1992). Thus, the biosynthetic pathway is built forward, where each step is advantageous to the survival of the ensemble of molecules in its time. Novel evolutionary changes are generally added to the existing systems without replacing them, as evidenced by biological conservatism, thus enabling one to detect and follow important stages in evolution.

The use of biological conservatism as a tool for the study of the evolution of biochemical reaction pathways was phrased by Bernal in 1960 (cited by Hartman, 1975) as follows: "The pattern . . . is one of stages of increasing inner complexity, following one another in order of time, each one including in itself structure and processes evolved at the lower levels."

Case studies

Some of the most interesting cases relevant to the present section are exemplified in the following section, followed by a discussion of their implications with regard to central schools of thought in the search for the origin of life.

RNA preceded DNA

Various biochemical reactions are thought to have been preserved over hundreds of millions, perhaps billions of years. Consider, for instance, the following features:

1. The synthesis of deoxyribonucleotides in the cells of all living organisms is carried out by reduction of preexisting ribonucleotides and catalyzed by a specific enzyme.
2. It is known that protein synthesis can sometimes take place without DNA, but never in the absence of RNA.
3. ATP, not a deoxyadenosine triphosphate, plays a central role in the cell as an energy carrier.
4. Catalytic cofactors such as NAD, NADP, FAD, and coenzyme A are conjugated molecules containing adenosine, not a deoxyadenosine moiety (see, for instance, de Duve, 1993; Maurel, 1992a, 1992b; Sievers et al., 1994).

Remembering also that the RNA molecules, tRNA and mRNA, carry out more biological functions than DNA, the most likely interpretation of these observations is that RNA preceded DNA in evolution. According to this interpretation, the order in which compounds are synthesized in contemporary metabolic cycles reflects the evolutionary process of their acquisition by early "living" entities. Presumably the molecules involved in the reducing reaction of nucleotides—RNA and various enzymes (proteins)—already existed as such or in the form of unknown precursors even in the remote history of biology, before the evolution of DNA.

Intermediary metabolites as precursors of amino acids

In spite of the diversity of the pathways for the biosynthesis of amino acids, they share a common denominator: Their carbon skeleton is derived from intermediates (the substances produced by a given metabolic step to be utilized by another one) of glycolysis, the pentose phosphate glycolysis, or the citric acid cycle (Stryer, 1995). In other words, the precursors of amino acids are the intermediary metabolites of the major metabolic cycles (chapter 9). According to Granick (1957), the order of events in those metabolic cycles reflects the history of the introduction of amino acids into the evolving metabolism of a very early stage of the origin of life; this may be interpreted as an indication that some ancient matabolic cycles preceded the amino acids in the first "living" chemical entities (Degani and Halman, 1967; Bishop et al., 1997).

Clues from the central role of sugars in contemporary biosynthesis Theoretical calculations (Weber, 1997) have established that redox disproportionation of carbon is the primary energy source driving amino acids and lipid biosynthesis from glucose, accounting for 96% and 84% of the total energy of lipid and amino acid biosynthesis, respectively. In contrast, nucleotide biosynthesis depends almost entirely on ATP for energy. Weber's conclusion is thus that in view of the central role of sugars as an indispensable source of energy for contemporary biosynthesis, and its ability to drive not only biotic but also abiotic syntheses, the evolution of sugar disproportionation reactions in the origin of metabolism and the origin of life seems likely. It is noted, however, that Weber's calculations serve only as a hint, with no information regarding a prebiotic mechanism by which those reactions would take place. Additional features of such a mechanism would include catalysis and activated species of the involved molecules, as well as an adequate environment.

The antiquity of the citric acid cycle

The central role of the citric acid cycle (the Krebs cycle) in cell metabolism suggests that it is an ancient pathway (see also Waddell et al., 1987; Bishop et al., 1997). At a first sight, and in view of its assumed antiquity, there is an apparent discrepancy between the evidence regarding the lack of free oxygen in the primordial atmosphere of the prebiotic earth, on the one hand, and the use of oxygen by the citric acid cycle, on the other hand (see also Waddell et al., 1987). A possible clue for the solution of this apparent discrepancy is that the initial steps of this cycle—the synthesis of citrate, its isomerization into isocitrate, and the decarboxylation to α-ketoglutarate—all evolved long before the appearance of oxygen in the earth's atmosphere (see Stryer, 1995; Lehninger et al., 1993). The oxidative attributes of the citric acid cycle were established during later periods, when free oxygen became available in the atmosphere (see Keefe, Lazcano, and Miller, 1995).

Hartman (1975) suggested applying Bernal's strategy to the citric acid cycle, which had been known for its central role in metabolism. His conjecture was that the citric acid cycle was followed by the building blocks of the major constituents of the cell—nucleotides, amino acids, lipids, and carbohydrates. A different evolutionary pathway for the citric acid cycle was discussed by Zubay (1996). The connection between the citric acid cycle and the genetic code is discussed in chapter 17.

The reductive citric acid cycle (RCC)

The reductive citric acid cycle (chapter 9) was suggested by Wächtershäuser (1988, 1992a) and by Buchanan and his collaborators (see Buchanan, 1992) as the key for the understanding of the early evolution of metabolism. The possible prebiotic role of this cycle will be discussed in chapter 21.

Granick, Morowitz, and Elitzur's postulates: The molecular level of Haeckel's biogenetic law

The interpretation of the metabolic pathways as "historical documents" for the reconstruction of the evolutionary pathways of life is based on the supposition that the order in which compounds are synthesized in certain metabolic pathways reflects the order in which reactions came into play in prebiotic chemistry. This realization culminated in 1950 in the following postulate: "Biosynthesis recapitulates biogenesis." More recently, Morowitz (1992), focusing on intermediary metabolism rephrased this postulate as follows: "Intermediary metabolism recapitulates prebiotic chemistry."

Both Granick and Morowitz's postulates are paraphrases on the "biogenetic law" postulated in 1866 by Haeckel (chapter 4), and known in his most succinct form as "Ontogeny recapitulates phylogeny." This postulate, which was published some seven years after the publication of Darwin's evolution theory, states that the developmental history of an individual creature recapitulates the evolutionary history of the group of creatures to which this individual belongs. Granick's and Morowitz's postulates refer to the molecular level of biology, which reflects a central feature of modern biology. It is general enough to include also the popular hypothesis that RNA preceded DNA in evolution.

Elitzur (1995), in a discussion of the thermodynamics of evolution, proposed the following generalization of Haeckel's biogenetic law: "Evolutionary progress involves internalization." In his words: "Processes that have initially shaped the species, using individual organisms as their dynamic constituents, later become internal mechanisms within the individual organism itself." Citing Lorenz, Elitzur accepts his notion that "Every organism . . . constitutes a 'hypothesis' about the environment, put to test of survival. Organisms that survive are those that constitute hypotheses that were confirmed upon their interaction with reality. These confirmed hypotheses are then added to the species' phylogenetic heritage." The inherent information of functioning structures about their environment was discussed by Dennett (1996, p. 197).

Calibration of molecular-biological models by geochemical data

Protein clocks

Yarus (1993) estimated the age of the genetic code and its machinery by using ancient microfossils of known age. It turned out that a major problem in the search into the deep roots of the phylogenetic tree is its time scale. Estimates of the age of the last common ancestor, including Yarus's estimate, range from 1.3 to 3.5 billion years.

A novel attempt to explore this problem has recently been advanced by R. F. Doolittle and his group (1996). These researchers studied genes for 57 different metabolic enzymes representing 15 major phylogenetic groups that may be considered protein clocks (see also Morell, 1996; Mooers and Redfield, 1996). Their conclusions do not agree with the accepted estimate, according to which the divergence time of the major biological groupings, namely prokaryotes and eukaryotes, occurred 3.5 Ga ago (fig. 16.2). It should be noted that the researchers assumed a rather constant rate of amino acid substitution. This assumption is not warranted (see also Wächtershäuser, 1996, and personal communication) and should be further explored. The implications of this debate are important. In addition to the possibility that both dates are correct (Mooers and Redfield, 1996) or to the validity of the assumption of an approximately constant rate of changes (Doolittle et al., 1996), it is even possible that "we are the descendants of a completely different diversification." (Mooers and Redfield, 1996).

Compartmentation by means of phospholipids does not require biopolymers characterized by specific sequences

Extant cytomembrane systems There are no chemical bonds between the amphiphilic molecules of the cell membrane; in the bidimensional, liquidlike structures that are formed, individual molecules can flip-flop, rotate, and diffuse. Indeed, phospholipid membranes can be formed either spontaneously or with a little energy input, under appropriate conditions. These systems require no particular structural organization, and therefore, unlike in the case of nucleic acids and protein molecules, there are unstable sequences and a few limitations regarding the internal order in the organization of the amphiphilic molecules of which biological membranes are made. Thus, there is no obvious meaningful intrinsic structural information in their organization; sequencing of biological membranes is therefore meaningless, and back-extrapolation from extant cells in order to characterize the membranes of primordial living entities should be based on other principles. The incorporation of directed peptides and proteins into biological membranes is assumed to have taken place after the establishment of compartmentation based on lipid membranes.

Contemporary membranes are known to have many forms and compositions (de Duve, 1991; Petty, 1993). There are considerable differences between the simple cell membrane of prokaryotes, on the one hand, and the sophisticated cytomembrane of eukaryotes, with its irregular boundaries, protrusions, and folds, on the other. The eukaryotic cell may thus be viewed as a single complex system capable of establishing an efficient connection and matter exchange between its different parts. The eukaryotic membrane is assumed to have developed from the prokaryotic membrane (de Duve, 1991, p. 62). This brings one to the last common ancestor, which is a rather evolved living entity, as we have already seen. The question is thus, is it possible to characterize the earliest membranes of living entities?

Hypothetical features of early membranes The last common ancestor is considered by many researchers to be a population of cells characterized by extensive lateral gene exchange. The latter feature means an easy gene transfer between individual cells and is thought to have brought about a very efficient evolution of that population. Presumably the transfer of single genes or small gene clusters has been more efficient than that of complete genomes. And no less important, an efficient transfer depends on the properties of the cell membrane. Continuing this line of thought, lipid mem-

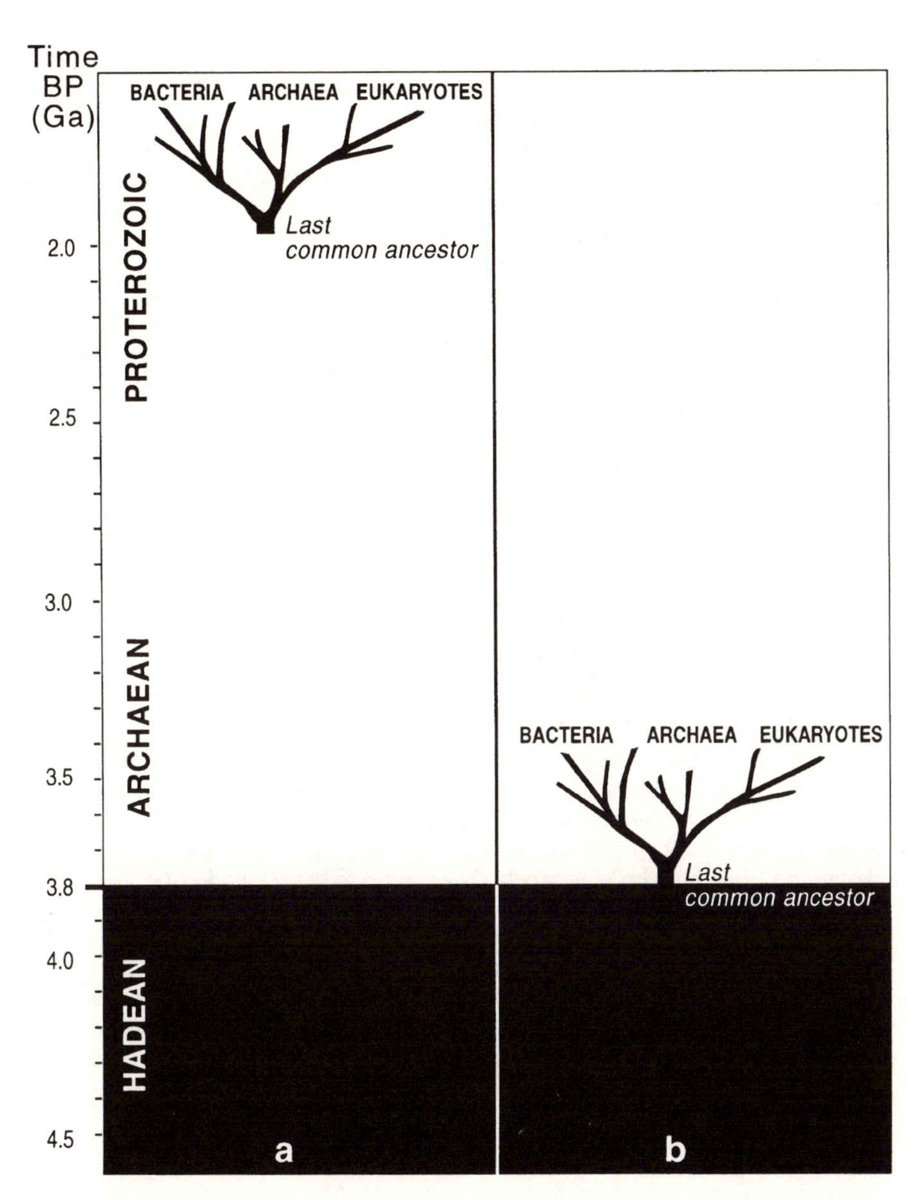

Fig. 16.2. The discrepancy between two approaches for the estimation of life history of our phylogenetic tree. **a.** Comparison of protein sequences (R. F. Doolittle et al., 1996) gives a phylogenetic tree where the last common ancestor lived about 1.8 Ga ago. **b.** The established calibration by microfossil age gives a phylogenetic age, the common ancestor of which lived about 3.5 Ga ago. (After Mooers and Redfield, 1996)

branes are not the only candidates to serve as primordial membranes: β-structures of alternating polypeptides were suggested by Brack and Orgel (1975), and porous protein layers were suggested by Zillig et al. (1992).

Bacteria and Archaea are characterized by different lipid membrane constituents (Zillig et al., 1992). The nature of the lipid layer of the last common ancestor is still unknown; as I already noted, it could well have been a protein layer. This could have facilitated the transfer of gene clusters between individual cells. Once the lipid membrane was established, further gene transfer would be unlikely (ibid.).

It is interesting to note that Wächtershäuser (chapter 21) suggested as early as 1988 that lipids were formed by early prebiotic synthesis of terpenoids (phosphorylated isoprenoids; see also Benner et al., 1989). It was later suggested by Ourisson and Nakatani (see Maddox, 1994) that these compounds were also components of the first cell membranes. Of these suggestions, only that by Wächtershäuser (chapter 21) is relevant to prebiotic evolution, dealing with the early stages of evolution before the appearance of cells.

Summary

The language of the origin of life is chemistry, encompassing elemental composition, chemical stability, and chirality. Probability considerations suggest that cells, organelles, metabolic cycles, and information-processing systems could not have arisen at once; the emergence of life is thus viewed as a step-by-step process. Moreover, under appropriate conditions the origin of life is inevitable. Clues from biology for the origin of life and its first stages are based on the principle of continuity and are thus reflected by conserved biochemical entities, functions, and processes, as well as sequences of the major biopolymers. These conserved attributes are thus repositories of evolutionary events. From the point of view of the history of the search into the evolution of life, Haeckel's biogenetic law is reflected at the molecular level of the cell, encompassing chemical features of metabolism and molecular biology. Accordingly, essential biomolecules of extant life and their functions are assumed to have already been central during the early stages of the origin-of-life processes. Furthermore, these conservative attributes of life also imply that various features of the primordial environment, most notably hydration and temperature regimes, may be inferred from extant biology.

17

Top-down Reconstruction of Processes and Early Evolutionary Stages without Specific Geochemical Considerations

The structure of a proposed first genetic system "should be simple as possible, but not simpler."

Albert Eschenmoser, "Chemistry of potentially prebiological natural products" (a paraphrase on a remark attributed to Einstein)

Early evolutionary stages that preceded the progenote

The vocabulary of origin-of-life researchers has been enriched recently by concepts and propositions such as "minimal cell, "chemoton," and various "worlds." The best known of the worlds is the "RNA-world." Not less important is the "ancient genetic code." One common denominator between these scientific propositions, which are rather different from one another, is that they represent early evolutionary stages in the origin-of-life processes, *all of which preceede the progenote.* A second common denominator between these propositions is that they are the results of back-extrapolations from extant biology to much earlier stages in the evolution of living entities, *essentially without specific geochemical considerations.* The distinction between top-down and bottom-up approaches is convenient but not always clear-cut; thus it is arbitrary to some extent.

Phenotypic versus hereditary origin of life

The essential attributes of living entities may be divided into *phenotypic* and *hereditary.* The first one is essentially metabolism, with its cycles of chemical reactions, driven by an extrinsic source of energy; the latter attribute has to do with template-directed syntheses of organic polymers. Based on the principle of biological continuity, it is assumed that this apparent duality is very old. Are there clues regarding the question of who came first? The two obvious possibilities are (see Maynard Smith and Szathmary, 1995, pp. 17–18, for a discussion) that the origin of life is either phenotypic or hereditary.

The origin of life is phenotypic

This is the "metabolism came first" postulate, where the metabolic systems had a kind of replication not based on template-directed synthesis. Only later on did these systems acquire the ability to synthesize the building blocks from which the replicating polymers were made.

The origin of life is hereditary

This is the "heredity came first" postulate, according to which replicating systems ("replicators") came first by means of natural selection: Those replicators that multiply most efficiently in their chemical environment would be selected; the ability to affect the chemical environment was acquired later, by the successors of these entities (Maynard Smith and Szathmáry, 1995).

These are the established patterns of thought, where the priority of either possibility is still the subject matter of controversy, as if they are mutually exclusive. There seems to exist, however, a third possibility: coevolution of phenotypic and hereditary features.

The origin of life is coevolution of phenotypic and hereditary features

Metabolizing and replicating systems coevolved, where primitive catalysts and templates "collaborated" in a synergetic fashion.

Processes and scenarios representing these three schools of thought are discussed later. Let us focus first on what Casti (1992) referred to as the "current flavor-of-the-month in the ORI [Origin of Life] business," which "seems to be a replicator-first theory involving the evolution of life from self-catalytic RNA."

The RNA-world theory and its main implications

> The finding of efficient RNA catalysts [during the early 1980s] was surprising because RNA is very different from traditional polypeptide enzymes.
>
> Thomas R. Cech, "The efficiency and versatility of catalytic RNA"

Woese, Crick, and Orgel have, quite independently, postulated already by the late 1960s that RNA was central to the early stages of the origin of life. The discovery of ribozymes by Cech and Altman and their collaborators during the early 1980s has highlighted a long-standing postulate regarding the fundamental role of RNA in the origin of life. It gave the theoretical considerations important experimental support by demonstrating catalytic activity of extant RNA. As Crick (1993) noted, none of these prominent scientists suggested that relics of catalytic activities would be found in extant RNA molecules.

The RNA-world theory that ensued (Gilbert, 1986) postulated a central role of RNA in the basic functions of primitive creatures, a role that preceded the "invention" of DNA and template-directed peptide synthesis. In the words of Waldrop

(1992), the RNA world is characterized as "a pre-DNA realm populated by organisms that stored genetic information in RNA, catalyzed chemical reactions with RNA, and carried out all other necessities of life with RNA, and RNA alone." This theory was translated into that part of the phylogenetic tree which preceded the last common ancestor. A new flood of speculations, research proposals, scientific studies, and new fashions and controversies was thus initiated.

As Joyce and Orgel (1993) have noted, the RNA-world theory was used and interpreted in different ways by various authors, therefore it would be impossible to define it rigorously. There are, however, three basic assumptions on which this theory stands: (1) At some time in the evolutionary history of our phylogenetic tree, RNA molecules served as carriers of genetic information. (2) Replication of RNA molecules was based on the Watson-Crick base-pairing rules. (3) Genetic (template-directed) proteins were not involved as catalysts in the chemical reactions of the RNA world.

For different points of view the reader is referred to Benner et al., 1993; de Duve, 1993; James and Ellington, 1995; Joyce, 1989; Lahav, 1993; Lazcano et al., 1992; Maizels and Weiner, 1993; Maurel, 1992a, 1992b; Orgel, 1994; Schuster, 1993, 1995; Wächtershäuser, 1992a, 1994b; Weiner and Maizels, 1987, 1991.

Involvement of RNA molecules in central processes

The involvement of RNA molecules in the central processes of information manipulation in extant cells is of great interest to biochemists, even though the relevance of these reactions to the origin of life is not clear. The reactions that attracted the attention of scientists are central in the maintenance and expression of genetic information in eukaryotic cells, namely, DNA replication, transcription, mRNA splicing, and translation (Cech, 1993a, 1993b).

Relevance to the origin of life

The indication (Noller et al., 1992; Noller, 1993) that peptidyl transferase reaction of peptide synthesis may be catalyzed by RNA has yet to be substantiated experimentally with regard to prebiotic reaction. Thus, the demonstration (Wilson and Szostak, 1995) that ribozymes can promote a rather broad range of catalytic activities may not be directly applicable to the origin-of-life problem per se. The reason is that these ribozymes are too large to be considered prebiotic (see also M. J. Moore, 1995). Similarly, the ribozyme capable of peptidyl transferase activity (Zhang and Cech, 1997) is not prebiotically plausible.

Relics of the RNA world?

Small circular pathogenic RNAs of plants have been suggested as relics of the RNA world in a precellular evolution (Diener, 1989). This issue is still controversial (see Chela-Flores, 1994).

Prebiotic RNA

The RNA-world hypothesis was suggested as an integral part of the prebiotic-soup paradigm, which also includes activated nucleotides (see Wächtershäuser, 1994a,

1994b). It does not postulate a chemical pathway for the prebiotic synthesis of RNA molecules. It is widely held that modern RNA is probably different from the hypothetical prebiotic RNA (for more information see Cohen, 1995; Connell and Christian, 1993; Joyce and Orgel, 1993; Kolb et al., 1994; Lazcano, 1994b; Schwartz, 1993).

According to Schuster (1993), the appearance of RNA in the chemical evolution era was late, therefore its introduction into the emerging biology was very early. Moreover, it seems likely that the first RNA or RNA-like molecules to have emerged were small. The smallest true ribozyme described so far is the trinucleotide UUU, which catalyzes, in the presence of Mn^{++}, a specific cleavage between G and A (Kazakov and Altman, 1992).

Catalytic repertoire of ribozymes

The catalytic repertoire of the first ribozymes discovered independently by Cech and Altman was rather dull; it consisted of little more than cutting and joining preexisting RNA. However, within a few years this repertoire increased steadily to include the joining together of oligonucleotides, as well as drawing energy for the reaction from a triphosphate group. No less interesting are the indications obtained by Noller and his group with regard to the possible involvement of RNA in the peptide-bond formation in the ribosome (Noller et al., 1992; see also Orgel, 1994; Lohse and Szostak, 1996; Zhang and Cech, 1997). These catalytic activities, however, were performed by rather long RNA molecules. Since the prebiotic synthesis of such large RNA molecules is still questionable, the prebiotic significance of such reactions has yet to be established.

Several research groups (Décout and Maurel, 1993; Décout et al., 1995; Maurel, 1992a, 1992b; Robertson and Miller, 1995a; 1995b) addressed the chemical reasons for the small repertoire of catalytic activities known so far in ribozymes. This limitation may be the result of the small number of functional groups available in RNA molecules—hydroxyls, phosphates, and the amino and imidazole moieties of the bases (fig. 17.1). For instance, in order to explore this problem Décout and Maurel (1993) and Décout et al. (1995) synthesized N^6-Ribosyladenine and its derivatives under plausible prebiotic conditions and studied their catalytic activities in peptidic and phosphodiester backbone processing.

Robertson and Miller (1995a; see also Robertson and Miller 1995b) carried out laboratory experiments, under potentially prebiotic conditions, in which they showed the plausibility of the formation of "5-substituted uracils with the side chains of most of the 20 amino acids in proteins" (see fig. 17.1). These results suggest that if indeed uracil was present on the primitive earth, then various derivatives of this molecules could have been formed in the prebiotic environment. These substituted uracils could have served as constituents of ribozymes in the RNA world, catalyzing reactions that, at later stages, became the domain of protein activity. Thus, ribozymes with such substituted uracils could have served as a bridge between the RNA world and the DNA-protein world.

Involvement of RNA in peptide synthesis The catalytic complex of the ribosome is made of RNA and protein molecules. The direct role of ribosomal RNA in the protein synthesis has been established by Samaha et al. (1995). This topic will be discussed in the context of a biogeochemical scenario in chapter 23.

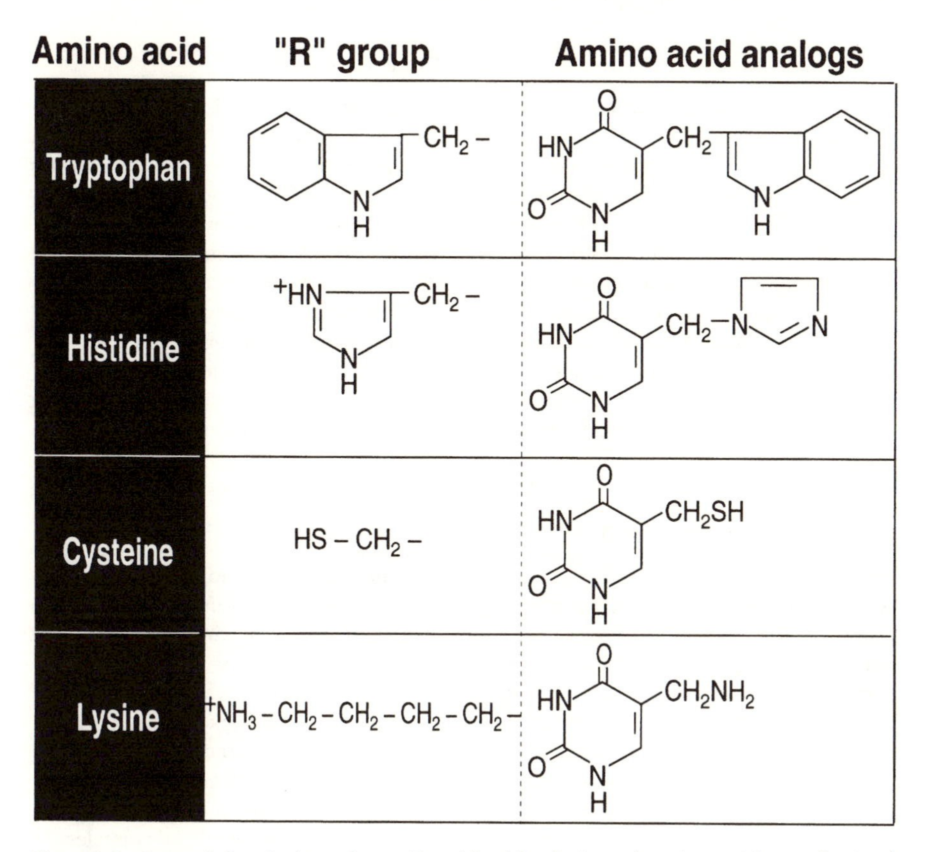

Fig. 17.1. Several 5-substituted uracils with side chains of amino acids synthesized under plausible prebiotic conditions. The "R" group is seen in the general structure of an amino acid, namely, $RCNH_2HCOOH$. (Adapted from Robertson and Miller, 1995a)

Artificial evolution of RNA molecules in the laboratory

Several groups recently succeeded in conducting what might be called "artificial evolution" or "directed molecular evolution" (Joyce, 1992; Jaeger, 1997). According to this approach, Darwinian evolution is applied on a molecular scale to large macromolecular populations through cycles of selections, amplifications, and mutations, thus directing and enhancing the evolution of the desired molecules. As a result of these developments, the repertoire of ribozyme catalysis has increased to include novel reaction mechanisms as well as reaction types. Examples to these works are mainly from the groups of Szostak (for instance Green and Szostak, 1992; Lorsch and Szostak, 1994; Sassanfar and Szostak, 1993) and Joyce (Beaudry and Joyce, 1992; Lehman and Joyce, 1993). These reactions, however, are not directly relevant to the origin of life, both because the sizes of these RNA strands are considerably larger than the presumed RNA-like molecules that may be considered prebiotically plausible and because the selection and amplification mechanisms of these reactions are not prebiotic.

The way from the RNA world: Transition to the DNA world

The presently known catalytic versatility of RNA molecules, and the apparent elegant solution of the "chicken-and-egg" paradox they provide, make this hypothesis very attractive. The next stage (stages?) in the evolution of living entities, according to the RNA-world theory, is connected to the evolution of DNA and the "invention" of the translation apparatus—the ribosome and the genetic code, by which template-directed (coded) peptides could be synthesized. The DNA could have evolved from RNA; being more chemically stable than RNA made it a more efficient candidate for the role of preserving the blueprint of the cell. Obviously, the evolutionary pathway from the RNA world had to go through a transitionary stage during which template-directed protein synthesis gradually emerged, concomitant with the evolution of the first metabolic pathways. It is accepted by most scientists that "the greater structural variety of amino acids allowed better catalytic properties in protein enzymes than in those composed of RNA" (Westheimer, 1987).

The "breakthrough organism"

Benner et al. (1989) have further applied parsimonious considerations regarding conserved features in extant organisms in their search for the pre-common-ancestor era. Using data on metabolic pathways, chemical structure, and enzymatic reaction mechanisms, and applying the principle of biological continuity, they were able to reconstruct various properties of hypothetical ancient organisms that preceded the progenote. Their approach, which essentially represents back-extrapolation from contemporary organisms, assumes an RNA world as the most ancient era they can characterize. In order to go from the RNA world to the present biology, they assume a transition stage in which a biology based on nucleotide-protein chemistry gradually evolved. Their scenario (Benner et al., 1989) thus consists of three episodes, as follows (fig. 17.2):

1. The RNA-world.
2. The transition stage characterized by a hypothetical construct named a "breakthrough organism," which is "the first organism to contain a genetically encoded messenger RNA that directed the synthesis of a protein selectable for its catalytic activity." This organism probably used DNA in the storage of genetic information.
3. The emergence of the progenote.

An interesting feature of the scenario suggested by this group is the stage in which translation was acquired. In contrast to most models, in which translation was assembled in a primitive environment characterized by primitive metabolism (see also chapter 23), Benner et al. (1993) suggest that the translation mechanism was established in a relatively complex environment provided by the breakthrough organism.

The origin of the amphiphilic constituents of cellular membranes should be looked for in the metabolic pathways by which they were synthesized, as exemplified by Benner et al. (1989). Obviously, the membrane proteins of the kinds common in extant organisms cannot be older than the first template-directed protein synthesis.

At present, this scenario seems rather vulnerable, as I will explain later. However, the back-extrapolation is based on scientific disciplines such as biochemistry, molecular biology, and enzymology and lays before the scientist what Benner and his as-

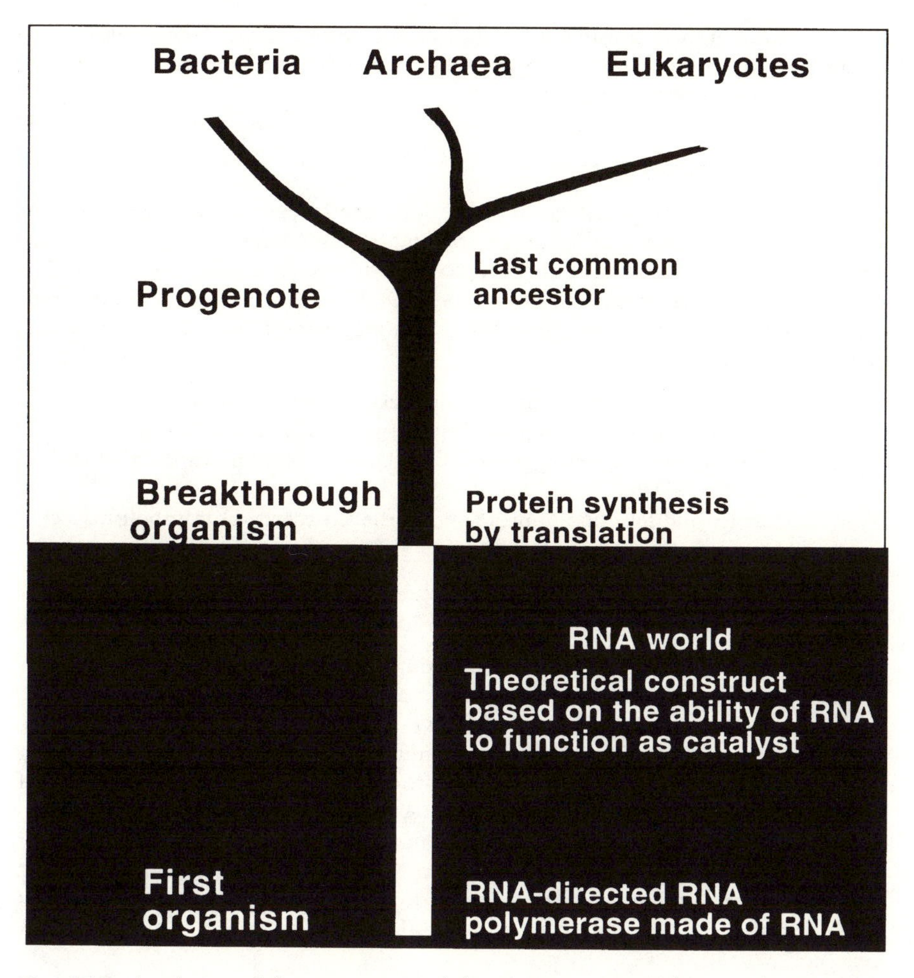

Fig. 17.2. A scheme of the emergence of the progenote according to Benner et al., 1989.

sociates consider "a palimpsest of the RNA world," which they were able to partly decipher (a palimpsest is a parchment that has been inscribed two or more times; the previous texts are imperfectly erased and partially legible). The problem is, of course, the interpretation of these observations. A partial list of the main traits assigned to the breakthrough organism according to Benner et al., (1989) are as follows:

1 RNA cofactors.
2. Reactions catalyzed: redox reactions, transmethylation, carbon-carbon bond formation, energy metabolism based on phosphate esters.
3. Use of DNA to store genetic information.
4. Biosynthesis of porphyrins.
5. Lack of synthesis of fatty acids, and use of terpenes as the major lipid component.

It is noted that these considerations deal with the progenote and the breakthrough organism and are therefore evolutionally far away from the hypothetical origin-of-life processes in which we are interested.

Additional transitory stages

Additional indications regarding the need to assume a hypothetical organism as an intermediate stage between the RNA world and the last common ancestor were also suggested by García-Meza et al. (1994), according to whom this creature was a rather complex cell, similar to modern prokaryotes. De Duve (1993) discussed the evolutionary pathways that followed the RNA world, suggesting a "multiplicity of distinct steps," as follows:

1. Formation of a primordial peptide-synthesizing machinery.
2. Progressive development of a primitive genetic code in this machinery.
3. Partitioning into competing protocells, where the protein-synthesizing machinery was selected.
4. Emergence and selection of proteins and the emergence of metabolism.

Was the last common ancestor characterized by an RNA genome?

Several researchers tried to calculate the minimal gene set needed to sustain the existence of a modern cell, where the calculations are based on sequenced genes of some of the smallest known bacteria. In a recent work, Mushegian and Koonin (1996; see also Maniloff, 1996) suggested that the minimal set consists of 256 genes. Taking into account the primitive entities suggested for the very early stages of the origin of life (chapters 20–23), this figure demonstrates the huge gap between a minimal extant cell and the presumably much smaller number of ensembles of oligomers that characterized the earliest stages of the origin of life. From our standpoint the most interesting idea of these researchers has to do with their suggestion that "it seems likely that the last common ancestor of the three primary kingdoms had an RNA genome."

What preceded the RNA world?

> If a metabolically complex "RNA world" existed then molecular biologists should be able to find or create modern RNA strings similar to those that presumably populated it, ribozymes that can catalyzed a wide variety of metabolic reactions.
>
> Kenneth D. James and Andrew D. Ellington,
> "The search for missing links between self-replicating nucleic acids and the RNA world"

In contrast to some premature optimistic expectations, it seems that the RNA-world theory has not provided so far tangible clues regarding the *origin* of life. The scenarios suggested so far do not treat the very first stages of the origin of the RNA world (see also de Duve, 1993). Neither did they treat the "invention" of translation. In other words, the RNA-world theory does not have any suggestion pertaining to the origin-of-life processes (see Benner et al., 1989).

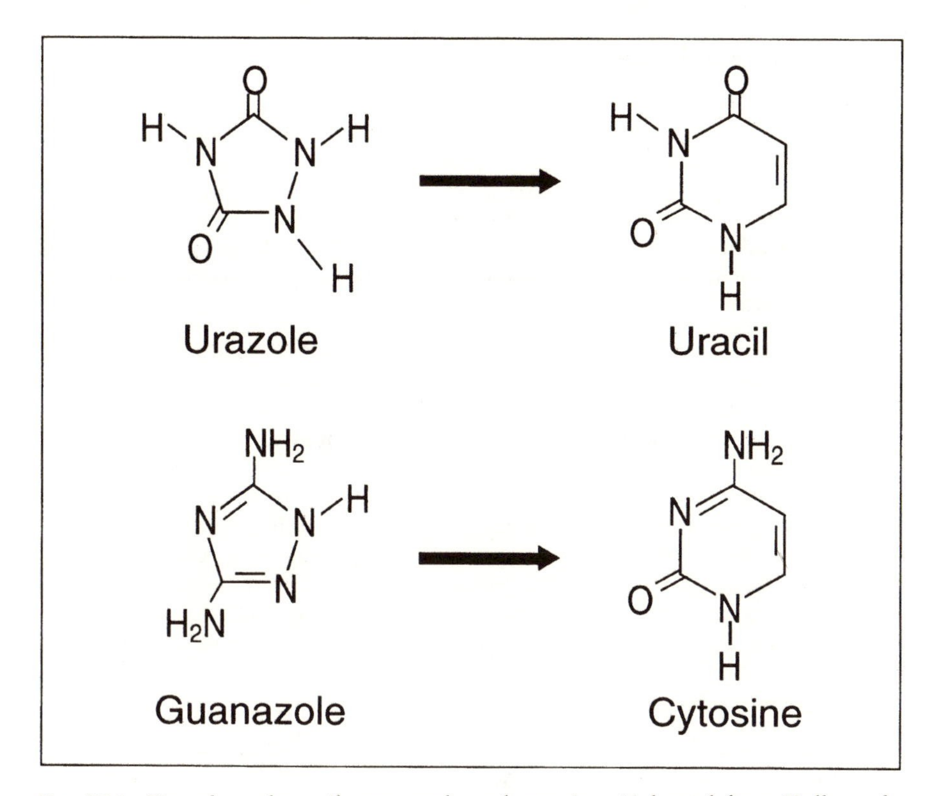

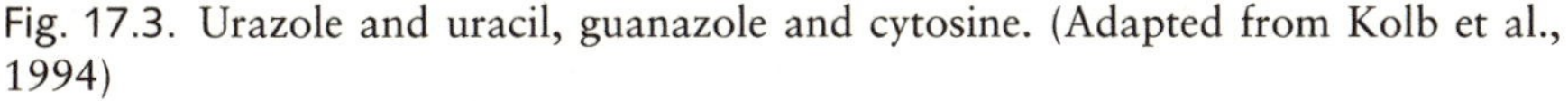

Fig. 17.3. Urazole and uracil, guanazole and cytosine. (Adapted from Kolb et al., 1994)

Possible precursors to RNA

According to the RNA-world theory, RNA preceded the origin of protein synthesis. As discussed by Joyce (1989), RNA itself is unlikely to have been the first genetic material. One possible solution to this problem would thus be to look for predecessors for RNA. For example, Kolb et al. (1994) suggested that urazole could have served as a precursor to uracil and that guanazole could have served as the precursor of cytosine (fig. 17.3). This suggestion, furthermore, has a biogeochemical flavor that will be discussed in Chapter 19.

Other kinds of possible precursors of RNA have to do mainly with solubility of phosphorus compounds and prebiotic phosphorilation of organic molecules. Recent studies suggest that prebiotic phosphorus compounds, such as phosphonic acids, are more plausible than the previously favored phosphate (phosphoric acid) as a prebiotic souce of phosphorus. These ideas support the suggestion that RNA predecessors were more easily synthesized than was expected under the assumption that the prebiotic phosphorus was phosphate. Moreover, these developments suggest that the backbone of the hypothetical predecessor of RNA was different from the extant sugar-phosphate chain. For more details see Arrhenius et al., 1997; Kolb et al., 1997; Maciá et al., 1997; and Schwartz, 1997a, 1997b.

A prebiotic genetic system

Arbitrary or incomplete as the description of life may be (chapter 11), it should include the attribute of a genetic system. Only a few attempts have been made, however, to suggest at least the general principles by which the very beginning of the genetic system emerged; one of these attempts is the aforementioned suggestion (Benner et al., 1993) that translation was acquired in a relatively complex system. This issue was partially addressed by Joyce and Orgel (1993, pp. 19–20), who considered two fundamental different kinds of transition from a simple genetic system to the RNA world, as follows:

1. A gradual transition from one kind of informational molecules to RNA molecules. Such a transition may be possible if the two systems are similar, such that the transition would not result in a kind of discontinuity related to their three-dimensional features. Candidate monomers are preferably characterized by synthetic routes simpler than those of nucleotides.
2. A "genetic takeover" process in which one self-replicating system "learns" how to synthesize and polymerize the constituents of the second system for its own selective advantage and is then taken over by it. Cairns-Smith (1982) favors the idea that this first genetic system was inorganic, presumably a clay mineral (chapter 20). He did not, however, discard the possibility that one organic system can replace another organic-genetic system.

The genotype/phenotype dichotomy

According to the RNA-world theory, the evolution of template-directed peptide synthesis may be perceived as the emergence of a genotype/phenotype dichotomy: Rather than one kind of chemical entity serving as both genotype and phenotype, these two attributes become uncoupled, though interrelated, as found in extant living forms (see Gordon, 1995). The alternative to this late decoupling of peptides and protonucleotides is the emergence of genotype/phenotype dichotomy with life itself, as suggested by the coevolution theory (Lahav, 1991; Lahav and Nir, 1997; see chapter 23).

Implied in both the RNA-world and Lahav's coevolution theories is either a late or concomitant evolution of metabolism, based on preformed primordial genetic and catalytic systems. The incorporation of metabolism into these scenarios has yet to be worked out. Similarly, compartmentalization could have evolved simultaneously with the evolution of template-directed synthesis, or later. As discussed in a later section, other schools have different suggestions, and this issue is still unresolved.

The chicken-and-egg paradox

This paradox stems from an attempt to back-extrapolate from extant molecular biology of information-processing and metabolism to the very beginning of these attributes in the hypothetical prebiotic evolution era. In contemporary living organisms, information processing is based on template-directed reactions performed by polynucleotides, where these reactions can proceed only with the help of the catalytic functions performed by biocatalysts, themselves the products of metabolic and template-directed reactions. Thus, life is equated with the simultaneous existence of two functions, namely, replication and metabolism. Since "in central biochemistry everything depends on everything" (Cairns-Smith, 1982, p. 94), the question, then, is

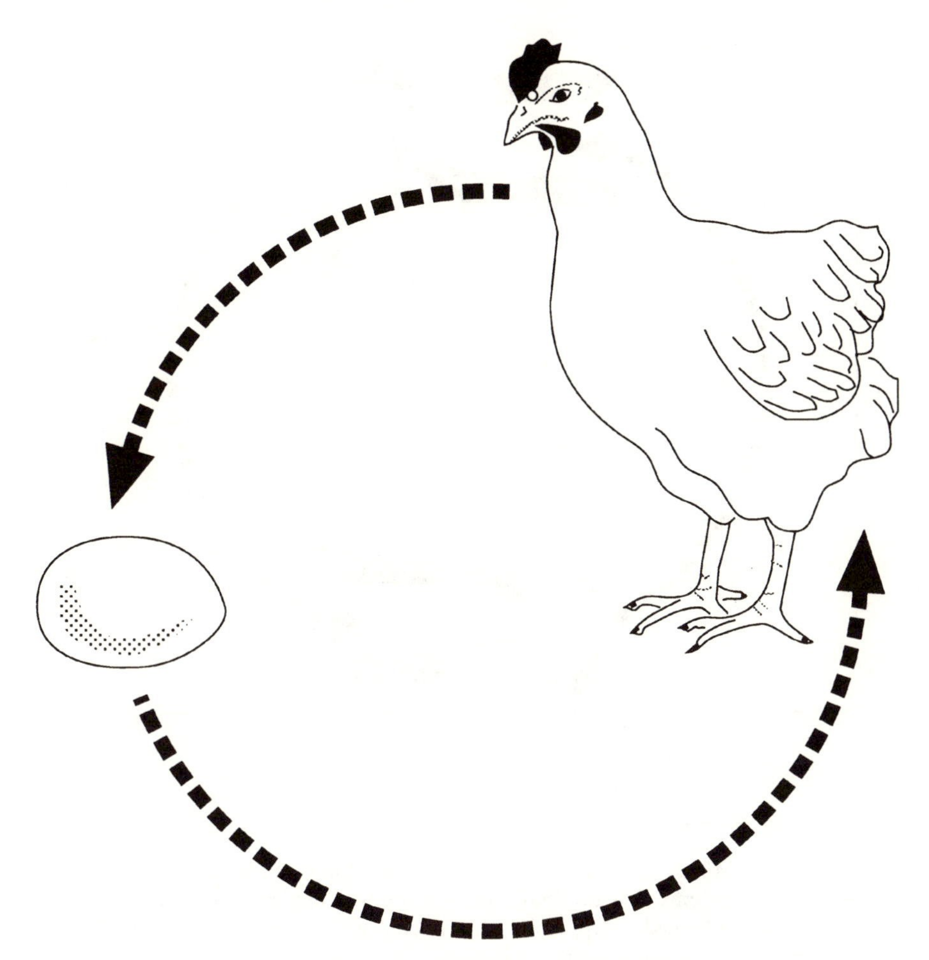

Fig. 17.4. The chicken-and-egg paradox.

which came first-replicating or metabolic functions and molecules (see Casti, 1992)? In terms of contemporary molecular biology the replicator is a polynucleotide, which serves as the informational molecule, and the metabolizer is a protein, which functions as a biocatalyst. The former is the genotype, symbolized by an egg, whereas the latter is the phenotype, symbolized by a chicken. The question is thus, which came first, the chicken or the egg (fig. 17.4)? As we shall see later this paradox may not be a paradox at all.

Similar to the chicken-and-egg paradox is the following paradox (the sixth order of the Mishna, the collection of Jewish oral laws): "Tongs were made of tongs"; thus, how were the first tongs made? (fig. 17.5).

Another chicken-and-egg paradox is discussed by Joyce and Orgel (1993). The chicken-and-egg paradoxes are thus destined to stimulate researchers and should be addressed by any hypothesis on the origin of life. Attempts to solve these paradoxes are likely to lean heavily on theoretical models, which may then be tested in the laboratory. Such a solution is yet to be found.

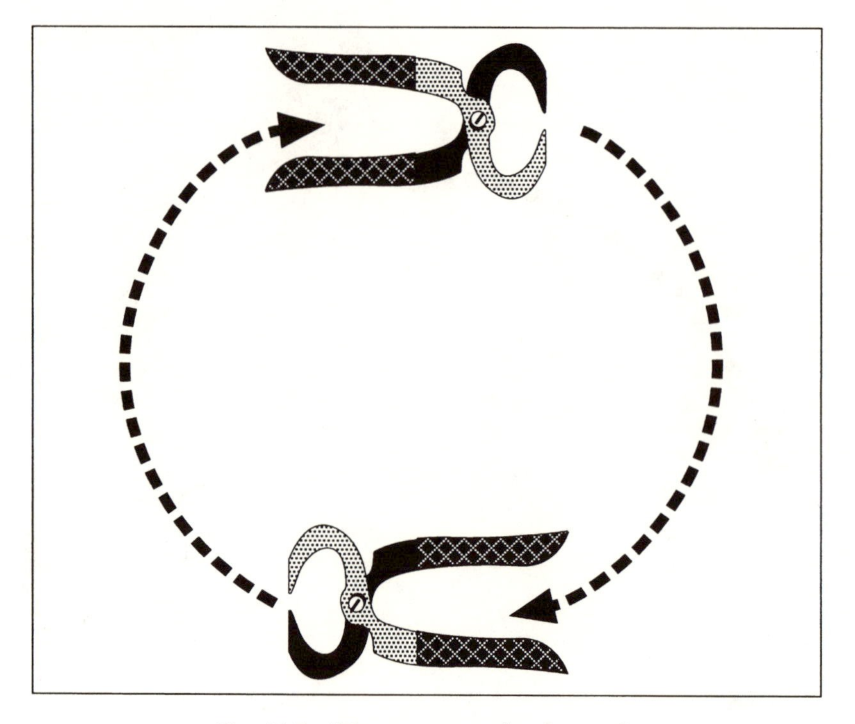

Fig. 17.5. "Tongs were made of tongs."

Minimum complexity

The RNA-world theory implies the popular assumption that life started with simple molecules and systems that gradually became more complex. Kauffman (1993, p. 295) recently challenged this presumed simplicity of the first living entities. This issue and the "minimal complexity" it implies is dicussed later.

The RNA world as a biological hypothesis without specific environment

The RNA-world connection with possible early specific environments on the surface of Earth is minimal—that is, it has to do with UV radiation (see Kolb et al., 1994) or with prebiotically-plausible conditions derived from the Oparin-Haldane-Miller-Urey paradigm. The main role of the environment in the RNA-world scenario is to supply building blocks and small oligomers of either RNA or RNA-like molecules. In addition to adequate temperatures, divalent ions such as Mg^{2+} or Zn^{2+} are assumed to be present; these metal ions are assumed to have been needed then as they are needed today; as catalysts.

Prebiotic plausibility, doubts, and reservations

Most researchers do not consider RNA molecules prebiotically plausible because of the problems encountered in its prebiotic synthesis. The experimental work carried out by Ferris and his collaborators (see Ferris, 1993a, 1994b; Ferris et al., 1996) may suggest a possible mechanism for the prebiotic synthesis of RNA. It should also be noted that the activation of the RNA building blocks in the replication reactions of RNA renders this experimental system prebiotically implausible because the activating agents used in laboratory experiments (Joyce and Orgel, 1993) are considered unstable in the prebiotic environments.

Weiss and Cherry (1993), while discussing the RNA world, raised a few fundamental questions that cannot be answered with certainty today; of special interest to us is the unclear relationship between RNA molecules and the primordial world. Is there a continuity between extant RNA molecules and the RNA-world molecules? Do they share the same ancestor? In other words, "it is uncertain whether specific mechanistic details . . . would have remained intact over the more than three billion years since the glory days of RNA catalysis."

Even more pessimistic with regard to the RNA world is the note made by Moore (1993. p. 131) while discussing the length of the "window of opportunity," the time interval in the history of the biosphere available for the RNA world. Moore expresses his doubts as follows: "when one starts thinking along these lines, one must consider the unthinkable, i.e. that the length of time the RNA-based organisms bestrode the earth might actually be zero." Regarding the possibility that the first ribosomes were made of RNA, he suggested that the assumption that the first ribosomes were made entirely of RNA is not plausible, adding, "Proteins must have come before ribosomes. After all, why would a device for making polypeptides evolve in an organism that had no use for protein? A protoribosome that contained both RNA and protein is thus entirely plausible."

The present confusion regarding the relevance of the RNA world to the origin of life has been recently discussed by Ellington (1994) and James and Ellington (1995).

Selection of ribose and its chemical stability

It is generally assumed that the prebiotic synthesis of ribose is the formose reaction (Miller, 1992; chapter 15). However, in this reaction numerous sugars are formed together with ribose. So far no plausible mechanism for the selection of ribose, out of the other sugars, has been suggested. In addition, the stability of ribose, like that of other sugars, is low. For instance, the ribose half-life at pH7 and 100°C was found experimentally to be 73 minutes. At the same pH, and at 0°C, the half-life was 44 years (Larralde et al., 1995). These half-lives are considered very low (compared to those of amino acids, which are in the range of hundreds of years) for a plausible molecular evolution process in which ribose is involved to take place. Indeed, some scientists have concluded that these selectivity and stability problems exclude ribose from a possible prebiotic involvement in molecular evolution. Thus, the existence of an RNA world according to Larralde et al. (1995), implies either the availability of a prebiotic source of ribose or the existence of RNA-like (pre-RNA) molecules with a different backbone. This, then, implies a "pre-RNA world" of a kind (Larralde et al., 1995). One possible solution could be the PNA world suggested by Nielsen and his collaborators (discussed later).

Homochirality (enantiomeric homogeneity) of the sugar moiety of nucleotides

Laboratory experiments with nonenzymatic replication of RNA strands (see Joyce 1987, 1989, for a review) pointed out to the problem of enantiomeric cross-inhibition, the poisoning of chain growth on a template following the addition of building blocks with enantiomers of ribose. The chiral purity of the building blocks of RNA, the sugar moiety, is crucial for their participation in the self-replication reactions and hence in the establishment of the RNA world. This conclusion does not tell whether the needed strong enantiomeric bias was exercised by the evolving "living" systems of the RNA world or whether it resulted from the fact that source of terrestrial chirality was extraterrestrial, as has been suggested by Bonner (1991, 1995), MacDermott (1993), Greenberg (1995), and Popa (1997). One possible solution for this problem will be described shortly, in the discussion of a specific template-directed reaction of RNA.

The problems with adenine

As noted earlier, the RNA-world theory is based on the Watson-Crick base pairing as the mechanism of template-directed reactions. Adenine plays essential role in all known living organisms in replication and numerous other reactions. However, its chemical properties cast doubts on its possible role in the very early stages of the origin of life. These chemical properties include its low yields in known syntheses under plausible prebiotic conditions, its susceptibility to hydrolysis and to reaction with a variety of prebiotically plausible electrophiles, and its ability to hydrogen-bond and function in any specific recognition scheme under the hypothesized conditions of the prebiotic soup (Shapiro, 1995). Similar to many other problems in this field of scientific endeavor, the role of adenine, or an adeninelike molecule, has yet to be resolved.

Template-directed reactions and the transition between the inanimate and the animate

Because of the centrality of template-directed syntheses in contemporary life, it is tempting to equate their emergence with the transition from chemistry to biology, arbitrary as such an association may be. For those who believe in the RNA world, self-replication is the beginning of both genetics and directed synthesis of catalysts (RNA molecules). The RNA molecules, or assemblies of these molecules that function as both genotypes and phenotypes, may be considered the first living creatures. The evolution of template-directed synthesis of peptides and proteins according to the RNA world is thus not a totally new "invention." Rather, it is a sophistication of an existing template-directed process, the basic tenets of which have already been in use in the RNA world. Therefore, template-directed protein synthesis could have evolved in a "living," thriving RNA world.

For those who prefer the coevolution of template-directed reactions of self-replication and peptide synthesis, the first chemical entities characterized by these two intimately connected processes may be considered the first living creatures.

Prospects

It is hard to predict whether the RNA-world theory is a short-lived mode. Its establishment among scientists has brought about novel theoretical and experimental works that have focused on some central aspects of the ancient forms of life. Whether or not this will be the only importance of the RNA world is unknown at present.

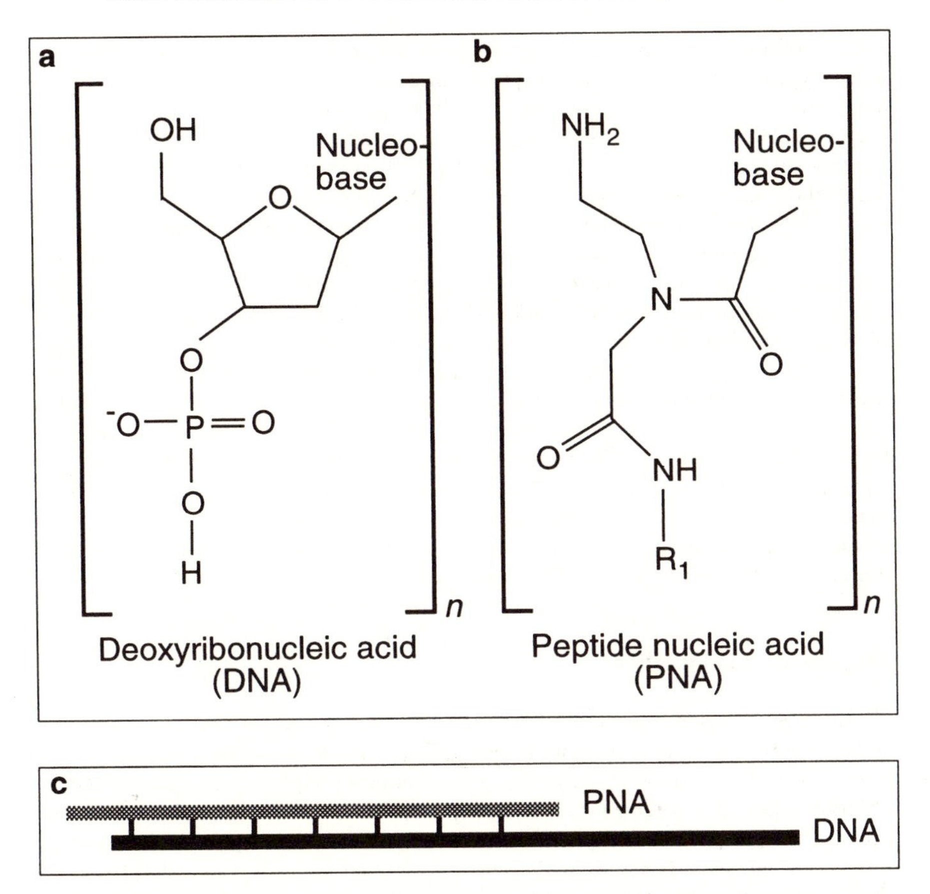

Fig. 17.6. Chemical structure of PNA and DNA. a. DNA. b. PNA. R_1 is either hydrogen atom or lysine amide for the PNAs investigated by Nielsen (1993) and Wittung et al. (1994). c. A scheme of binding of PNA to single-strand DNA. (Adapted from Nielsen, 1993)

The PNA/RNA world

In view of the difficulties in explaining prebiotic synthesis of RNA molecules, several research groups have looked for molecules that could have served as predecessors to the RNA molecules. Obviously, such molecules should be prebiotically feasible; no less important, they should have chemical properties that would enable information transfer with RNA molecules in the transition between the pre-RNA world and the RNA world. One such a candidate is a peptide nucleic acid (PNA), which consists of a peptide backbone to which nucleobases are attached (fig. 17.6; see also Wittung et al., 1994)

PNA was found to bind to oligo(deoxy)ribonucleotides according to the Watson-Crick base pairing rules. Chemically, the PNA bridges the gap between nucleic acids and proteins; PNA molecules can carry genetic information, where the backbone is made of peptides. According to Nielsen (1993), "it is conceivable that prebiotic mimicing conditions which produces nucleobase-amino acid conjugates can be found" in Miller-type experiments. In addition to serving as a bridge between two "worlds," it has been suggested that PNA served as the basis of an achiral genetic material. It has been speculated that when used as an adsorbed template (see chapter 20) on a chiral mineral surface, it is also able to act as a chiral template. It is thus fashionable to speak about a PNA world that preceded the RNA world. Recent experiments (Böhler et al., 1995) have shown that RNA oligonucleotides facilitate the enzyme-free synthesis of complementary PNA strands and vice versa. This supports Nielsen's suggestion that a transition between different genetic systems is plausible (see Piccirilli, 1995). However, more recent work (Schmidt et al., 1997) has shown that enantiomeric cross-inhibition also takes place in PNA-directed reactions.

The PNA/RNA world is an interesting hypothesis that seems not to violate the principle of biological continuity and has to be further explored. It should be noted, however, that the only peptide in the PNA world is that of the backbone of the informational molecules. In such a system the appearance of additional prebiotic, nontemplated peptides, including catalytic peptides, cannot be discarded. Thus, the PNA hypothesis should also take additional peptides into account. In the presence of catalytic peptides, entirely different scenarios can be envisaged that may prove to be more efficient than the "pure" PNA system.

Template-directed reactions

A template "provides instructions for the formation of a single product from a substrate or substrates which otherwise have the potential to assemble and react in a variety of ways" (Anderson et al., 1993). A biochemical template is characterized by its ability to direct the organization of various molecules in a specific order. The importance of template-directed replicating systems extends today beyond the classical biological systems of RNA and DNA. It includes nonbiological systems that have been developed in an attempt to explore a new kind of chemistry of organic polymers (Orgel, 1992; Rebek, 1994).

Template-directed polymerization—self-replication (self-copying) and peptide synthesis—is a central feature in biology and, by continuity, also during the early stages of biology. The evolutionary stage in which this system emerged is model-dependent and thus controversial. For instance, according to Wächtershäuser's (1992a) pyrite world (chapter 21), these systems emerged relatively late, whereas according to Lahav's (1991; Lahav and Nir, 1997) coevolution hypothesis, the emergence of these systems was a rather early event that may be equated with the origin of life, as discussed in chapter 23.

The very concept of prebiotic template-directed synthesis is based on the biological continuity principle and the simplification guideline. Generally speaking, the laboratory studies of these reactions were carried out in the framework of the heterotrophic paradigm, in the sense that the experimentalists focused on the stage of interactions among prefabricated building blocks and their oligomers and polymers, not on the mechanism of formation of these molecules. In such a system the most common assumption is a pool of dissolved organic molecules—a prebiotic soup.

The processes that have attracted scientists most have been the evolution of template-directed syntheses of oligomers and polymers from their building blocks, namely, replication of polynucleotides, on the one hand, and peptide synthesis, on the other. Implied in the focusing on template-directed synthesis is the assumption that a meaningful evolutionary process should start with a genetic system, primitive as it may be. This is not self-evident, however. As we shall see later, it has also been postulated that a primordial metabolism came before coded synthesis (see, for instance, Wong, 1975, 1980; Morowitz, 1992; Wächtershäuser, 1988). The coexistence of various processes of both schools is another possibility that has yet to be considered.

Self-replication and complementary-strand-synthesis of RNA without enzymes

RNA molecules have been adopted by several research groups as models for templates, on which the mononucleotides would be attached specifically by the known Watson-Crick complementarity rules and later would condense to form a complementary strand. This is a structural simplification of the contemporary replication scheme of polynucleotides in the cell.

Template-directed reactions were studied extensively by Orgel and his collaborators (see Orgel, 1992) in the context of prebiotic chemistry. The kinetics of these reactions was studied by Kanavarioti and White (1987), Kanavarioti (1992, 1994, 1997), Kanavarioti and Baird (1995), and Kanavarioti et al. (1993). In these "prebiotic" experiments, *no enzymes have been allowed*; such biopolymers have been assumed to be incompatible with the prebiotic picture of small, simple, and noncoded molecules. In these studies, the experimental system consisted of RNA molecules, and homochiral mononucleotides (with the optically active D-sugar, the building blocks of RNA). Using this approach, it has been found that whereas template-directed synthesis of complementary strands was relatively easy to demonstrate in the laboratory, self-replication of RNA strands defied all experimental efforts, except in a few specific cases. We shall focus for a moment on the template-directed reactions that were elaborated by Orgel and his collaborators.

The natural triphosphate activation is not effective

The many laboratory experiments carried out by Orgel's school have shown that template-directed condensation of mononucleotides on RNA strands in the absence of enzymes did not occur. Apparently, the universal natural activating agent of nucleotides, the triphosphate ion, is not active enough for the condensation reaction of the mononucleotides to take place on the RNA strand in the absence of a catalyst. A template-directed reaction did take place, however, in certain systems where the mononucleotide building blocks were preactivated by reactive chemical groups that are more effective than the triphosphate. The result was a complementary strand on the RNA template under study. The most powerful activating agent in these reactions were imidazolides such as phosphorimidazolide of adenine (ImPA) and guanosine 5′-phospho-2-methyl-imidazole (2-MeImPG) (fig. 17.7). Imidazole is considered by some researchers to be a reasonable prebiotic compound, since it can be readily prepared under prebiotic conditions (Miller, 1992). However, because of its high chemical reactivity, its accumulation in prebiotic environments such as the prebiotic sea is still not clear.

Example of a template-directed reaction experiment

The nonenzymatic template-directed synthesis was developed by Orgel's school (Inue and Orgel, 1983; Joyce, 1989). The reaction mixture consists of poly-C as a template

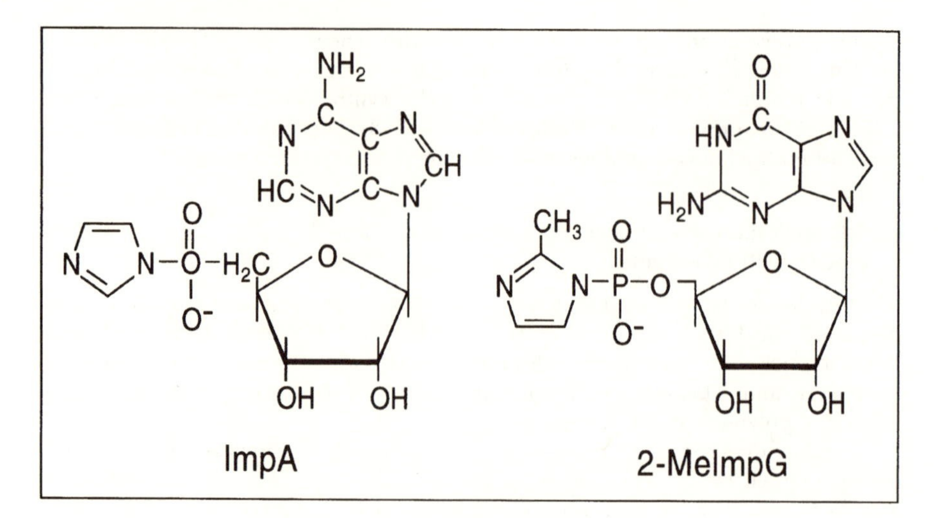

Fig. 17.7. Structure of ImpA and 2-MeImPG. (Adapted from Miller, 1992)

and activated Gs as mononucleotides, where the sugar moiety in these molecules is the homochiral (D) ribose. Na^+ and Mg^{++} are included in the buffered solution. The reaction mixture is incubated for one week at a low temperature of about 5°C (in order to enhance hydrogen bonding between the poly-C template and the activated G mononucleotides), and the concentration of the condensation product, oligo-G, is then measured by HPLC. In a typical experiment, the population of oligo-Gs ranges from dimers to oligomers of 40 and more building blocks. Equation 17.1 shows a polymerization of an activated purine ribonucleotide (ImP-guanine, ImPG) on a polypyridine template (polycytosine, poly-C). The template-directed reaction and the formation of a complementary strand of poly-G (the complex on the right side of the equation) proceeds efficiently according to the following equation (Orgel, 1986; Miller, 1992), where the complex on the right hand of equation 17.1 can be separated into its two components poly-C and poly-G:

$$\text{poly-C} + \text{ImPG} \rightarrow \text{poly-C:poly-G} \qquad (17.1)$$

This template-directed synthesis is only the first half of a self-replication reaction; in the second half of the reaction the template should be poly-G (or oligo-G) and the activated monomers, the C mononucleotides. However, the activated cytosine (ImPC) does not polymerize on the poly-G template, as shown schematically in equation 17.2, where X denotes that this reaction does not take place:

$$\text{poly-G} + \text{ImPC} \text{ —X}\rightarrow \text{poly-G:poly-C} \qquad (17.2)$$

Should such a reaction proceed under these experimental conditions, a full cycle of self-replication would be obtained according to this scheme, where *C and *G are activated C and G, poly-C and poly-G are the templates, and $(G)_n$ and $(C)_m$ are the polymers formed in each of these template-directed reactions:

$$n(^*G) \xrightarrow{\text{poly-C}} (G)_n \qquad (17.3)$$

$$m(^{*}\mathrm{C}) \xrightarrow{\text{poly-G}} (\mathrm{C})_m \tag{17.4}$$

As noted earlier, it was found that, except for specific systems, the second reaction (17.4) does not take place. The reason is probably the tendency of poly- and oligo-Gs to undergo self-structuring, where certain domains on the same poly-G strand are bound very strongly to each other. The result is that the C monomers are not able to hydrogen-bond to the template, therefore the complementary strand formation does not take place (see Joyce and Orgel, 1993). It is noted that polymerization of the activated form of A (i.e., ImPA) on a poly-U template does take place, but the polymerization of the activated form of U (i.e. ImPU) on a poly-A template does not. In line with these experiments, other templates were also used in complementary-strand formation reactions. These include the production of various phosphodiester linkages, as well as linear and cyclic nucleotide strands (Ertem and Ferris, 1996).

The self-replication reaction is still an experimental problem

A two-step reaction, that is, self-replication, was successfully performed only in a few specific cases with short palindromic templates (von Kiedrowski, 1986, 1993; Sievers and von Kidrowski, 1994; Sievers et al., 1994; Li and Nicolaou, 1994; Ferris, 1994a; see also Gordon et al., 1994). The reaction is still under study, and a scenario based on the Orgel school reaction "by which non-enzymatic self-replication of short RNA molecules could occur" was suggested by Kanavarioti (1992) but has not been tested so far.

A novel kind of RNA precursor was recently suggested by Schwartz (1997a, 1997b), based on the observation that a significant portion of the phosphorus in the Murchison meteorite is in the form of methyl and ethyl phosphonic acids (Cooper et al., 1992). The relevant RNA-like molecule is thus characterized by a backbone structure based on ribose-2,4-diphosphonic acid.

The chirality problem in template-directed reactions

The chirality problem of the template-directed reaction was circumvented (and postponed) by the use of homochiral nucleotides. Based on laboratory experiments, it has been established that nucleotides synthesized by prebiotic reactions are always racemic—that is, they contain equal concentrations of D and L enantiomers. Thus a better simulated prebiotic experiment should use D and L ribose rather than D-ribose alone. When the same kind of experiment was carried out with the racemic mixture of activated D-ribonucleotides, the template-directed reaction did not proceed, because of enantiomeric cross-inhibition (Joyce et al., 1984). A more complex system arises if one takes into account additional possible molecular configurations of the nucleotides under study. Thus, both α- and β- as well as anti- and syn- configurations of nucleotides are also produced in prebiotic synthesis, and additional kinds of cross-inhibition are expected in a prebiotically-plausible environment (Joyce, 1989; Joyce and Orgel, 1993). These problems have remained unsolved so far.

It is interesting to briefly discuss one approach that has been suggested in order to overcome this stumbling block of enantiomeric cross-inhibition. According to this suggestion (see Joyce et al., 1984; Joyce, 1987), a self-replicating molecule similar to RNA but nonchiral (Joyce, 1989) may be assumed. According to this suggestion, the backbone of the first templates would be based on prochiral nucleoside analogs characterized by open-ring molecules, such as glycerol. Being achiral, these templates would not suffer from enantiomeric cross-inhibition (see Lazcano, 1994b). Pioneering work along this line of thought was made by Schwartz and Orgel (1985), who showed that nucleoside analogues can be polymerized in template-directed reactions.

Some experimental support to the plausibility of this suggestion, together with more details on these problems can be found in Schwartz, 1993; Goodwin et al., 1994; Schwartz and van Vliet, 1994; van Vliet et al., 1994, 1995; Joyce and Orgel, 1993; Brack, 1993a; and James and Ellington, 1995. However, a more recent work (Schmidt et al., 1997; see above) suggests that "it now seems unlikely that the choice of a new template, whether chiral or achiral, will overcome enantiomeric cross-inhibition so generally as to permit template-directed replication of oligomers long enough to seed the direct emergenceof RNA from a solution of racemic ribonucleotides."

Simpler replicators are needed

One central challenge for origin-of-life research (James and Ellington, 1995) is to devise simple replicators, such as those elaborated by the von Kiedrowski or Nicolau labs, capable of undergoing further development and thus increasing their complexity toward the systems developed by the Szostak and Joyce labs. (The cited works are of great interest by their own merits: The former two researchers developed simplified systems capable of self-replicating; the latter two developed the technology for the selection of ribozymes for specific catalytic functions out of very large populations of RNA molecules. These works are outside the scope of the present book.)

James and Ellington (1995) also suggested a scenario for an evolutionary process of simple replicators; their scheme includes competition, accretion of additional sequence information, selection for catalytic properties, and diversification.

Catalyzed RNA polymerization in prebiotic processes

Back-extrapolating from biology, there are two major possible biocatalysts—ribozymes and peptides—as well as a presumed transition from small and simple to bigger and more complex organic molecules and assemblies thereof. Considerations of the prebiotic environment also implies the possible involvement of environmental entities such as metal cations, minerals, and water, and parameters such as temperature and hydration, in the processes under study. Surprisingly, very little experimental and theoretical work has been done on prebiotic catalyzed template-directed reactions, as reviewed briefly in the following section.

RNA-catalyzed polymerization Replication of RNA molecules catalyzed by ribozymes is in the framework of the RNA world. Recent work by Ekland and Bartel (1996) has shown that such catalytic reactions, "using the same reaction employed by protein enzymes that catalyze RNA polymerization, can be carried out in the laboratory." In their experiments these researchers used a template RNA and nucleoside triphosphates (pppC, pppA and pppG), as well as a ribozyme that extended an RNA primer by successive addition of mononucleotides. The added nucleotides were joined to the growing RNA chain by 3′,5′-phosphodiester linkages; the ribozyme showed a remarkable template fidelity. The authors conclude, "The progress made in deriving RNA polymerase from RNA sequences bodes well for the eventual demonstration for autocatalytic RNA replication." Indeed, autocatalytic RNA replication would be an important achievement; however, in view of the RNA-world theory's persistent problems in connection to the molecular evolution process, the relevance of such a feat to origin-of-life research has yet to be worked out. Furthermore, as will be discussed in chapter 23, it would be interesting to see whether the ribozyme in this reaction can be replaced by a catalytic peptide.

Polymerization catalyzed by small peptides It is surprising that very little effort has been dedicated to prebiotic reactions catalyzed by small peptides, which are considered prebiotically plausible and therefore relevant to replication processes. It is convenient to divide the catalytic polymarization reactions studied so far into three categories: RNA replication, RNA-template-directed peptide synthesis, and self-replication of peptides.

With regard to RNA replications very small peptides, up to tripeptides, were found ineffective in catalyzing Orgel's reaction of template-directed RNA polymerization (Orgel, personal communication).

As far as I know, RNA-template-directed peptide synthesis has not been studied so far.

With regard to self-replication of peptides, according to the central dogma of molecular biology, information always flows in the direction from the genotype to phenotype—from DNA to RNA to protein; proteins do not possess the ability to duplicate themselves. Theoretical suggestions regarding the ability of proteins to replicate by means of amino acid pairing were advanced in the early 1980s (Root-Bernstein, 1982, 1983); however, until very recently such replication reactions were not reported (Brack, 1994).

The latter postulate was recently challenged by Ghadiri and his associates (Lee et al., 1996; Peaff, 1996), who for the first time succeeded in showing experimentally self-replication of peptides. In their experiment a 32-residue α-helical peptide acted autocatalytically in templating its own synthesis in neutral, dilute aqueous solutions. As noted by Kauffman (1996), this discovery "may prove to be either a mere chemical curiosity, or seminal." Because of the implications of this work for origin-of-life research, it is expected that more work will be performed in the near future in order to further explore the intriguing question of peptide self-replication.

Another strategy seems to be needed

As noted earlier, small organic catalytic molecules have hardly been used in the template-directed and self-replication experiments I have described. Moreover, we shall see later that it is plausible to assume prebiotic environments that, rather than no enzymes, contain small organic catalysts. This approach may be helpful in the successful establishment of self-replication reactions, as discussed in chapter 23.

Early stages of the translation machinery and the genetic code: The top-down approach

The first speculation about the origin of the genetic code was suggested by Gamow (1954). In spite of a continuous effort by hundreds of scientists since then, the problem of the origin of the genetic code has not been solved as yet. In retrospect this is expected, in view of the complexity of the protein synthesis machine. Given such a complex system, containing more than a hundred components (Lazcano, 1994b), it is not surprising that Moras (1992) noted with much pessimism that "the absence of direct link between the anticodon loop and the site of aminoacylation suggests that the search for a simple stereochemical correlation between the three letter genetic code and the amino acid or the synthetase (associated with the idea of a second genetic code) is hopeless."

Analysis of the plethora of papers with titles suggesting their dedication to the cause of the origin of the genetic code shows that the overwhelming majority of these works actually deal with its evolution. This may be "because the authors assume that the origin of the genetic code is inevitable once they have created a scenario that provides the components of an informational molecule" (Yockey, 1992). Recalling that the hypothetical entity named progenote (chapter 10) already possessed a primordial translation machinery, the following discussion goes beyond the progenote to treat the evolutionary stages that preceded this entity.

The present section briefly reviews the main attempts to use the top-down approach and go backward in time in order to suggest possible ancient stages and scenarios for the evolution of the genetic code. These propositions, per definition, deal with neither the very first reactions of the molecular evolution processes, nor the primordial emergence of the genetic code. It is interesting to note that no specific environments have been proposed in the great majority of the hypotheses advanced so far in order to describe early stages of the genetic code. As will be discussed in chapter 23, the combination of both top-down and bottom-up approaches is more likely to come up with plausible scenarios for the origin of the genetic code and its machinery than the use of just one approach. In other words, taking into consideration the geochemical conditions in specific prebiotic environments seems to be necessary for the search into the origin of both life and the genetic code.

The peptide-synthesis machinery

The two molecules on which most researchers have focused are tRNA and aminoacyl-tRNA synthetase. The approaches used include conserved domains and phylogeneric studies, as well as structural features. The phylogenetic analysis of the molecule essentially involves the alignment of the molecules from many organisms by computer, the identification of conserved domains (molecular fossils), the construction of their phylogenetic trees, and the reconstruction of the order of appearance of central features of the protein-synthesis machinery during evolution time.

The structural studies are aimed at the establishment of the three-dimensional features of the molecules. By combining the molecular features of various approaches, it is possible to partially characterize early stages of the evolution of the protein-synthesis machinery.

The genetic code: Generally accepted assumptions and some controversial ones

In view of the enormous number of hypotheses and speculations regarding the evolution of the primordial genetic code and its machinery, their detailed review in the present book is not possible. For a succinct review of central attributes of the code, see Zubay, 1996. The following features are related to certain aspects of the *emergence* of the code; moreover, they are accepted by most researchers.

Early number of amino acids involved

The early arsenal of amino acids was small, perhaps as few as four or even two (see de Duve, 1991; Hartman, 1995a, 1995b; Kuhn and Waser, 1994a, 1994b). According to Béland and Allen (1994), the primitive genetic code had only 20 separate words.

The commaless code

The first code is likely to have used the commaless triplet codon (where messages are readable only in one frame, such that no frame-shifted codons are allowed) rather than the binary codon. The reason for that assumption, which is essentially related to the principle of continuity, was phrased by Crick (1968) as follows: "A change in codon size necessarily makes nonsense of all previous messages and would almost certainly be lethal."

The genetic code is neither universal nor a frozen accident

Both the universality of the genetic code and the concept that it is a frozen accident have been critically discussed by various researchers (see de Duve, 1991; Jukes and Osawa, 1993; Maynard Smith and Szathmáry, 1995, for reviews). Based on statistical analysis of the sensitivity of the different properties of amino acids to point mutations in the DNA, several authors have suggested that the genetic code did not originate through a frozen accident. Furthermore, phylogenetic studies of some aminoacyl-tRNA synthetases suggest that "the entire code was not completely assigned at the time of the diversion of bacteria from nucleated cells" (Cedergren and Miramontes, 1996). Thus, the recent discoveries of several nonuniversal genetic codes indicate that it is variable rather than frozen. According to Ueda and Watanabe (1993), "the code has appeared frozen because of its extremely high rigidity in comparison to other biological systems." Moreover, the latter authors suggest that this rigidity may be explained by the recognition mechanisms of tRNAs by aminoacyl-tRNA synthetase.

Thus, the genetic code is not identical ("universal") in all biological systems. However, since most known organisms do have an identical code, the latter is generally referred to as the standard genetic code. Codon assignments could change and thus could also evolve (Jukes and Osawa, 1993). These changes, however, are very slow; for a long time it was assumed that the code was "frozen" at an early evolutionary stage, when changes would have been disruptive for life. This is the "frozen accident" hypothesis suggested by Crick (1968).

Our challenge is, of course, to explore the *formation* of the genetic code. Obviously, except for the general guidelines of primordial simplicity, the features cited earlier cannot be used directly in the deciphering of this problem.

Main approaches in the study of early stages of the genetic code

> With only four nucleotides and a long evolutionary history, searching for vestiges of the genetic code within the tRNA structure is like searching for a small needle in a big haystack.
>
> Paul R. Schimmel, "Origin of genetic code"

We will now briefly discuss the main approaches used by researchers, as well as their main findings and interpretations, in order to explore the early stages of the code and its machinery. They include direct interaction, matchmakers, correlations, and phylogenetic considerations.

Direct base–amino acids interactions

The use of the principle of biological continuity is best exemplified by the suggestion (Orgel, 1989) that the attachment of amino acid to an RNA adaptor by the 3′-termini of the RNA was the first step in the evolution of the peptide formation machinery. The RNA is thus the predecessor of tRNA (pre-tRNA, proto-tRNA), and the ester bond between the 3′-end of the RNA and the amino acid is continuous to the ester bond between the corresponding chemical moieties of the extant translation system. Additional stages are discussed by Orgel. It should be noted that this suggestion does not provide a mechanism for the primordial translation process and the environment in which it functions (Lazcano, 1994b; Lacey et al., 1992).

One of the earliest lines of research has been to characterize the direct interaction between amino acids and nucleotides (see, for instance, Hopfield, 1978; Shimizu, 1982). Indeed, "an understanding of the interaction of amino acids with oligonucleotides is generally recognized to be a necessary prerequisite to any attempt to explain the origin of the genetic code" (Zieboll and Orgel, 1994). Since the contemporary machinery of translation does not need such direct interactions (because the tRNA interposes between the amino acid and its codon), these studies reflect an attempt to back-extrapolate toward primordial interactions, which supposedly could have taken place before the advent of the large and sophisticated tRNAs and ribosome.

Early studies on particular base–amino acid affinities or a stereochemical fit between these two molecular species were disappointing. These kinds of early works thus seemed to suggest that the code was established by accidental associations that were later selected to be frozen; accordingly, the code was considered a frozen accident (Taylor and Coates, 1989).

An interesting observation in this context is the specific binding of the amino acid arginine onto the guanosine cofactor-binding site of the group I intron (Yarus, 1988, 1993; Yarus and Christian, 1989; see also Calnan et al., 1991; Noller, 1993). According to Yarus, the origin of the arginine codon might have occurred via direct amino acid–RNA recognition. A similar conclusion was reached by Zieboll and Orgel (1994).

Reciprocal information transfer between oligoribonucleotides and oligopeptides was suggested by Zhang and Egli (1994). Three types of complementary Watson-Crick-type hydrogen bonds are proposed between the interacting molecules. The proposal deals with structural aspects of information transfer between the two different polymers but does not address the mechanism of evolution of the genetic code.

Matchmakers

Rather than counting on fitness and direct attachment of amino acids and tRNAs, it may be possible to bring together these two molecular species by means of a third party. Based on phylogenetic analysis of aminoacyl-tRNA synthetases, Nagel and R. F. Doolittle (1995) proposed a "matchmaker" (presumably RNA), which can bring together the amino acids with their tRNAs; the introduction of the contemporary aminoacyl-tRNA synthetases was thus a gradual evolutionary process. (For additional aspects of this theory, see Wood, 1991.)

Correlations

Because the origin of the genetic code is a problem with no obvious place where to begin (Orgel, 1986, 1989), many workers have tried to look for correlations between various chemical properties of amino acids, on the one hand, and various properties of the genetic code and its machinery, on the other. The list of amino acid properties includes polarity and hydrophobicity-hydrophilicity (Blalock and Smith, 1984; Di Giulio, 1996a; Lacey and Mullins, 1983; Lacey et al., 1985; Lacey and Staves, 1990; Staves et al., 1988; Taylor and Coates, 1989), plausibility of prebiotic synthesis, side chains, and metabolic pathways of synthesis (Wong, 1975, 1976, 1980; Taylor and Coates, 1989).

Phylogenetical and functional considerations

The following topics exemplify the problems that have been investigated by various researchers.

Ribosomal peptide-bond formation So far, no experimental evidence for catalyzing peptide-bond formation between activated amino acids by prebiotically plausible ribozymes has been published. However, many researchers find such a possibility feasible (see Lazcano, 1994b, 1994c; Noller et al., 1992; Lazcano, 1994a; Zhang and Cech, 1997).

Evolution of ribosomes from independent fragments Several lines of evidence, as well as analogy to the last common ancestor and its rationale, led Lacey and Staves (1990) to focus on the constituent of the ribosomal RNA known as 5S rRNA. According to these researchers, "contemporary tRNAs and 5S rRNAs had a common ancestor." Their hypothetical entity, which served as a universal translator, was about 120 nucleotides long; it "gave rise to 5S rRNA and to a universal tRNA from which the tRNAs for specific amino acids arose through variable processing. Contemporary 5S rRNA evolved through mutations which were dictated by its present functions." The assumption that in an earlier evolutionary stage, such as the RNA world, 5S rRNA was made of a collection of independent smaller RNA fragments was discussed by Y.-H. Lee et al. (1993). Similarly, the absence of sequence homology between two components of the ribosome, namely 16S and 23S ribosomal RNA, corroborates with independent evolutionary origin of two RNA pieces (Noller, 1993).

Primordial tRNA, amino-acyl synthetase, and related origin of mRNA and tRNA Hopfield (1978) suggested that the origin of the genetic code is associated with the structure of primordial tRNA. Eigen and Winkler-Oswatitsch (1981a, 1981b, 1983) analyzed the relationships among primordial RNAs and concluded that primitive tRNAs had served as both mRNAs and tRNA adaptors before starting their specialization. This idea of a related origin is an important feature of Lahav's (1991) co-evolution hypothesis on the emergence of the translation machinery, which will be discussed later in the present chapter, as well as in chapter 23.

Another approach relevant to the early RNA world was suggested by Dick and

Schamel (1995), according to whom tRNA-like molecules, ribozymic charging catalysts, and some small and large subunit rRNA evolved from the same ancestral RNA molecule. They further suggested that "generation and evolution of tRNA were coupled to the evolution of synthetases, ribosomal RNA, and introns from the beginning," where the original tRNA precursor hairpins functioned as "replication and recombination control elements." Early stages of the ribosome are also discussed by Gordon (1995).

Aminoacyl-tRNA synthetases In extant biology, aminoacyl-tRNA synthetases (aaRSs) are the enzymes responsible for the correct attachment of amino acids onto their corresponding tRNAs (fig. 17.8). As such, they perform a crucial step in the translation of a sequence of nucleotides into a sequence of amino acids. It is thus of great interest to explore their evolutionary root, as well as their primordial structure and functions.

Sequence analysis and comparative studies of crystal structures of aaRSs complexed with cognate tRNAs (for reviews, see Härtlein and Cusack 1995; Nicholas and McClain, 1995; Ferreira and Cavalcanti, 1997; Rodin and Ohno, 1995) revealed two distinct classes, I and II, of 10 synthetases each. It is recalled that there is one synthetase for each amino acid, and each synthetase attaches its amino acid to all of the tRNA isoacceptors, which contain the anticodon corresponding to the amino acid. Because of lack of similarity, independent origin of these two classes was formerly suggested by most researchers. However, assuming independent origin created other problems as discussed by Rodin and Ohno (1995; see also later the discussion on Wong's coevolution theory). This discrepancy seems to have been solved recently by Rodin and Ohno (1995), according to whom the two classes of aaRSs served as complementary strands of the same nucleic acid and emerged synchronously. This conclusion may be helpful in the elucidation of the origin of the genetic code and its machinery.

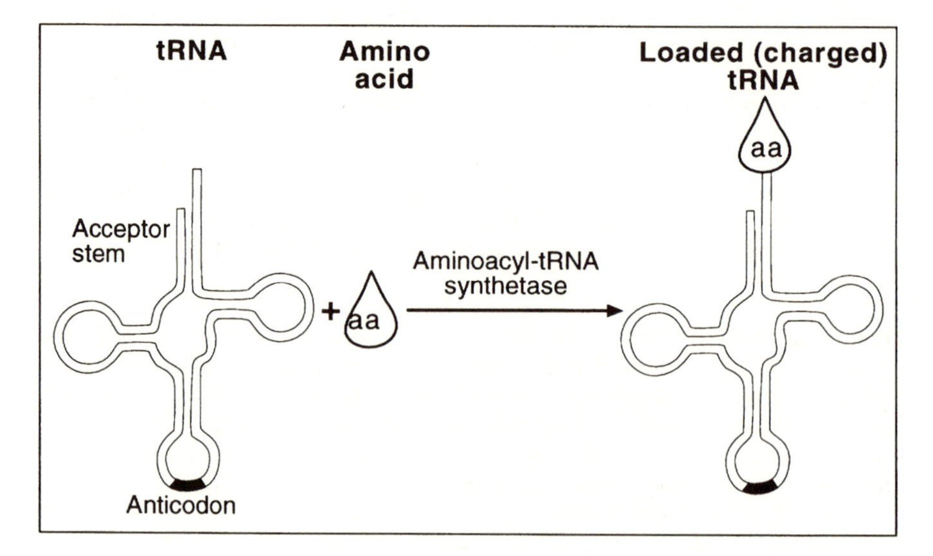

Fig. 17.8. A scheme of transfer RNA (tRNA) and the loading reaction by amino acid.

tRNA identity and the structure of early synthetases Early work on this topic was carried out by Schulman (Saks et al., 1994). One of the most sophisticated experiments designed to back-extrapolate from contemporary ribosomes to ancient translation systems was carried out by Schimmel and his group (for recent reviews, see Arnez and Moras, 1997; Buechter and Schimmel, 1993; Schimmel et al., 1993; Schimmel and Henderson, 1994; Schimmel and de Pouplana, 1995). Schimmel focused on the so-called tRNA identity, the structural features of tRNA that are recognized by aaRS. These tRNA identity features include the anticodon as well as other elements in tRNA molecules. Thus, in several cases, these identity features are based mainly on the anticodon or include the anticodon as a major component. Evidently, this coincidence between the primary (classical) genetic code and the tRNA identity is too simple for biology. Indeed, there are cases in which the anticodon has little or nothing to do with the tRNA identity. For instance, a number of aaRSs recognize two or more distinct tRNAs bearing different anticodons; this is explained by additional identity elements on the tRNA molecules. Thus, the identity of alanine-specific tRNA depends exclusively on the presence of a G3-U70 wobble base pair in the acceptor stem of the molecule (Francklyn and Schimmel, 1989; Hou and Schimmel, 1988; Fig. 17.9).

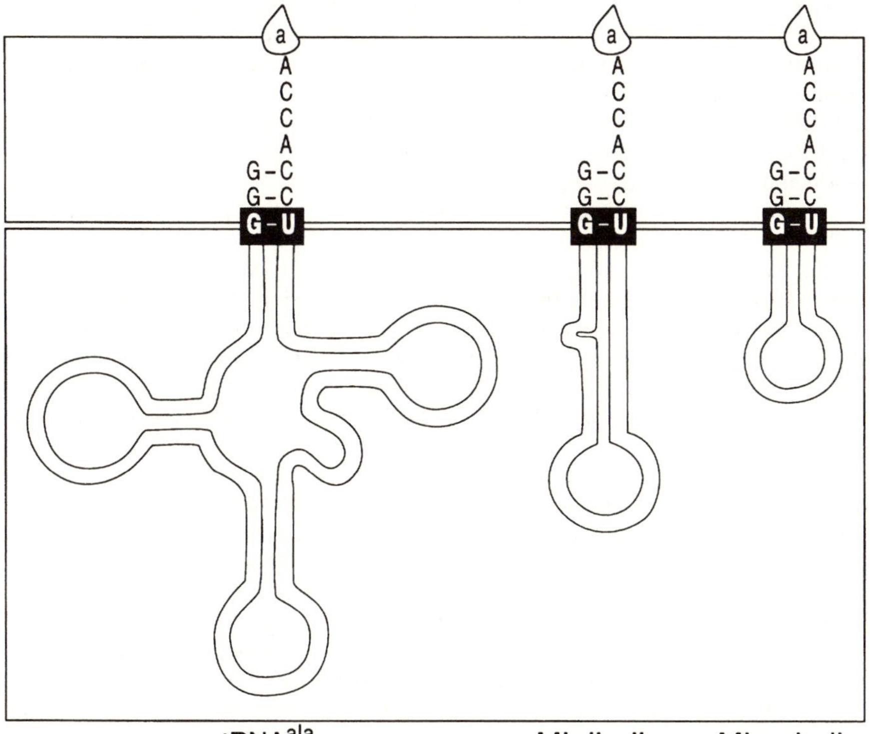

Fig. 17.9. Scheme of the structure of *E. coli* alanine tRNA loaded by alanine (a). The identity of the alanine-specific tRNA depends exclusively on the presence of a G3-U70 wobble base pair in the acceptor stem of the molecule. Also shown are loaded minihelix and microhelix with the same identity base pair, which can also be aminoacylated. (Modified from Hou et al., 1989)

Schimmel and his associates elegantly manipulated their experimental system by using minihelices and microhelices that lacked the anticodon moiety but possessed the characteristic identity marks in their acceptor stems. It turned out that when the anticodon was not necessary for the identification of the tRNA, the synthetase was able to charge the minihelix based solely on the identity features along the acceptor stem. These results prompted de Duve to suggest the controversial concept of a "second genetic code," namely, a built-in identity feature of the tRNA-aaRS relationships, which he named "paracodon."

Schimmel et al. (1993) studied aaRSs, which are very old proteins. The picture emerging from these works has to do with the determinants of the recognition elements in both tRNAs and aaRSs and is related to early stages of the evolution of the translation apparatus. The recognition elements on the acceptor stem of the tRNA molecule were found to be close to the attached amino acid; presumably these recognition elements preceded the anticodon, suggesting a smaller proto-tRNA in earlier evolutionary stages (Schimmel et al., 1993; Schimmel and Schmidt, 1995).

All the enzymes of the synthetase group are made of discrete domains. One such domain, which is found in all synthetases, carry out the main functions of the synthetases: catalizing amino acid activation, binding the acceptor stem of tRNA, and transferring the amino acid to its binding site on the tRNA. This domain, which is considered a core synthetase, is probably the conserved portion of the primitive enzyme. Its size did not have to extend more than 1 to 2 nm beyond the amino acid attachment site (at the 3′ end of the primitive tRNA), in which the identity elements of the tRNA resided (Buechter and Schimmel, 1993). For those who adopt the RNA-world hypothesis, the recognition of amino acids is the connection between the RNA world and template-directed proteins.

In another recent study on the evolution of aaRSs and the genetic code, Härtlein and Cusack (1995) suggested that synthetases "are not the oldest protein enzymes, but survived as RNA enzymes during the early period of evolution of protein catalysts." According to this view, the division of synthetases into two classes is connected to the replacement of primordial RNA synthetases by the more adapted protein synthetases. If this is true, then "the genetic code was essentially frozen before the protein synthetases that we know today came into existence." We have seen earlier that according to Rodin and Ohno (1995), the two synthetase groups share the same primordial nucleic acid.

Are the acceptor-stem and anticodon-codon domains related? It is recalled that the three-dimensional structure of tRNA is a rather flat L-shaped molecule, typically comprising 75–93 nucleotides (see Schimmel and Henderson, 1994). The two regions on which we focus our discussion are the 3′ CCA end, to which an amino acid is loaded, and the anticodon trinucleotide sequence, which, according to the algorithm of the genetic code, corresponds to the loaded amino acid; the separation distance of these two regions is at least 7 nm. Based on various lines of evidence, it has been suggested that these two domains—the acceptor stem-loop, on the one hand, and the anticodon loop, on the other—as well as their flanking nucleotides existed some time in the evolutionary history of the tRNA as separate entities, either as discrete domains or as discrete molecules (Schimmel and Henderson, 1994).

Attempts to relate acceptor-stem and anticodon-codon domains are of special importance in the invention of scenarios for the beginning of the translation machine and the genetic code.

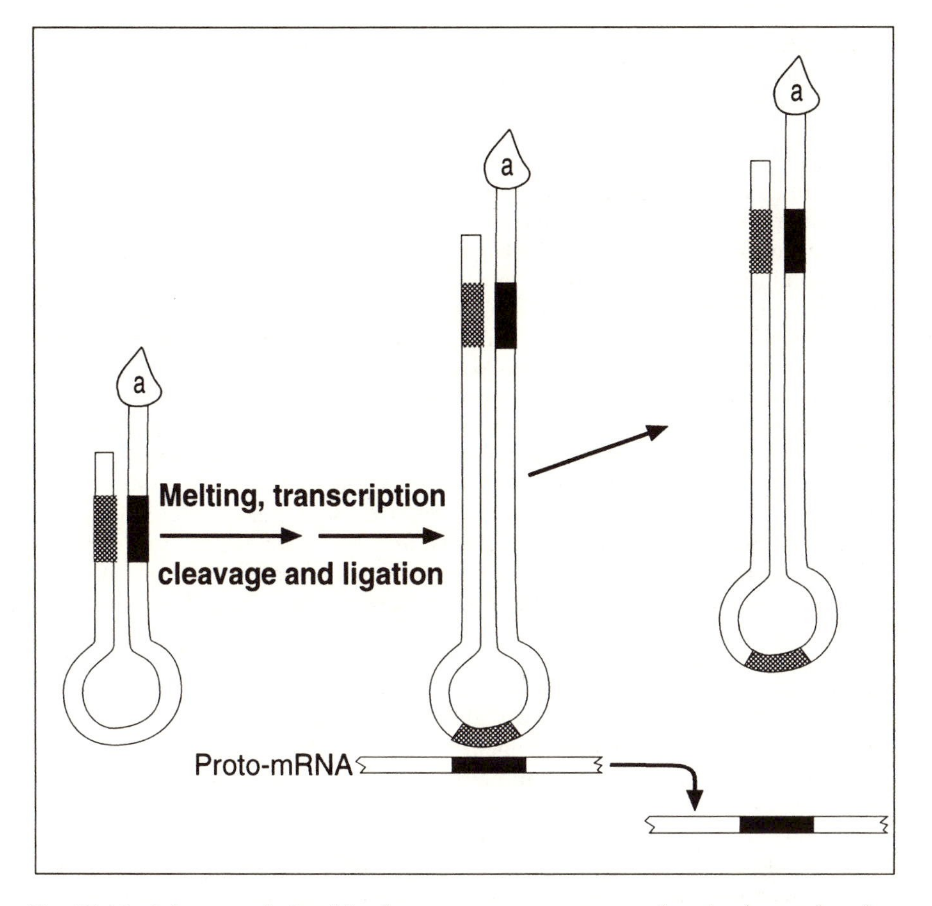

Fig. 17.10. Inherent relationships between acceptor-stem and anticodon-codon domains according to Möller and Janssen (1990).

Möller and Janssen's model Möller and Janssen (1990, 1992) carried out a statistical study according to which the 3-4-5 positions on the acceptor stem of certain extant tRNAs are conserved remnants of a prototype genetic code. These are the tRNAs of the primordial amino acids: alanine, glycine, aspartic acid, and valine. These amino acids are considered by some researchers more ancient constituents of biology than the other amino acids, since they are found among the products of the Miller-type experiments (chapter 15). Möller and Janssen further suggested a mechanism relating the sequence of these 3-4-5 positions on an acceptor stem to the anticodon of the corresponding tRNA (fig. 17.10). Moreover, their mechanism involves the formation of a small, separate domain, which comprises the relevant codon, the complementary domain of the above anticodon; this domain is thus a part of the proto-mRNA. These recognition domains have a special role in yet another hypothesis (see chapter 23).

More recently, Rodin et al. (1996) further elaborated on Möller and Janssen's search for relationships in tRNA sequences, searching for vestiges of the original codon-anticodon pair in the 1–2-3 positions. An important feature of their approach was to use constructed consensus acceptor stems of tRNAs corresponding to each of the 20 amino acids by examining 1,268 known tRNA sequences. The model suggested by Rodin et al. (1996) is based on the RNA-world theory. In order to avoid the chicken-and-egg argument, they had to assume direct interaction between anticodon and amino acid at the 3′ half of their proto-tRNA. Moreover, they hypothesize that "the genetic code could have existed already in the RNA world, possibly by direct aminoacylation of anticodons helped by RNA synthetase-like activity inherent in group I introns of tRNA genes." The latter hypothesis is rather dubious, since the role of peptides in the RNA world is not clear and its benefits are not obvious (P. B. Moore, 1993). Nevertheless, Rodin and his associates note that internal sequence periodicity suggests that "tRNA most likely evolved from very short hairpins . . . so that unavoidably, at some critical step of their elongation, the recognition of anticodon and that of the acceptor stem by aaRS came to be coupled" (Rodin et al., 1996). It was found that the structure of the consensus image of the acceptor domains was a double stranded palindrome 11 base pairs long, with complementary triplets in the center. This structure could have served as the precursor of tRNA; the original anticodon-codon pair can thus be identified at the 1–2-3 positions of the present tRNA acceptors. Despite all its beauty, this model does not address the origin of template-directed protein synthesis, nor does it address the origin of the genetic code.

The prebiotic "proto-tRNA cycle" According to Lahav (1991), those proto-RNA strands which can form the secondary structure of proto-tRNA can form a "proto-tRNA cycle" (fig. 17.11) in which building blocks of proto-RNA are hydrogen-bonded to their complementary domains on proto-tRNAs. Upon condensation of these building blocks, their dissociation from their "mother templates," and ligation, they form proto-mRNAs. Each proto-mRNA consists of domains that correspond to the complementary domains of their mother proto-tRNAs.

The replicating (copying) process under study is considered a most rudimentary stage in the evolution of such processes. Due to the crudeness of this primordial replication mechanism, the first indication for the existence of self-copying would be the formation of a population of "families" of strands possessing domains of both similar and complementary sequences. Moreover, the most primitive replicating process is likely to be characterized by information transfer of *domains* rather than *whole strands* (chapter 10). Can such domains be considered the predecessors of exon-intron systems (see Lahav, 1989)?

This model is a part of a biogeochemical scenario. It is presented in the present context because it has to do with the acceptor-stem and anticodon-codon domains. It will be further elaborated as a biogeochemical model in chapter 23.

Primordial stages in the recognition between synthetases, tRNAs, and amino acids tRNA sequences were used to test hypotheses regarding the order in which amino acids were assigned to the genetic code, as well as the nature of the primordial recognition system and its evolution. Recent evidence suggests that the first synthetases were RNAs rather than proteins (Saks and Sampson, 1995). The expansion of the genetic code, however, probably required peptides, which replaced those RNA synthetases before all the codon assignments were completed (Saks and Sampson, 1995).

Additional aspects of the early stages of coded protein syntheses can be found in Wetzel, 1995, and Rogers and Söll, 1995.

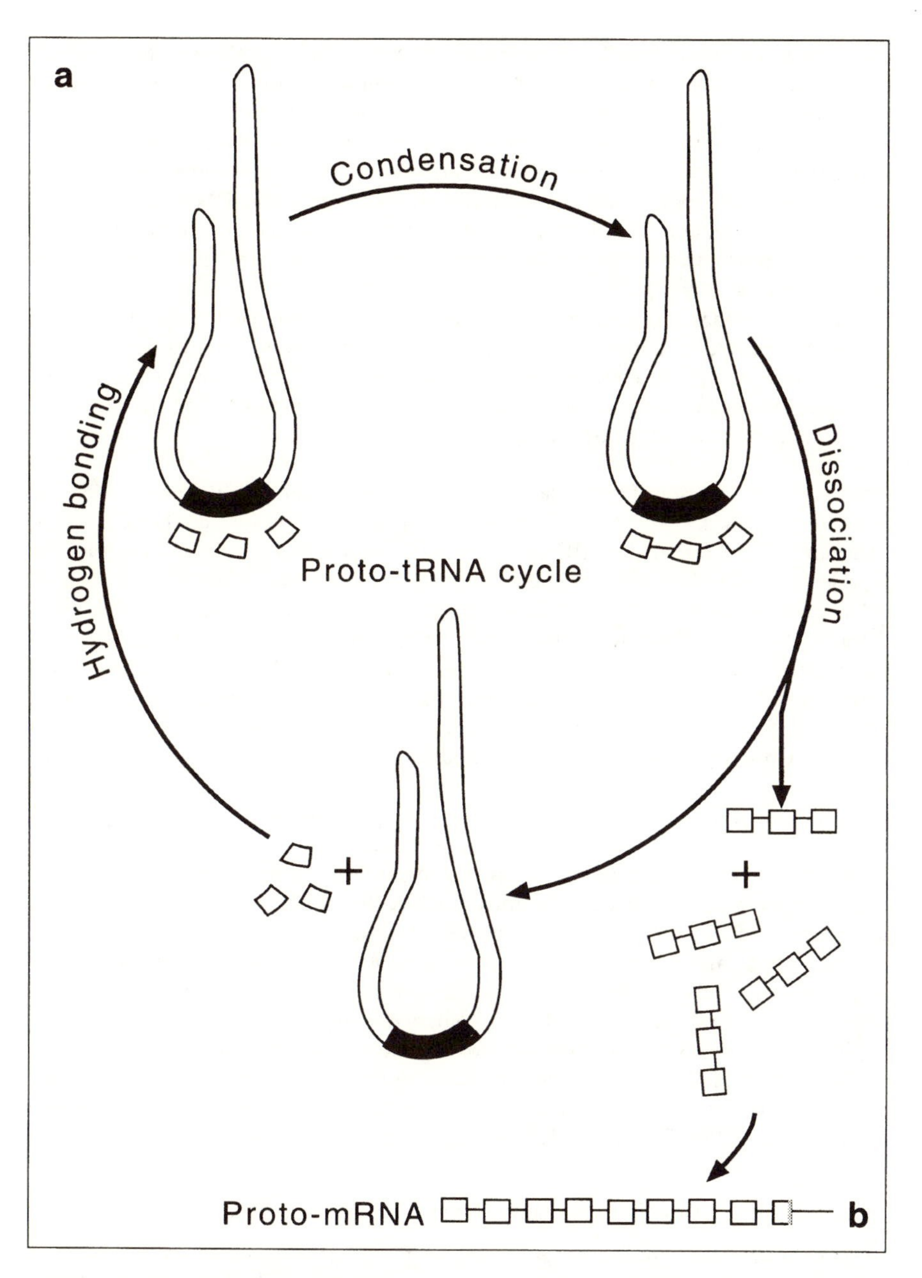

Fig. 17.11. The "prebiotic proto-tRNA cycle": development of codon-anticodon relationships according to Lahav (1991). a. Complementary domain buildup by hydrogen-bonding of building blocks, their condensation to form a complementary domain, and the dissocoation of this complementary domain from its "mother template." b. Ligation of a number of complementary domains to form a proto-mRNA. (Following Lahav, 1991)

The coding coenzyme handles hypothesis This hypothesis addresses the origin of the genetic code but not the origin of protein synthesis by translation. According to Szathmáry (1993; see also Maynard Smith and Szathmáry, 1995), useful coding arose before translation, in a metabolically complex RNA world. In this system, protocells were formed spontaneously and contained ribozymes that catalyzed a variety of reactions, including oxidation/reduction and transesterification. DNA was invented before translation, and, based on earlier suggestions, photosynthesis was carried out with the help of tetrapyrrols; the membranes of the vesicles were made of terpenes (see Benner et al., 1989).

Some functional groups are assumed to have characterized the RNA building blocks. Moreover, the existence of cofactors is a mechanism to increase the catalytic capabilities of the ribozymes. Amino acids bonded to nucleotides increase the range of their activities and specificities. A nucleotide can thus serve as a "handle" to which such cofactors are attached by base pairing.

Was CoA a predecessor of tRNA-like molecule? Several authors (Reanney, 1977; see Di Giulio, 1996a, 1997) proposed a model for the origin of protein synthesis in which a variant of CoA rather than tRNA served in the loading of amino acids in the first peptide syntheses. According to these authors, this stage was followed by a stage of tRNA loading, therefore the transition between the thioester world (de Duve, 1991; chapter 19) and the RNA world could have taken place without violating the principle of continuity; presumably the early stages of interactions between tRNA-like molecules loaded with amino acids could have taken place in the absence of a template (Orgel, 1989; Di Giulio, 1996b). This suggestion corroborates with the prebiotic plausibility of CoA synthesis found by Miller and Schlesinger (1993a, 1993b; chapter 15).

Hypothetical primordial mechanisms for template-directed peptide synthesis Simplified protein synthesis systems that presumably could serve as predecessors to extant ribosomes were suggested by various researchers. In their simplest form these systems consist of strands of a primordial RNA (RNA-like molecule, proto-RNA) and amino acids. Two kinds of such strands are needed, namely, a proto-mRNA and a proto-tRNA capable of adopting a secondary structure similar to that of extant tRNA. Amino acids can be attached to the 3′-termini of proto-tRNAs (see figure. 17.8). Peptide-bond formation takes place between two amino acids loaded on two neighboring proto-tRNAs hydrogen bonded to their complementary sites on a proto-mRNA strand. This system is a simplification of the extant ribosome. For more details the reader is referred to Crick et al., 1976; Kuhn, 1976; Kuhn and Waser, 1994a, 1994b; Orgel, 1989; Wood, 1991; Lahav, 1991; Lahav and Nir, 1997.

Scenarios for the evolution of early stages of the genetic code

Wong's coevolution hypothesis An interesting attempt to relate the biosynthesis of amino acids and the evolution of the genetic code was made by Wong (1975, 1976; see also Amirnovin, 1997); his theory for the coevolution of the genetic code and the biosynthesic apparatus postulates that the genetic code reflects the prebiotic biosynthetic pathways of amino acids, which can be recognized in extant cells. The first amino acids to be coded were those amino acids that were most abundant in the prebiotic soup. During later stages these amino acids served as precursors for additional amino acids. The prebiotic metabolic pathways gradually became enzymatized and eventually reached their present evolutionary stage.

It is noted that Wong follows the Oparin-Urey-Miller paradigm (the prebiotic soup), and his scenario is thus heterotrophic (see also Edwards, 1996; Eriani et al., 1995; Rodin and Ohno, 1995).

One line of evidence in favor of Wong's coevolution theory comes from a statistical analysis of the phylogeny of tRNA molecules (Di Giulio, 1994b, 1995, 1997). In this study, ancestral-consensus tRNA constructed from diverse organisms was analyzed; it was found that of the 15 ancestral tRNA sequences under study, three have a common sequence, GAGCGC, in certain corresponding nucleotide positions. The three are $tRNA^{Thr}$, $tRNA^{Ile}$, and $tRNA^{Met}$; the common sequence is interpreted as an indication for a related origin of the three tRNAs. Indeed, the three amino acids threonine, isoleucine, and methionine are closely related biosynthetically. Thus, the mechanism proposed by Wong's coevolution theory seems to be detectable in the phylogeny of tRNA molecules. Moreover, "the traces of this mechanism seem to have also been found in an equivalent analysis conducted on the trees of aminoacyl-tRNA synthetases" (Di Giulio, 1994b).

One interesting discrepancy in Wong's theory seems to have been solved recently. This has to do with Wong's postulates that there are two major groups of amino acids, according to their biosynthetic pathways. However, the existence of two distinct aminoacyl-tRNA synthertases, each containing 10 synthetases, created a problem: Each of the groups of amino acids postulated by Wong's coevolution theory contained amino acids belonging to the two aaRS groups. Moreover, the emergence of two independent protein synthesis systems using reduced sets of amino acids seems unlikely. It was recently suggested by Rodin and Ohno (1995) that the problem may be solved by assuming that the two groups of synthetases emerged synchronously as complementary strands from the same primordial nucleic acid.

Taylor and Coates's contribution Following Wong, Taylor and Coates (1989) further elaborated the correlation between the three codon bases of the genetic code, the synthetic pathway of the amino acids in the intermediary metabolite pattern of the citric acid cycle, and certain physicochemical features of corresponding amino acids. In spite of a few exceptions, the general pattern of these relationships is that the most hydrophobic amino acids are characterized by U in their midbase codon, whereas the most hydrophilic amino acids are characterized by A in their midbase codon. Moreover, "with the exception of G, the first base is generally invariant within a synthetic pathway. G-coded amino acids show a different order, being found only at the head of the synthetic pathways." According to the authors, "the apparently systematic nature of these relationships has profound implications for the origin of the genetic code."

It should be noted that these observations and correlations, as well as other ones that have not been mentioned here, do not give tangible clues regarding the mechanism of formation of the genetic code. Indeed, Morowitz (1992) commented that these correlations are derived from the precursors of the syntheses of these amino acids, and the logic of these relationships has yet to be found. Similarly, Cedergren and Miramontes (1997) noted that "the co-evolutionary theory [according to Wong] does not address the origin of the code, but rather proposes a plausible pathway from prehistoric code to the present-day version."

First experimental clues It is interesting to note that Morowitz, Chang, and Peterson (Morowitz, 1992; Morowitz et al., 1995) demonstrated that the Wong-Taylor-Coates coevolution theory can be partially tested in the laboratory. The experiment they performed was the synthesis of glutamic acid in this reaction:

$$NH_3 + \alpha\text{-ketoglutaric acid} \rightarrow \text{glutamic acid} \qquad (17.5)$$

This reaction was carried out in the absence of enzymes, under rather mild conditions. This facile reaction, which deals with a possible prebiotic synthetic pathway for amino acids, may help in bridging the gap between prebiotic chemistry and the metabolic chart. More recently, a non-enzymatic transamination reaction between amino acids and keto acids was reported by Bishop et al. (1997).

As I mentioned earlier, several authors (see Di Giulio, 1994a, 1994b and literature reviewed therein) established that tRNA sequences contain phylogenetic information and can be used in the study of the evolution of the genetic code. Of special relevance to Wong's hypothesis is the possibility suggested by di Giulio (1994a, 1994b) that the mechanism proposed by Wong "is still detectable in the phylogeny of RNA molecules."

The reductive carboxylic acid cycle The biosynthetic pathways of amino acids that were addressed by Wong, Taylor, and Coates are apparently not as old as the ones originating with the reductive carboxylic acid cycle (chapter 9). Thus, the relationships between the primordial genetic code and the biosynthetic pathways of amino acids may be extended to what is presumed to be the most ancient metabolic cycle—the reductive carboxylic acid cycle. This subject calls for further study. Additional discussion of this cycle is given in chapter 21 in the context of the pyrite-world theory.

Metabolite channeling and the origin of the code This theory (Edwards, 1996) is based on Wong's theory as well as on Wächtershäuser's pyrite world (chapter 21). It will be discussed in the context of biogeochemical theories (chapter 20).

Did the original code begin with glycine (GG), alanine (GC), arginine (CG), and proline (CC)? According to Hartman (1995a), "The evolutionary record of the genetic code is probably contained in the structure and sequences of the aminoacyl-tRNA synthetases. That evolutionary record is now being uncovered in the amino acid sequences and the crystalographic structures of the twenty synthetases." The suggested scheme starts with a simplified system involving only two bases: guanine (G) and cytosine (C). These would code for glycine (GG), alanine (GC), arginine (CG), and proline (CC), which correspond to the two first bases of the extant genetic code for each of the four amino acids under study. The third position of the codon can be occupied by any of the bases. The model also outlines the central evolutionary stages through which the genetic code further evolved, based on the structure and sequence of the aminoacyl-tRNA synthetases.

The double strand coding This phenomenon of double-strand coding (fig. 17.12) implies that the genetic code is adapted to sense and antisense reading of the information encoded in the sequences of DNA and RNA strands (Blalock and Smith, 1984; Blalock, 1990; Konecny et al., 1993, 1995). It is not yet clear at what evolutionary stage of the genetic code the double-strand coding emerged.

It was noted by Zull and Smith (1990) that the sense-antisense relationship is not likely to be fortuitous and may be related to the evolution of the code redundancy; it may have been established at a very early time, where "the evolution of the specific redundancy of the codons could be related to selective pressure toward retention of information in the antisense strands of the genetic material."

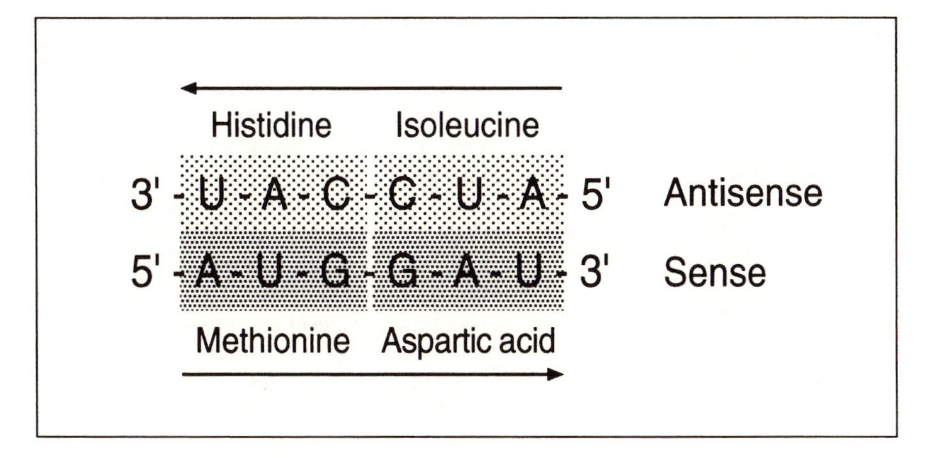

Fig. 17.12. The sense-antisense transformation. (Following Konecny et al., 1993)

The 20-word genetic code According to Béland and Allen (1994), the primitive genetic code consisted of only 20 separated words and was characterized also by sense and antisense reading.

A primordial genetic information started with a coding for seven amino acids and one stop signal The starting point of this theory (Jiménez-Sánchez, 1995) is a rather advanced stage of molecular evolution—the RNA world. In such a system the genetic information began with RNA molecules made of two letters, A and U, which are proposed to be "grouped in eight triplets coding for seven amino acids and one stop signal." Further evolution then brought about the gradual introduction of two more letters, G and C, thus producing the contemporary stage of 64 codons, with its higher informational capability.

Primordial translation apparatus in a system of primordial metabolic cycles Based on the apparent linkage between aaRSs, on the one hand, and the three essential cofactors in metabolism—ATP, CoA, and biotin—on the other, Delarue (1995) suggested a primitive mechanism for a translation apparatus. The origin of the genetic code, however, took place "after the apparition of some functional (statistical) proteins." Template-free systems are assumed to be the predecessors of the primordial translation apparatus, which is reconstructed by assuming continuity between the primordial system and the two classes of extant aaRSs.

The first genetically encoded proteins are suggested to have been nonrandom alternating copolymers, where the translation mechanism proceeds along the guidelines suggested by Crick et al. (1976). The mechanism of the primitive translation apparatus is a rather complex, two-step "molecular motor," flipping back and forth between two states, where the energy needed in each step is transferred by ATP. The transition between the two steps involves the loading of the CCA end of the tRNA by amino acids, and movement of the tRNA molecule along the mRNA strand. The scenario also includes additional stages of the evolution of the genetic code, peptide elongation, peptides' organization in metabolic cycles, and peptide-catalyzed RNA polymerization.

Chemoton, minimal cell, and protocell: Hypothetical simplest "living" cellular entities

Having started the present journey from the progenote toward the remote past of the hypothesized origin of life, our next stop would be the hypothetical cellular entities that have been suggested by different researchers for the earliest stages of living cellular organisms. It should be noted that these hypothetical entities are much more primitive than the "minimal gene set" for cellular life suggested recently by Mushegian and Koonin (1996). However, they are also more advanced than the hypothetical system in which the very beginning of life took place. Thus, Morowitz et al. (1988) define biogenesis as "that continuum of chemical processes which . . . led from an essentially random mixture of inorganic and organic substances in the prebiotic environment to the first assemblies recognized as living cells. Along this continuum there must have been a minimum protocell, which we define as an entity thermodynamically separated from the environment and able to replicate using available nutrient molecules and energy sources." De Duve (1991, pp. 58, 99) suggests that the "first primitive protocell" should be considered the progenote itself. Aspects of transition processes leading to life were recently discussed by Szathmáry and Maynard Smith (1997).

The following suggestions regarding primitive forms of cells are based on the top-down approach. The reconstruction of protocells according to the bottom-up approach will be discussed in chapter 19.

The chemoton

The chemoton is an abstract construct based on top-down extrapolation from extant cells. It was conceived by Ganti (1987), who phrased his approach as follows: "We have to look for the subsystems of the simplest living systems, then to construct the abstract models of the soft [chemical] minimal systems displaying the qualitative properties of the individual subsystems. Finally we have to unite the subsystems into a single, functionally uniform system. All these well done, we must obtain the abstract model of the simplest living system." The abstract model consists of three subsystems: a chemical motor (i.e., an autocatalytic cycle for metabolism), a fluid bilayer membrane, and the copying system (i.e., a self-replicating template macromolecule). By uniting these three subsystems, one gets the *chemoton* . This abstract entity was suggested by Ganti as "the minimal system of life" (see Szathmáry, 1994)

The minimal cell (minimum protocell)

The minimal cell consists of the minimal number of components of a living entity needed to fulfill the autopoietic criteria of system-logical requirements (see also Oró and Lazcano, 1984; Fleischaker, 1988, 1990a, 1990b; Mingers, 1995). In its general phrasing (see Varela, 1994, and references therein) the autopoietic point of view is rather abstract and not easy to translate into laboratory experimental program. Three kinds of minimal cells have been considered by Fleischaker (1988): the minimal extant cell, the minimal Terran cell, and the minimal Universal cell. Though no experiment has been published so far on the synthesis of a minimal cell in the laboratory, the autopoietic criteria for its function and composition have been discussed by Fleischaker (1990a), who concluded that this entity must exhibit all the following operations:

1. Self-assembly of its molecular components.
2. Capture of energy and uphill energy transduction.
3. Incorporation and transportation of matter throughout the system.
4. Production of all cell components by transformations within the cell.
5. Replacement of all cell components by transformations within the system.

This definition is both universal and minimal (Fleischaker, 1990a).

The protocell

"Protocell" is a general term that has been used rather loosely by different authors. The list of properties a protocell must have is as follows (Morowitz, 1992): separability from its environment, capability of accreting matter from its environment and using available compounds, and ability to transduce energy into a useful form and to be a member of a population of similar but not identical organisms. In addition, this entity must be on its evolutionary pathway to the beginning of the phylogenetic tree, the last universal ancestor; this, according to Morowitz, is "experimentally testable."

It is noted that Morowitz's list can be applied to very early living entities.

Is there a lower size limit for the simplest cellular living entities?

The controversial term "nanobacteria" has recently been applied to very small objects observed in geological specimens from Mars and Earth (see McKay et al., 1996). The validity of the interpretation of these small objects as fossils of living organisms has been questioned, however, and this debate is still going on (Morowitz, 1992, 1996; Maniloff, 1996, 1997; Nealson, 1997; Psenner and Loferer, 1997; Folk, 1997). A central argument in this debate is exemplified in the calculated number of atoms in an oval cell with a radious of 0.1μm, which is considered close to the theoretical minimum radius of a living cell: This number is about 100 million. According to these estimates, smaller entities would not be large enough to maintain the minimal activities of a living cell. Thus, taking into account the functions of such minimal cells, the cells' size should be considerable. In the context of the origin of life, a central problem would be, then, to bridge the seemingly huge gap between the minimal cells, on the one hand, and the primordial assemblies of organic molecules in the transition between inanimate and animate, on the other. This topic will be discussed further in chapter 23.

The origin of the first cellular entities is still obscure

The three hypothetical entities—the minimal cell, the chemoton, and the protocell—do not tell us much about their origin. They are top-down constructs that demonstrate a certain organizational level of "minimal life," with hardly an indication as to the route from the inanimate to the animate. Moreover, no specific assumptions with regard to the prebiotic environment are made; the most important information about the environment in which the living entities were presumably formed is that it was hydrated, at least part of the time. We are left with the problem of the *origin* of living entities, continuing our back-extrapolation and guided by the principle of biological continuity. Several aspects of the continuity of this transition have been discussed by Küppers (1990, p. 134). Aspects of function and organization in the origin-of-life process were discussed by Blomberg (1997).

Summary

The most popular scenario to date that is based on back-extrapolation from biology to the early history of life is the RNA-world theory. This hypothetical stage that preceded the last common ancestor is based on the ability of extant RNA molecules to function as both catalysts and templates; as such, it is a top-down theoretical construct that implies an elegant solution for the chicken-and-egg paradox. However, it is accepted today by most researchers that the RNA world, if ever existed, cannot explain the transition of inanimate to animate matter and thus must have been preceded by more ancient molecular-evolution systems.

Hypotheses dealing with transitions from a pre-RNA world to the RNA world and from the RNA world to the DNA world have been suggested by various researchers. Central to the studies of the role of RNA in the origin-of-life process are template-directed reactions of complementary strand formation and self-replication. So far, except for specific cases that cannot be considered prebiotically plausible, self-replication of RNA molecules without enzymes was not observed in laboratory experiments.

The top-down approach was also applied to the early stages of the translation machinery and the genetic code. The plethora of hypotheses regarding this topic include a connection between the genetic code and the synthesis of amino acids in primordial metabolic cycles, formation of tRNA and its recognition domains, the number of primordial amino acids and protonucleotides, the code as a "frozen accident," early synthetases, and the possible involvement of CoA in the translation machinery. Practically all of these scenarios deal with early stages in the evolution of the genetic code; none of them suggests a mechanism for the emergence of the genetic code.

At a higher level of organization, the top-down approach was applied to the early stages of the ribosome and to what is considered to be the simplest living cellular entities. These latter theoretical constructs—the chemoton, the minimal cell and the protocell—are evolutionally advanced chemical entities that are not directly related to the transition from inanimate to animate.

18

Bottom-up Reconstruction Without Specific Biogeochemical Conditions

> The chemist strives to explain the inanimate world by reference to mechanistic laws. The historian strives to understand the world of human culture by reference to a fabric of plans and purposes. . . . Nowhere is this encounter in sharper focus than in the problem of the origin of life.
>
> Günter Wächtershäuser, "The origin of life and its methodological challenge"

Using the bottom-up approach, we now consider the reconstruction of the second stage of the origin-of-life process by beginning with the building blocks of the central biopolymers. Thus, given the initial conditions of concentrations of amino acids, nucleotides, or their predecessors (extant biomolecules and accessory chemicals in an aqueous system), is it possible to initiate and maintain processes and scenarios of their organization into chemical entities characterized by central attributes of living organisms?

Error threshold and hypercycles

With the bottom-up approach, it is time now to focus on additional aspects of the hypothetical molecular evolution process, through which molecular replication and coded catalysts could have emerged.

The "error threshold" of a copying process

The phrase "error threshold" is derived from the theoretical description of the mutation-selection processes of an RNA molecule serving as a master copy. In a population of RNA molecules and their building blocks, the master copy molecule cannot grow beyond a certain error threshold without suffering a deterioration in the fidelity of the copying process. For instance, a rough estimate of the maximum length of polynucleotide molecules, assuming "prebiotic accuracy" in a replication reaction, is 50 to 100 nucleotides (see Maynard Smith and Szathmary, 1995; Kauffman, 1993), since it is generally accepted that the copying fidelity was low at the early evolution stages.

Such theoretical considerations may be relevant to a certain stage of prebiotic evolution, but not necessarily to the very early stages of molecular evolution, which are the focus of this book. Moreover, they do not address the mechanism of the replication and molecular evolution processes.

The hypercycle

The theoretical construct called the hypercycle was developed first by Eigen (1971) and Eigen and Schuster (1977), who argued that catalytic feedback of a biochemical system has to be organized in a closed loop, a hypercycle (see also Wills, 1994). A hypercycle may be defined as follows (Jantsch, 1980, p. 32): "A hypercycle is a closed circle of transformatory or catalytic process in which one or more participants act as autocatalysts." In other words, the hypercycle is a closed loop of chemical reactions with a catalytic feedback. In this hypercycle the first member specifically replicates the second member, which specifically replicates the third, and so on. The last member of this cycle should catalyze the first member, and so forth.

Despite all its beauty and elegance, the hypercycle is not directly relevant to the origin of life because it does not answer the question: "How did the first hypercycle emerge in the first place?" Moreover, it does not address the molecular evolution process in which the "replicators" evolved.

What would be conserved if "the tape were played twice"?

Inspired by Gould (1989, p. 347) it was Fontana and Buss (1994), in a paper with the title given in this subhead, who asked, "If we had the option of observing a control earth, would we observe, say, the evolution of *Homo sapiens* or the evolution of something unambiguously identifiable as a metazoan or even something akin to eukaryote?" The question focuses on the history of life, which, in their words, is "the product of both contingency and necessity." In a theoretical study dealing with the possible recurrence of certain biological features, these authors found that among the features that would be retained if "the tape were played twice" are hypercycles of self-reproducing objects. The hypercycle thus seems to be a general principle of molecular organization.

For further information on this controversial issue the reader is referred to two prominent origin-of-lifers, de Duve (1995a, 1995b, 1996) and Morowitz (1992).

Double-origin hypotheses

Following Von Neumann, Dyson (1985) has expanded the idea that replication and metabolism, although so inherently linked in all cellular functions and structures, are logically separable. Accordingly, he suggested combining Eigen's approach, which treats essentially template-directed reactions, with Oparin's approach, which focuses on metabolism. Using the symbiosis theory of Lynn Margulis (see Margulis, 1993b) as a manifestation of the double origin of the Eukarya, he suggested that Eigen and Oparin's propositions be combined into a unified double-origin theory. This idea was also extended to Cairns-Smith's proposition of "clay life," according to which biology was preceded by a form of "mineral life" (chapter 20).

Dyson states, "I happen to prefer the Oparin theory, not because I think it is necessarily right but because it is unfashionable" (1985. p. 33). He calls his model "a Toy Model of the Oparin Theory." The model (see Lifson, 1997, for criticism) starts with a population of compartments (droplets) containing amino acids and peptides. Some of the amino acids and peptides are active and may undergo a variety of interactions, which, according to the underlying assumptions, would turn them into an "autocatalytic set" of peptides. The dynamics of the system may bring about a transition from a lower "dead" state to an upper "live" state. The transition from the low to the upper state is possible, under plausible conditions, if the number of amino acids in a droplet is between 2,000 and 20,000.

The incorporation of nucleic acids into the scenario is a late stage of evolution, which is hardly dealt with by Dyson. The autocatalytic sets under study would generate a large number of by-products, including, perhaps, nucleic acids. The latter molecules are "parasitic" on the autocatalytic pathway, but later become symbionts and thus are incorporated into the evolving "living" entities (Maynard Smith and Szathmáry, 1995, p. 72). Dyson's hypothesis is an interesting attempt to describe in an abstract way an evolving system with a minimum of chemical mechanisms. It is an intellectual exercise that may serve as a starting point for experimentalists to translate his model into a chemical reality. It is also relevant to Kaufman's hypothesis, which is described in the next section.

Emergence of self-reproducing systems of catalytic polymers

The approach of a "minimum chemical mechanism" was developed by S. A. Kauffman during the 1980s and culminated in the publication of a book in 1993. At the outset of his discussion Kauffman defines the problem he wants to attack by focusing on self-reproducing systems characterized also by metabolism.

Initial complexity rather than simplicity

It should be noted that the mere survivability and multiplication of RNA or RNA-like molecules is evolutionarily meaningful if it can lead to more complex systems in which coded catalysts would help these molecules to survive and multiply. Thus, the selection of both RNA molecules and catalysts should be related to the survivability of their own organizational system. The organizational method that may lead both to surpass the error threshold and the establishment of a new organizational level involves linking a set of templates in a closed cycle with both autocatalytic and hypercyclic attributes (Kauffman, 1993, p. 359). Following Kauffman, Lancet and his group (Lancet et al., 1994; Segré et al., 1998) developed their own computer model based on energized precursor monomers and characterized by mutual catalysis in sets of random oligomers. According to them, this model can serve as a tool for investigating the dynamics of self-organization mechanisms in populations of interacting molecules.

The molecular arsenal of Kauffman's proposition is a collection of polymer catalysts such as peptides and RNA-like polymers. Upon increasing the concentration of these polymer catalysts beyond a critical complexity threshold, a high probability for the formation of a subsystem of polymers is reached, the members of which are catalyzed by other members of the subsystem. Such sets of autocatalytic, self-organized polymers are able to reproduce collectively. Life, according to Kauffman, started as a set of minimal size comprising peptides or RNA catalysts and functioning collectively in a coordinated, autocatalytic manner. The critical complexity threshold is still rather high. According to his estimate, the number of peptides needed for an autocatalytic set is 20,000. The catalytic set, with its catalytic diversification, forms a catalytic closure, which is a kind of self-sufficient molecular population or an "evolutionary autarchy." The complexity of smaller system is not enough to achieve catalytic closure.

The critical threshold complexity is related to Dyson's question, "Why life is so complicated?" (1985, p. 60).

Evolution without a genome and metabolic web

If the formation of all molecules in a catalytic set is performed by catalytic processes of the catalytic set's molecules, template-replication is not needed anymore for the primordial stage of evolution under consideration. As a result, a replication of arbitrary RNA (or RNA-like) sequences can be performed in such a system without difficulty. Given the problems involved in the prebiotic replication of arbitrary RNA molecules, such an alternative synthetic route looks quite attractive. Similarly, such a theoretical system might be capable of self-reproducing peptides (p. 340). It is recalled that a possible hint toward prebiotic self-reproduction of peptides was recently reported by D. H. Lee et al. (1996; chapter 17). The relevance of this reaction to the origin of life has yet to be established.

Kauffman extends his idea also to the initiation and evolution of metabolism. Since life began at a rather high level of complexity, utilization of various energy sources, coupling of endergonic and exergonic reactions, and metabolic functions would begin accordingly (p. 355).

Complexity

The term "complexity" in biology, including the problematics of its uses, has been addressed recently by several authors in connection with evolution (for different points of view see Maynard Smith and Szathmáry, 1995; Mingers, 1995; Yockey, 1992); it is used here with the intuitive understanding implied in Kauffman's postulate that all free-living organisms are characterized by a minimal level of complexity (1993, p. 294). Thus, it is assumed that as a catalytic set of molecules is built up, its complexity also increases. Indeed, it is generally accepted intuitively that by whatever definition of life, a minimal complexity is needed for life to emerge.

Are there simpler models?

The challenge posed by Kauffman is not his "complexity first" postulate and its concomitant "complexity threshold"; this guideline is implied, in fact, in several attempts to describe the origin of central reactions and molecular evolution processes. It is the size of his catalytic set and its functions, the concept of "crystalization" of life, and the plausibility of his model that should be examined.

Consider a small volume of a prebiotic solution in which organic molecules start their accumulation and interactions. Kauffman's catalytic set may not be the only possibility to initiate reaction pathways, cycles, and processes that may evolve into "living" entities. Even in the specific case where a population of molecules needed to serve as a catalytic set is obtained abruptly and Kauffman's catalytic set may be initiated accordingly, "shortcuts" may also be formed, thus bypassing potential catalytic sets of a large size. In other words, in the case of both gradual and abrupt buildups of this molecular population, other reaction pathways and processes may be likely to evolve and to form reaction feedback loops: interactions between amino acids, peptides, RNA-like building blocks and strands, as well as other system constituents, would start from the beginning and take place all the time. And thus, before functioning as a large and complicated catalytic set, the molecules under consideration may initiate and function in other, more parsimonial processes related to the origin of life, forming a "detour" through simpler pathways. The threshold complexity is thus a function of the system composition, constituent composition, and operating conditions. The challenge is to hypothesize the simplest system capable of forming feedback loops and functioning as autocatalyst, in an environment teeming with many other constituents with their own reactions and processes, and then to carry out the needed experiments in the laboratory.

The great majority of prebiotic scenarios suggested so far are based on a gradual accumulation of organic molecules and the concomitant interactions that are relevant to the origin of life. Several such scenarios, characterized by a small number of relevant constituents as well as various prebiotic biogeochemical features, are described in more detail in chapters 21 and 23.

In concluding this section I should note that certain mathematical derivations in Kauffman's theory have recently been challenged by Lifson (1997), thus adding one more controversial issue to the long list of debates characterizing this scientific field. On the other hand, the importance of Kauffman's theory has been acknowledged by various researchers and philosophers, as discussed, for instance, by Dennett (1996, pp. 220–227). For more information and comments, see Kauffman, 1993; and Maynard Smith and Szathmáry, 1995.

Populations of ribotides and the beginning of the genetic code according to Ferreira and Coutinho

The model developed by Ferreira and Coutinho (1993) is divided into two stages: the establishment of a population of self-replicating oligoribotides and the transition to peptide-assisted interactions. The first stage is based on the RNA-world concept, where two small random oligomers are condensed, catalyzed by a third fragment. The starting point is a mixture of ribotidelike molecules with a phosphate group in a position denoted as 5′. This group can interact with another reactive moiety denoted 3′, consisting of a free OH group. The condensation reaction between these two reactive groups is described by a Michaelis-Menten model, and the resulting ribotide population is calculated.

In the second stage, randomly synthesized peptides with one predominant amino acid are introduced, increasing the growth rate of some of the oligomers thus formed. In the resulting system, a favorable codon-anticodon–amino acid relationship is established, though a detailed mechanism is not specified. The origin of the genetic code and the mechanism of its emergence are still open questions.

Summary

The bottom-up reconstruction without specific biogeochemical conditions starts with the building blocks of the major biopolymers and explores the organization of these monomers into biologically meaningful structures systems, and functions. The present chapter encompasses central topics dealing with catalyzed replication of informational molecules and the evolution of complex systems capable of functioning as feedback systems. These include Eigen's "hypercycle" and Kauffman's "catalytic sets," which deal with catalytic feedback of a biochemical system organized in a closed loop, and Dyson's "double origin" model. Though these models and theories do not deal with the earliest hypothetical events of the origin of life, they form a general theoretical basis that may be applicable to the earliest interactions and processes characterizing the transition from inanimate to animate.

19

Bottom-Up Biogeochemical Reconstruction

Starting from Organic Scratch in the Absence of Minerals

> Any attempt to establish prebiotic chemistry or prebiotic physics is bound to be rather arbitrary. The only thing we can do is to propose a large number of models from which the geologist and the biologist will have to pick out those models which fit into a consistent picture.
>
> Aharon Katzir-Katchalsky, cited in Noam Lahav and David H. White, "A possible role of fluctuating clay-water systems in the production of ordered prebiotic oligomers"

Biogeochemistry: The role of the environment

The biogeochemical approach is based on specific environments, without which the reactions undergone by organic oligomers and polymers would not be understood in the context of the molecular evolution process (Miller and Orgel, 1974). The present chapter focuses on the study of prebiotic geochemical environments, including primordial energy sources and some important reactions and processes of prebiotically plausible organic molecules. The establishment of hypothetical pathways and scenarios based on laboratory experiments as well as on computer-modeling simulations of the origin-of-life processes will be discussed in the following chapters.

Our multiaspect challenge

The multidimensional mosaic called "the search into the origin of life" may be presented in various (necessarily schematic) ways. And just as it is impossible to define life on the basis of one aspect or criterion, so it is also difficult—in fact, impossible—to portray the origin of life according to just one aspect or point of view. Moreover, given the present lack of a central paradigm for the origin of life, it is expected that every self-respected researcher would have his or her own preferences, biases, and didactics in the presentation of the numerous scientific propositions dealing with reactions, processes, and scenarios of this complex issue. Indeed, the very abundance of hypotheses and speculations is an inherent attribute of this field of scientific endeavor.

The list of the main biogeochemical origin-of-life aspects dealt with in the present and next chapters focuses on the following interrelated topics: energy sources, plausibility of the molecules involved, trophic methods, amphiphilic membranes and models of protocells, the primordial genetic system, and template-directed reactions. The possible involvement of minerals in prebiotic processes will be discussed in the following chapters.

Examples of prebiotically plausibile energy sources and central molecules

> The coupling between energy liberating and energy requiring chemical reactions is a basic property of life as is the capacity of every living cell to exchange energy and matter with its environment.
>
> Herric Baltscheffsky, "Chemical origin and early evolution of biological energy conversion"

Energy flow is the driving force of all biological activities, as well as the origin of life. Our problem is to characterize simple energetic processes that were capable of driving prebiotically plausible reactions in the uphill direction and support their increase of complexity. Thus, we focus now on the very early stages of molecular evolution, where the organic building blocks were "manipulated" by their environment into chemical pathways that led to the emergence of the first "living" entities. Accordingly, the energy sources of that stage are involved in the buildup of monomers, oligomers, and polymers (see R. F. Fox, 1988, p. 153; Deamer and Fleischaker, 1994).

Relevance of prebiotic energy sources

The relevant list of energy sources includes radiant energy (solar radiation), chemical reactions (inorganic and organic), and heat (including shock waves). Manipulation of energy deals with energy liberation, energy coupling, and energy-rich compounds (Baltscheffsky and Baltscheffsky, 1994). The latter compounds often contain atoms and groups such as those shown in fig. 19.1. Of the energy sources listed earlier, shock-wave energy is neither specific nor regular and thus is less interesting in the biogeochemical context of this discussion. Thermal energy, on the other hand, with the concomitant dehydration, is relevant to the condensation reactions of the prebiotic era, in spite of being nonspecific.

Radiant energy and chemical energy are used by extant organisms, therefore it is tempting to apply the principle of continuity regarding these energy sources in the search for the beginning of energy conversion. A primordial mechanism for the formation of proton gradient in protocells is discussed in the following section. In this discussion we shall focus on the predecessor of ATP (adenosine triphosphate), a central molecule for contemporary energy conversion.

Pyrophosphate (PPi): Could it have served as a primitive energy donor?

In contemporary cells the phosphate-bond energy in ATP is used to drive chemical reactions (see Westheimer, 1987, for a discussion on why nature chose phosphate).

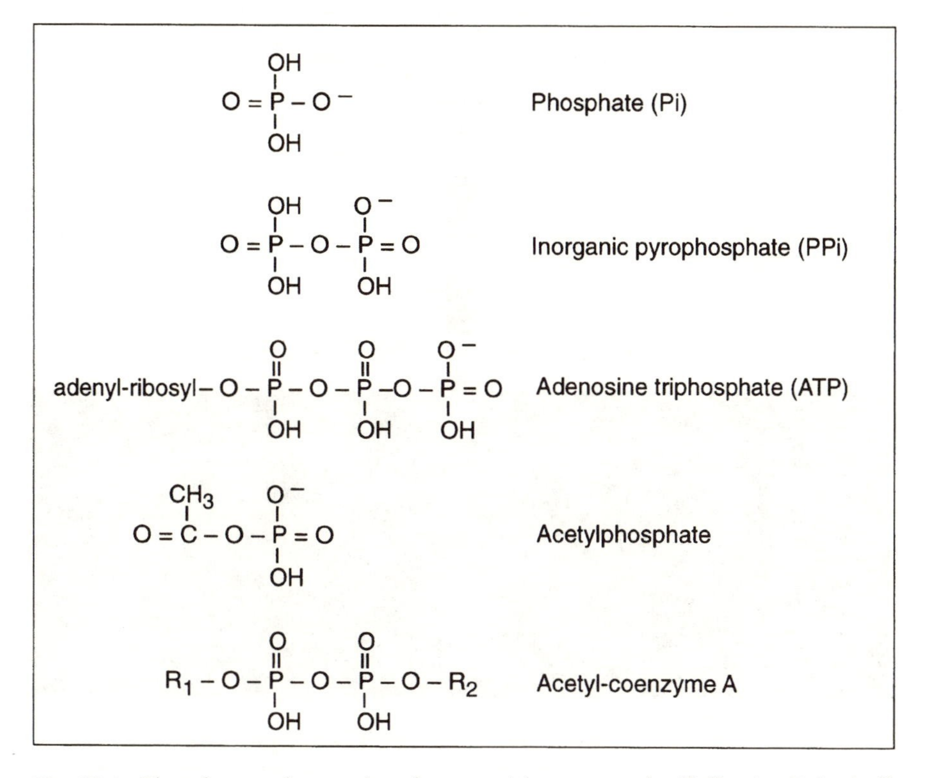

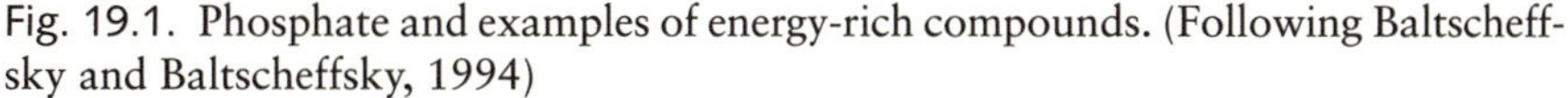
Fig. 19.1. Phosphate and examples of energy-rich compounds. (Following Baltscheffsky and Baltscheffsky, 1994)

This energy is associated with the pyrophosphate moiety, as discussed in chapter 9. In our attempt to simplify reactions and processes, it now seems natural to ask whether the inorganic pyrophosphate can also be used as a primordial energy-rich molecule. The first answer to this question was given by Lipmann (1965), the founder of bioenergetics: The inorganic pyrophosphate could have preceded ATP as an energy donor. The most prominent proponent of this theory since then has been Herric Baltscheffsky (1997), who coined the expression "the pyrophosphate world" (fig. 19.2). Indeed, it has been known for a long time that heating and drying phosphate solutions brings about the formation of pyrophosphate (equation 19.1).

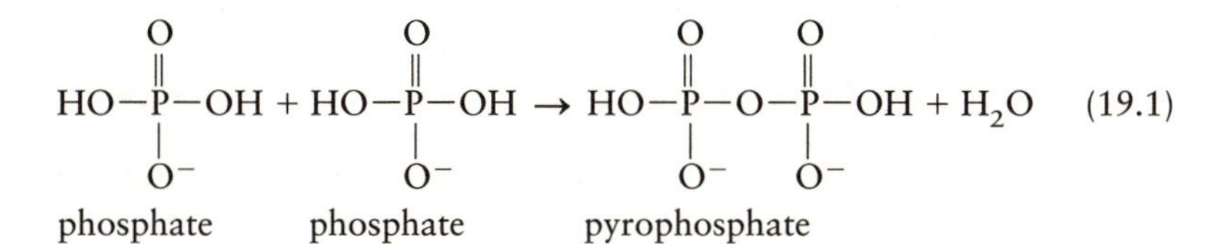

$$\underset{\text{phosphate}}{\mathrm{HO{-}P(=O)(O^-){-}OH}} + \underset{\text{phosphate}}{\mathrm{HO{-}P(=O)(O^-){-}OH}} \rightarrow \underset{\text{pyrophosphate}}{\mathrm{HO{-}P(=O)(O^-){-}O{-}P(=O)(O^-){-}OH}} + H_2O \quad (19.1)$$

Miller and Paris (1964) showed the formation of pyrophosphate from hydroxylapatite and potassium cyanate. The formation of pyrophosphate on Mg phosphate was studied by Hermes-Lima and Vieyra (1992; see references therein), where it was

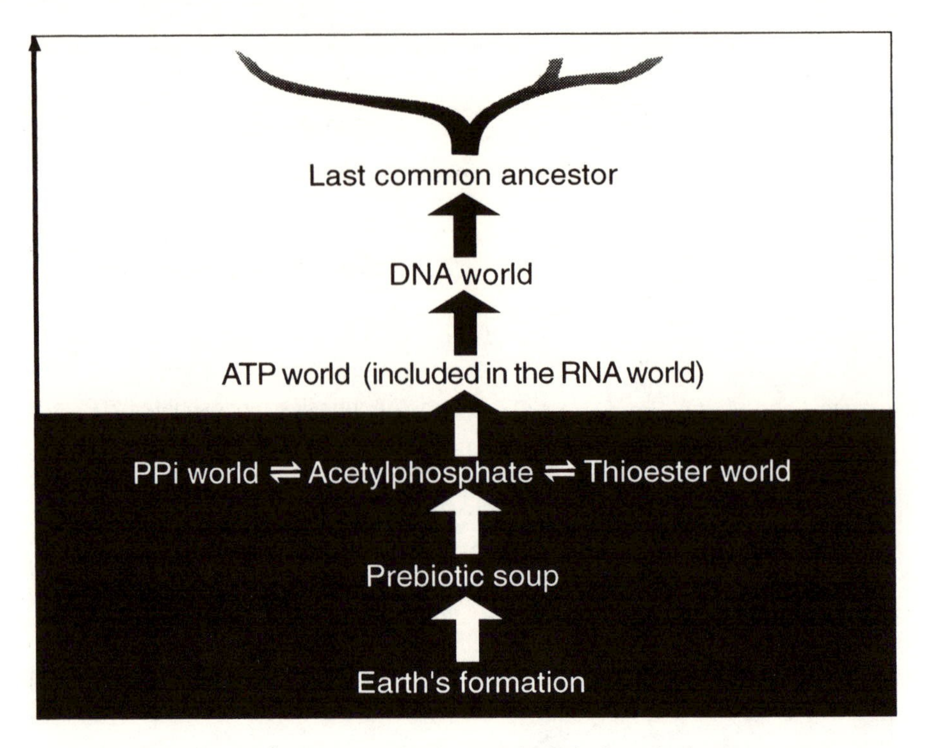

Fig. 19.2. Transition of energy-conversion methods before the last common ancestor. Not drawn to scale. (Modified from Baltscheffsky and Baltscheffsky, 1994)

shown that the reaction is catalyzed by the solid phase (mineral catalysis is discussed in chapter 20).

The problem of the prebiotic source of this inorganic compound was investigated by Yamagata et al. (1991). In their laboratory experiment Yamagata and his associates simulated the magmatic synthesis of the water-soluble pyrophosphate. According to these researchers, this magmatic pyrophosphate is the only identified source of pyrophosphate on the primordial earth. Moreover, this is a "large-scale production of pyrophosphate on the primitive Earth" (Yamagata et al., 1991).

Keefe and Miller (1995, 1996a; see also Lazcano and Miller, 1996) investigated various potentially prebiotic processes in which condensed phosphates and cyclic metaphosphates are formed or concentrated. Although a number of the processes under study gave substantial yields of pyrophosphate and trimetaphosphate, the authors concluded that these reactions are not robust enough to have formed high phosphate concentration in the prebiotic ocean. Their conclusion is therefore that the involvement of phosphates in the origin-of-life processes, including the production of phosphate esters, the involvement of phosphate in the first genetic material, and the function of phosphoanhydrides as prebiotic energy source, is implausible. Therefore, according to these authors, a reasonable approach to solving the problem of a prebiotic phosphate source would be to look for other, as-yet-undiscovered mechanisms for the concentration and synthesis of polyphosphates.

A central argument used by Keefe and Miller is the average phosphate concentration in the primordial ocean water. It should be remembered that the low average phosphate concentration these investigators found is of little relevance to the problem under study (Baltscheffsky, 1997). One reason for this is that the origin of life is likely to have emerged in specific niches characterized by higher solute concentrations (presumably including also phosphates), such as fluctuating systems undergoing wetting-drying processes. Another reason is that the transition of soluble phosphate into apatite is very slow. Under the conditions of continuous recycling of phosphate on the primordial earth, as suggested by Yamagata et al. (1991; see also Tiedemann, 1997), relatively high concentrations of supersaturated phosphate solutions (with respect to apatite) are likely to have been present in various prebiotic environments.

It is interesting to note that a *late* incorporation of P into biology was suggested also by Hartman (1992b), based on biochemical and molecular-biological arguments. The relevance of Hartman's considerations to the geochemical arguments raised by Keefe and Miller is hard to evaluate at present (see also Schwartz 1997a, 1997b).

Transition of energy-conversion methods

Using the present buzz-word "world," Baltscheffsky (1993) described the contemporary biology that uses ATP as "ATP world." Similarly, the primordial evolutionary stage that preceded the ATP world has been hypothesized to be based on PPi and is thus called the "PPi world." The energy-rich pyrophosphate is prebiotically plausible (according to Baltscheffski, Yamagata, Arrhenius, Mojzsis, and others, but not according to Miller and his associates) and could have served as the first molecule involved in energy manipulation. Moreover, its function in energy-transfer processes is independent of the energy source (i.e., radiant or chemical).

So far, no indication has been obtained for a possible pathway for energy conversion from PPi to the extant energy conversion based on ATP. While this possibility still remains in the realm of future discoveries, it is noted that the ATP could have served as a central molecule in the metabolism of the RNA world (Baltscheffsky, 1993). Baltscheffsky and Baltscheffsky (1994) suggested that molecules of acetylphosphate (fig. 19.2) could have served as a link between the inorganic PPi and the organic thioester in prebiotic and early biotic metabolism, according to the following scheme:

$$\text{PPi} \rightleftharpoons \text{Acetylphosphate} \rightleftharpoons \text{Acetyl-S-R} \tag{19.2}$$

These reactions occur in living cells, therefore the scheme corroborates with the principle of continuity. The general scheme of the transition between the prenucleotide world (which includes the PPi world, acetylphosphate, and the thioester world) and the nucleotide world (which includes the ATP world and the RNA world) describes the energy-conversion processes during the chemical evolution era. These processes are an integral part of the transition from the inanimate to the animate (fig. 19.2). The thioester world belongs to the present section, but because it is also related to minerals involved in redox reactions, it will be discussed in chapter 21.

The triose model

A different approach to possible sources of energy in the prebiotic era was suggested by Weber (1987). According to his model, glyceraldehyde could have acted as a source of both energy and monomers. The central constituents of this model are two trioses, glyceraldehyde and dihydroxyacetone, that are assumed to arise abiotically from formaldehyde. As shown by Weber (see also de Duve, 1991, pp. 165–166), this system can produce polyglyceric acid, which then functions as a catalyst in polymerization processes and, perhaps, in information transfer processes. The prebiotic significance of these reactions is their ability to "unite the origin of metabolism and the origin of polymer synthesis in a single process" (Weber, 1987). The similarity between this system and de Duve's thioester world is discussed in de Duve (1991).

Damaging energy sources might affect molecular evolution processes

An interesting example for biogeochemical plausibility considerations is the suggestion (Kolb et al., 1994) that urazole and guanazole served as predecessors to the uracil and cytosine bases of extant nucleic acids (Fig. 17.3), respectively: The two suggested precursors are transparent in the ultraviolet (UV) region of the spectrum, whereas uracil absorbs UV light in this region. Assuming that urazole and guanazole were constituents of predecessors of nucleic acids in the prebiotic era, this implies that these two molecules would be favored in the early origin-of-life stages, in the absence of an ozone layer on the early earth (chapters 16, 17). It is recalled that the UV flux at the surface of the early earth was likely to be much higher than at present, due to the ozone layer in the latter case. In Darwinian terms, "the ultraviolet transparency of urazole and guanazole would be a strong selective advantage for such nucleic acids, unless there were strong ultraviolet absorbers in the environment to protect ultraviolet-absorbing nucleic acids" (Kolb et al., 1994). A favored group of plausible UV absorbers is minerals, as mentioned in chapter 20.

The beginning of the two trophic methods

The suggested approaches are divided into two, namely, heterotrophic and autotrophic scenarios. The use of this terminology is based on the classification of extant organisms according to their source of organic carbon and their source of energy (chapter 6). Corroborating with the use of the principle of continuity, this nomenclature has been extended to the prebiotic era in order to characterize the synthetic processes of the prebiotic organic molecules that served as the building blocks of biology. According to Wächtershäuser (1992a, 1992b), the division between these two ways of life is a most fundamental distinction between producers and consumers (parasites).

The extension of extant trophic methods to the origin-of-life processes is thus a fundamental assumption in the search for the origin of the trophic method. This extension, however, introduces some difficulties, since it is not always sharp-edged where the origin of life is concerned. With the hypothetical primordial processes under study, the existence of living entities in a specific system is a matter of arbitrary definition of life. However, the heterotrophs are defined as living organisms using organic molecules that were preformed, either directly or indirectly, by autotrophs. Thus, it is necessary also to include, in the category of nutrients, organic molecules preformed inorganically. And if we do so, the division between heterotrophic and autotrophic beginning of life may not be clear-cut, as we will discuss shortly.

Prebiotic heterotrophy

According to the biogeochemical paradigm of Oparin-Haldane-Urey-Miller, the building blocks of biochemistry were formed spontaneously on the prebiotic earth from inorganic molecules, by various energy sources and mechanisms. When these molecules reached the primordial sea, the prebiotic soup was formed. The essence of the latter concept may be extended to include small prebiotic bodies of water such as puddles and even atmospheric water droplets and bubble-aerosol-droplets at the ocean-atmosphere interface (see chapter 22). Moreover, when a more general definition of the heterotrophic paradigm is used, the sources of the preformed organic molecules also include extraterrestrial infall.

In each of these prebiotic environments, the supply of the building blocks is likely to have extended over some time. Meanwhile, some of the organic molecules reacted with other constituents of their environment and grew up into assemblies of oligomers and polymers, which evolved gradually into the first living entities. These primordial entities did not have to synthesize their "nutrients" de novo; the latter ones, in the form of reduced carbon compounds, were abundant in the prebiotic environment. Thus, the first living creatures were heterotrophic; their ability to synthesize the needed nutrients evolved during later stages.

At least in principle, the first stages of the establishment of heterotrophic assemblies of molecules could have taken place in the absence of efficient means of compartmentation, such as membranes. For instance, a small droplet of solution hovering in the atmosphere may function temporarily as a compartment under certain conditions. However, the prebiotic synthesis (and/or extraterrestrial infall) of amphiphiles that would form primitive cell membranes early in this era seems a more plausible assumption.

Prebiotic autotrophy

According to the prebiotic autotrophy paradigm, CO_2 fixation and the building blocks of biochemistry, which eventually evolved into "living" entities, were initially formed at those sites where both the energy and the raw materials for these syntheses were available; the products of these syntheses became involved spontaneously in reaction cycles that were initiated by their formation. Further interactions of these molecules resulted in their evolution into more complex entities. The energy cost of the two trophic methods is discussed further in de Duve, 1991.

What was the first trophic method?

The attribution of either heterotrophy or autotrophy to the origin of life is problematic at present. Assuming a last common ancestor for the two trophic methods, there seem to be only two possibilities: (1) One of the two trophic methods was primordial, and the second one evolved later. (2) The two trophic methods evolved simultaneously.

The established (though implied) assumption is that the first living organisms were based on just one trophic method. Most of the scenarios suggested so far for various domains of the entire origin-of-life scenario are based on a prebiotic soup of a kind and thus belong to the heterotrophic camp. However, the autotrophic scenario suggested by Wächtershäuser (1988, 1990, 1992b; Edwards, 1996) has been acknowledged by many researchers as a viable and important proposition (chapter 21). As a matter of fact, for the last few years this has been one of the hottest controversies among researchers in the origin-of-life field. Thus, at the present state of our knowledge, the preference of either camp is to a large extent arbitrary.

Primordial "collaboration" between the two trophic methods?

According to Wächtershäuser (1992a, 1992b) the primordial trophic method could be either autotrophic or heterotrophic, since the two methods are mutually exclusive. Is it possible to conceive a scenario in which the borderline between the two trophic methods is not clear-cut? Is it possible to conceive a scenario for the origin of primordial living entities based on central attributes of the two trophic methods?

Consider, for instance, a scenario where small lipid vesicles are synthesized spontaneously under plausible prebiotic conditions (Morowitz, 1992). Under appropriate conditions these vesicles encapsulate chromophors that may initiate a primitive reaction cycle related to metabolism. If such a vesicle would evolve into an autophototrophic living entity, its origin may be arbitrarily considered a "collaboration" between molecules synthesized by the two trophic methods. Thus, in the twilight zone of the origin of the photoautotrophic method under study, the definition of the two trophic methods becomes arbitrary, like the definitions of life and its beginning (Olomucki, 1993; see chapter 7).

In the example under consideration, a combination of the two trophic methods may be posited, since the establishment of the lipid vesicles precedes the CO_2 fixation reactions and the emergence of life. In this case, *prebiotic autotrophy begins with prebiotic heterotrophy,* unless the first organic molecules involved in the origin of an autotroph are arbitrarily considered "living."

If, however, the trophic method is designated only for chemical entities defined as living, then the trophic method is relevant only when the (arbitrarily defined) living state is reached.

Heterotrophic systems are the beginning of many scenarios

Soon after the publication of the Miller-Urey experiment and the excitement it induced in the scientific community, it was hypothesized (and hoped) that all the central building blocks of biochemistry could be synthesized under prebiotic conditions. This hope has not yet been materialized, however. One major reason has been the changes that the phrase "prebiotically feasible" has undergone as prebiotic research intensified. The best example for this statement has to do with the models of the prebiotic atmosphere, which have changed from highly reducing, according to Urey and Miller, to mildly reducing, according to later models (Kasting 1933a, 1993b). Moreover, the dominating model of a prebiotic ocean as the major environment for molecular evolution, with the conceptual biases attached to it, was gradually replaced by a variety of environment candidates, each with its specific reactions, processes, and scenarios.

Why 20 coded protein amino acids?

An example of biogeochemical considerations in which both biological and geochemical features are introduced has to do with the selection of amino acids during the early stages of biology. The geochemical considerations are connected with the primordial ocean and the heterotrophic scenario. The number of known amino acids far exceeds the number of those acids which are used in biology; the number of possible amino acids is even much larger. The question is thus, "Why and how were these amino acids selected during the emergence of life?" These questions were addressed by Weber and Miller (1981), who examined the following attributes of their organic molecules in the presumed prebiotic arena:

1. Availability of amino acids in the primitive ocean.
2. Function of the amino acids in proteins.
3. Stability of amino acids and peptides.
4. Stability of amino acids to racemization.
5. Stability of the amino acid–tRNA complex.

Their conclusion was that the amino acids used in biology are the most suitable candidates, as judged by the combination of the various criteria. At the end of their discussion, Weber and Miller (1981) made an interesting speculation: "If life were to arise on another planet, we would expect that the catalysts would be poly-α-amino acids and that about 75% of the amino acids would be the same as on the earth." And because this speculation could not be tested at that time, they added, justifiably, that "the idea . . . can be tested when an independent origin and evolution of life is found on another planet." More about the cosmogeochemistry of amino acids can be found in Bada, 1991.

In a more recent work, Béland and Allen (1994) discussed the evolution of the genetic code and suggested a primitive genetic code with only 20 separate words, as an explanation for the 20 coded amino acids in extant life (chapter 17).

The second stage of a heterotrophic scenario

The next stage of the heterotrophic scenario has to answer the question, "What could the building blocks (formed in the first stage) do?" One answer to this question was given in the form of a sudden thermal jump designed to achieve organic polymers that, morphologically, resemble extant cells. The other answer is the "Lego-like approach."

Thermal-jump reactions: Could life have started by a sudden event?

According to this approach, a sudden increase in the temperature of mixtures of amino acids may bring about polymerization and formation of cell-like colloidal particles. S. W. Fox and his coworkers (see Fox and Dose, 1977; R. F. Fox, 1988; S. Fox, 1988; and chapters 15, 22) used thermal polymerization of amino acids to synthesize colloidal microspheres that, under the microscope, resemble living organisms such as yeast cells. Their reaction mixture consisted of an excess of aspartic and glutamic acid. The thermal treatment, normally ranging from 120°C to 180°C, was followed by washing with water and resulted in the polymers listed earlier. Chemically, the microspheres are made of a proteinlike substance called "proteinoid" (proteinlike), in which many of the bonds are not peptide bonds (chemical interreactions with the carboxyl or amino groups that are not involved in the normal peptide bond formation).

The biogeochemical scenario is the formation of microspheres as a result of washing the proteinoids from the hot rocky environment where they were formed to the sea. The microspheres thus served as metastable environments in which chemical processes could have taken place, protected to some extent from the harsh environment. Evolutionary processes taking place inside the microspheres are helped by the properties of the proteinoids.

Fox's hypothesis is controversial today, as it was when first published. It is adequate to cite Ferris (1989), who wrote: "Fox' theory has little validity. Consequently, the other claims . . . —on the significance of the budding of microspheres, their implied sexuality, the "membrane" potentials and so on—are meaningless in the context of the origin of life." More information, and critical comments, may be found in R. F. Fox, 1988; Brack, 1993b; and Yockey, 1992.

The "Lego-like" methodology

According to this approach, the second stage of a heterotrophic system may remind one of Lego play: The building blocks have to be used and manipulated so as to simulate an evolutionary process. It is recalled that an evolutionary process can proceed by both small and large steps (chapter 10); thus, the evolution rate may be considerably directed and managed. In order to accomplish this, one has to devise prebiotically feasible scenarios for the emergence of oligomers and polymers, reaction cycles, and functions that are compatible with the principle of biological continuity. At some point in the oligomerization processes, template-directed synthesis would be invented and would open up many more evolutionary options, as discussed in chapter 23.

Minimal cells and amphiphiles, monolayers, micelles, bilayers, and liposomes (vesicles)

It is instructive to continue the present discussion by focusing on some of the main experimental and theoretical aspects of the minimal cell, starting from the preformed constituents of the cell membrane and ending in the hypothetical divergence of the evolutionary pathways into autotrophs and heterotrophs.

Concentrating mechanisms of organic compounds

Organic compounds on the primordial earth presumably found their way from the sites of their formation, or their infall from extraterrestrial sources, to the terrestrial environment, where water-soluble compounds would be dissolved in the waters of oceans, lakes, and puddles. The relatively low density hydrocarbons and their derivatives, however, would accumulate at air-water interfaces, forming prebiotic oil slicks (see Deamer et al., 1994; Morowitz, 1992). These oil slicks would be of various sizes, depending on the sizes of the water surfaces; wind would concentrate them considerably. Thus, whereas the average concentration of organic amphiphiles on the earth's surface would be very small, their local concentrations could be much higher. Under appropriate conditions, membranes would form, followed by the formation of vesicles capable of encapsulating various organic molecules (fig. 19.3). Development of chemiosmotic potential between the encapsulated solution and the ambient solution can drive "nutrient" transport in and out of the vesicles.

Being exposed to solar radiation, some of the encapsulated organic compounds

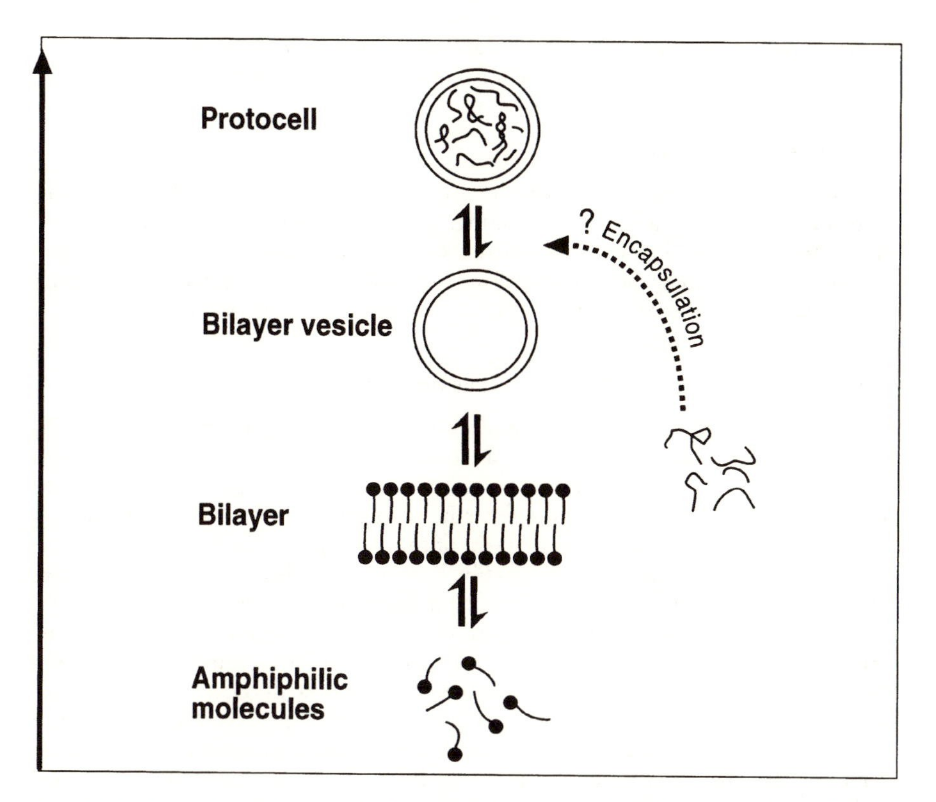

Fig. 19.3. Hypothetical stages in the transition of amphiphilic molecules to bilayer vesicles and protocellular systems capable of encapsulating organic monomers and oligomers and reaching the stage of cellular life. (Following Deamer et al., 1994)

would undergo photochemical reactions that may trigger the beginning of chemical processes, thus leading to the evolution of photoautotrophs, as described later.

Early membranes and their hypothesized appearance

The importance of membranes in the very first steps of the evolution of living entities was already stressed by Oparin (1957; see also Lazcano et al., 1992). A prebiotic model of such a membrane was suggested only much later, however, by Deamer and Oro (1980). Because most molecular evolution research was traditionally focused on peptides and nucleotides, and because the specific incorporation of membranes into these models was not clear, the study of the role of membranes in the origin of life was delayed compared with that of biopolymers. This may have influenced the hypotheses on the primordial entities involved in origin-of-life processes. Today the importance of membranes in the study of the origin of life is recognized by most researchers (Lazcano et al., 1992).

The time of appearance of membranes is model-dependent, varying from "membrane first" to "membranes somewhat later." The latter possibility is realized, for instance, in White's autogen (chapters 23) and in Wächtershäuser's pyrite world (chapter 21), where the primordial compartmentation mechanism is adsorption.

Predecessors of minimal cells

The minimal cells discussed in chapter 17 are rather advanced chemical entities; thus, each of them is a landmark, not a beginning. In order to explore the "membrane first" thesis, one has to look for model reactions where membranes are formed spontaneously under prebiotic conditions. For more details the reader is referred to Lazcano et al., 1992, and Deamer, 1994, 1997).

Difficulties in self-assembly of primitive membranes

As shown in fig. 19.3, amphiphilic molecules can self-assemble into bilayers, thus forming oriented aggregates, which can form bilayer vesicles. This is a physico-chemical process, where no covalent bonds are involved. The next stage in the pathway leading to a protocell of a kind is the incorporation of other organic molecules into the vesicles. One possible mechanism of this process is discussed later.

The bilayer structures are not readily formed from short ($< C_{12}$) molecules, whereas long-chain hydrocarbons are difficult to synthesize under plausible prebiotic conditions. Long-chain amphiphiles were not observed in at least one kind of carbonaceous meteorite. Moreover, it may be that amphiphilic compouds of the kind that can form lipid vesicles were rare on the primordial earth (Deamer et al., 1994). For instance, in spite of the relatively high concentrations of surface-active compounds in the Murchison carbonaceous meteorite, the likely contribution of this kind of meteorites (carbonaceous condrites) to the infall of organic matter to Earth seems negligible (see Chyba and Sagan, 1992; Deamer, 1994). Thus, the formation of membrane structures of the kind needed for the earliest cells is an unsolved problem.

Scientific research does not stop here

It would be a mistake to stop the research at this point and wait until, for instance, a source of prebiotic long-chain amphiphiles were discovered. Rather, scientists prefer to circumvent the present obstacle. An example of such circumvention would be to focus on the next stage of this research, namely, to assume a source of prebiotic amphiphiles capable of membrane formation. Using the appropriate fatty acids, which are not prebiotically plausible, simple bilayered liposomic structures can be prepared, and their formation process, as well as properties, can be studied. Such experiments include encapsulation of various compounds and the study of various reactions inside the vesicles (Chakrabarti and Deamer, 1994; Chakrabarti et al., 1994; Deamer, 1994; Deamer and Harang, 1990), as we shall see in a moment.

Encapsulation of large molecules by lipid bilayers

In order to enclose a large foreign molecule, the lipid bilayer must be reversibly broken at first, allowing the entry of this large molecule, and then resealed. It seems that the properties of lipid-bilayers are tailor-made just for this task: When a liposome breaks and large molecules leak inside, it would seal, under appropriate conditions, thereby encapsulating these molecules (Chakrabarti et al., 1994; Deamer, 1994).

Of the several methods of encapsulation—mechanical, osmotic, and hydration-

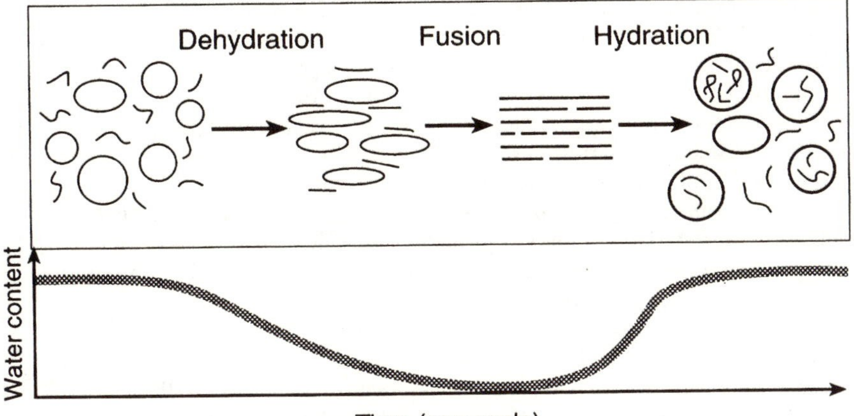

Fig. 19.4. Encapsulation of molecules according to the hydration-dehydration method. (Adapted from Deamer et al., 1994)

dehydration—the latter seems to be most prebiotic because it simulates processes that are likely to have taken place on the primordial earth (see Lahav, 1994, for a review). The mechanism of encapsulation is as follows (Deamer et al., 1994): When liposomes are dehydrated, they tend to fuse into multilayered structures. If solutes are present, they are sandwiched in between these layers. Upon rehydration, the lipid layers form vesicles containing the entrapped solutes (fig. 19.4).

Membranes could have had additional functions. For example, they have been implicated in catalytic reactions. One of the functions suggested for lipid membranes is related to their hydrophobicity-hydrophilicity features. According to Hartman (1995b) membranes could have functioned as the earliest aminoacyl-tRNA synthetases.

Polypeptides as membranes

According to Brack and Orgel (1975; Brack, 1994), alternating hydrophobic-hydrophilic polypeptides can form β sheets that would form bilayers. Such membranes would be porous and may have been glued to hold together by the polynucleotides that were involved in their synthesis (Orgel, 1987).

Experimental and theoretical aspects of synthetic minimal cells

Liposomes containing large molecules that can serve as models for minimal cells have recently been prepared by several research groups, as we will see in later sections.

Several attempts to synthesize systems that researchers claimed to be models of minimal cells were reported (Luisi and Varela, 1989; Luisi et al., 1994). Luisi et al. (1994) have developed the "core-and-shell reproduction"—a process by which the shell reproduction of spherically bounded systems (micelles or vesicles) proceeds simultaneously with the replication of nucleic acids hosted inside the micelles or vesicles. The vesicles are made of linear fatty acids, whereas the template-directed systems are based on either the polynucleotide phosphorilase-catalyzed synthesis of poly(A) from ADP, or the enzyme Q_β replicase, which is able to catalyze the synthesis of an RNA template. According to Luisi and his collaborators, "the template self-reproduction of linear sequences of oligonucleotides and the autopoietic shell reproduction of micelles and vesicles" can be viewed "as the first examples of autopoietic core-and-shell reproduction, which represent the simplest possible models for cellular life (as defined within the limiting frame of autopoiesis)." This is only the first step in the simulation of the formation of a synthetic minimal cell.

This experimental approach introduces an important feature of synthetic minimal cells, namely, that they encompass a range of complexities of their components. For instance, the same vesicle may host different replicating entities, but all of them are "core-and-shell reproduction."

A simpler system was prepared by Chakrabarti et al. (1994), in which an RNA polymerase was entrapped by a protocell. In this model, the lipidic vesicles was prepared without substrate. The substrate, adenosine diphosphate (ADP) molecules, was provided externally, serving as building blocks for the long-chain RNA polymer that was synthesized inside the vesicle. Having been formed under such conditions, various reactions are expected inside liposomes. These include the establishment of electrochemical potential between the inner environment and the ambient solution, and formation of proton gradient, photochemical reactions of guest molecules (Deamer, 1997), and template-directed syntheses.

Permeability of lipid membranes

One of the problems encountered by students of models of early membrane is the slow rate of diffusion across lipid bilayers. There are at least two possible ways to increase the influx rate of solutes into such model cells: to change the composition of these model cells, and to introduce transport mechanisms that facilitate the fluxes, including the formation of transmembrane pH gradient (Chakrabarti and Deamer, 1994; Deamer, 1997).

The permeability of small water-soluble molecules through the walls of lipidic liposomes is low. In the absence of sophisticated proteins, which are not considered prebiotically plausible by most researchers, additional transport mechanisms should be invoked. These include a primitive carrier-mediated diffusion, and facilitated diffusion (see Lazcano et al., 1992; Stilwell, 1980).

Prebiotic protocells with bioenergetic functions

Replication, thermodynamic isolation in the prebiotic environment, and the ability to use nutrient molecules and energy sources are central attributes of a minimum protocell (Morowitz, 1992, p. 103; chapter 17). Deamer and his collaborators (1994) have recently outlined a series of steps that would lead to the formation of a protocell with bioenergetic functions, as follows:

1. Formation of amphiphiles capable of self-assembly into monolayers, micelles, and bilayer vesicles, in aqueous environments.
2. Growth of the vesicles, for instance by energized addition of amphiphiles, followed by incorporation into the vesicles by partitioning mechanisms.
3. Capture of catalytic and replicating macromolecules by an appropriate mechanism, such as environmental hydration-dehydration cycles.
4. Capture of macromolecules such as pigments capable of absorbing light energy, where the light energy is used to form chemiosmotic proton gradient.

5. Acquisition of some energized process for transport of potential nutrients. A plausible source of the required energy is a transmembrane proton gradient capable of accumulating "nutrients" from the environment.

The future experimental challenges of the study of the earliest membranes in the origin-of-life process were described recently (Deamer et al., 1994) as follows: (1) synthesis of long hydrocarbons (12 or more carbons); (2) addition of both ionic and polar features to the hydrocarbon chains.

Emergence of the first protocells according to Morowitz

For organic molecules at the surface of the primordial sea, solar energy is the major prebiotic source of energy. Moreover, organic molecules, either produced in the prebiotic atmosphere or in the form of extra-terrestrial infall, sooner or later would reach this surface. The combination of an energy source and a variety of organic molecules can be manipulated, theoretically, into photoautotrophic protocells, as suggested by Morowitz (1992). Although the exact mechanism of this process is not known, the appearance of closed vesicles, with the partially isolated solution volume, polar interior and exterior, and a nonpolar membrane are a landmark in the origin-of-life process.

Among the dissolved molecules inside these vesicles were some chromophores—molecules that absorb light. With these chromophores, the spontaneously formed vesicles can convert light energy to electrical potential energy, thus serving as energy transduction devices (see also Deamer and Harang, 1990). In this way, protocells were formed, characterized by the basic elements of bioenergetics. Once an electrical potential was maintained across the membrane, various reactions could have occurred, including oxidation-reduction, and acid- or base-driven reactions. The chemical energy needed for the primordial metabolic processes involves phosphate-bond and reducing energy; thus, according to this scenario, phosphate was recruited into bichemistry at an early stage.

Gradually these protocells developed a metabolism based on CO_2, AH_2 (in extant cells AH_2 is NADH + H^+), and P—O—P. Nitrogen was introduced into the system by the entry of ammonia (NH_3) into the metabolism; ammonia reacted with keto acids (chapter 17), themselves intermediates of the primordial metabolism. The amino acids thus produced can form small peptides that can be attached to the surface of the protocell membranes, thus enhancing their catalytic specificity. It is noted that at the present stage the peptides are non-template-directed.

The self-replicating metabolic protocells represent a population of "living" entities characterized by "Darwinian mode of competition," where beneficial mutations are selected and novel biochemical complexities eventually lead to the last common ancestor. Coded genetic information is introduced into the system only during the latter transition from the protocells to the universal ancestor. The introduction of the genetic code is thus a *late* event, according to Morowitz; his scenario belongs to the "metabolism first" school. It does not offer a mechanism for the emergence of template-directed syntheses and the genetic code.

Heterotrophic minimal cells and the origin of life

Conceptually, the minimal cells listed earlier are made of preformed components. Vesicles with the features described earlier may be considered primordial in the sense that their formation would depend mainly on the presence of plausible concentrations of prebiotic amphiphiles under adequate environmental conditions.

The incorporation of another central attribute of the minimal cell—self-replication—is more difficult to conceive. This problem is of special importance, since it is hard to visualize the emergence of template-directed syntheses without the presence of a compartment of a kind. Addressing the problem of the collaboration between vesicle and self-copying processes, Luisi et al. (1994) noted: "The two processes of core-and-shell reproduction . . . are independent from one another, and it would be instead more interesting if they would be chemically linked with each other." Moreover, the synthetic models of minimal cells and related systems demonstrate certain features of these entities *but do not simulate their evolution.* Thus, the presently known synthetic self-reproducing vesicles are still far from the hypothetical minimal cells that presumably evolved spontaneously somewhere on Earth, billions of years ago.

Summary

The role of the environment is a central topic of the search for the origin of life. The present discussion of the bottom-up approach focuses on the main origin-of-life processes and scenarios in which minerals are not directly involved. These encompass energy sources for the primordial reactions, the primordial trophic methods, plausibility of some of the main molecules involved in the molecular evolution processes, and the prebiotic reactions of amphiphiles, including bilayer vesicle formation and organic molecules encapsulation, cellularization processes, and an autotrophic "metabolism first" scenario.

20

Bottom-up Biogeochemical Reconstruction

Minerals Functioning as Scaffolds, Adsorbents, Catalysts, and Information Carriers

In order to clarify the concept of organic being, let us look at mineral bodies. Fixed and unshakeable in their various elementary parts, they appear in combinations which, though formed according to laws, present neither order nor limits.

Goethe, cited in Jacob Lorch, "The charisma of crystals in biology"

The living organism has already been compared with a crystal, and the comparison is, *mutatis mutandis* [allowing for necessary changes], justifiable.

August Friedrich Leopold Weismann, the father of the theory of germ-plasm, cited in Jacob Lorch, "The charisma of crystals in biology"

Why minerals?

Minerals preceded life: Were they involved in the *origin* of life? Minerals have always been present on the earth's surface, since its accretion, some 4.6 Ga ago. Thus, if the beginning of life indeed took place on our planet, then various minerals not only witnessed the very first stages of this processes but could also have been involved in it.

The biblical stories and various old myths and folklores, such as the Golem of Rabbi Yehudah-Liwa from Prague, involved clays in the creation of life. The omnipresence of clays, and the ease with which living figurines can be modeled out of them, may have been among the reasons for the development of these very ancient traditions and folklore. But whatever the reasons for the importance of minerals and crystals in these old traditions may be, minerals have also been linked to life in biology, the scientific study of life, as Lorch (1975) discussed in his paper entitled "The charisma of crystals in biology."

The involvement of minerals, including clays, in the origin of life was suggested more recently by a number of scientists. Their hypotheses are based on both mineral ubiquity on the earth's surface and on the minerals' chemical and physicochemical properties. The first modern attempt to invoke minerals in the origin of life was made by Schwab in 1934 (cited by Bonner, 1991). The problem under study was the origin of chirality in living creatures, and Schwab suggested that asymmetric crystals, such as quartz, were involved in catalytic processes that gave rise to the first asymmetry of biological molecules.

The involvement of minerals in the origin of life was also suggested by the geochemist Goldschmidt in 1945 (published in 1951; cited by Cairns-Smith et al., 1992) and, quite independently, several years later, by the biophysicist Bernal (chapter 5). The first one, however, went unnoticed, whereas the latter was very influential. For more information on historical aspects see Kamminga, 1982.

Mineral candidates

The kinds of minerals on the prebiotic earth are not known with certainty. However, it has been assumed by various researchers that many different minerals could have existed in the prebiotic era. Moreover, various prebiotic environments have been proposed, each with its own minerals. Among the host of minerals suggested for various prebiotic functions associated with the origin of life, the most popular ones have been layer silicates, mainly the montmorillonite and kaolinite clay minerals. The complete list includes silica; alumina; quartz; zeolites; oxides/hydroxides of Fe, Mg, Ca, Co, Cu, Mn, Ni, Zn, and Al; phosphates; carbonates; and iron sulfide and cyanides (see Cairns-Smith et al., 1992; Kamaluddin et al., 1994; Lahav, 1994; Negrón-Mendoza et al., 1996, for reviews). Regolith grains, without additional mineralogical specifications, were also suggested (Nussinov and Maron, 1990; Nussinov et al., 1997).

In addition to this list of stable minerals, soluble minerals such as chlorides and sulfates have also been proposed (Lahav and Chang, 1982). These minerals are considered part-time solids during the dehydration period of an environmental system fluctuating between wet and dry states. Mineral salt solutions were explored by Schwendinger and Rode (1989).

Functions suggested for minerals

Most of the functions suggested so far are based on surface interactions between minerals and organic molecules. It is generally accepted that minerals also can protect organic molecules from hard UV photons capable of disrupting many kinds of covalent bonds (see Kolb et al., 1994, chapter 19).

The main functions suggested for minerals can be divided into the following groups: (1) adsorption and host structures, including serving as scaffolds for adsorbed templates and mineral-induced reactions in double-layered hydroxides; (2) primordial membrane; (3) catalysis; (4) symmetry breaking; (5) information transfer; (6) redox reactions; (7) energy source.

The following discussion focuses mainly on topics 1–5. Topics 1, 2, and 3 are also discussed in chapter 23. Topics 6 and 7 are discussed in chapter 21.

Adsorption and heterocatalysis in prebiotic processes

Adsorption reactions per se are of limited interest in the present context, where we focus on the organizational principles and mechanisms by which the fundamental attributes of life could have been formed during the chemical evolution era. Catalytical activity of minerals is associated with adsorption of the reactants from the solution onto the solid surfaces. A great deal of experimental work has been done on adsorption and catalysis of minerals in organic reactions, in relation to prebiotic condensation reactions and the formation of organic oligomers and polymers. The problem with condensation reactions is that although the adsorption of their products (oligomers and polymers) is affected by minerals, there is no evidence that mineral surfaces can direct the sequence of organic building blocks of a polymer in the way an organic template does. Remembering that template-directed synthesis is a central attribute of the origin-of-life process, it is obvious that non-template-directed oligomers and polymers are of limited interest in the present discussion. More details can be found in recent reviews (Bujdak and Rode, 1995; Chang, 1993; Ferris, 1993a, 1993b; Gedulin and Arrhenius, 1994; Hartman, 1995a, 1995b; Lahav, 1994; Mauzerall, 1992; Pitsch et al., 1995).

Adsorption—the simplest mechanism of compartmentation

The most primitive form of compartmentation was suggested by H. D. White (1980) in his autogen scenario and by Lahav and White (1980) in their adsorbed-template model. Wächtershäuser (1988, 1992a, 1992b) further developed this idea and discussed its physicochemical attributes and its role in primordial molecular selection in his "surface metabolism" scenario. According to this mechanism, the separation of the chemical entities under study from their environment is carried out not by a physical boundary, such as a membrane, but by an adsorbing surface. This mechanism has yet to be worked out experimentally, but it does not contradict the biological continuity principle.

Thus, the adsorption of chemical entities undergoing chemical evolution processes can be considered a primordial mechanism of compartmentation, which was available in certain prebiotic environments when the first organic building blocks of biochemistry were prebiotically synthesized.

Porous structures as possible means of compartmentation

Several authors suggested small pores, such as in rocks or clay aggregates, as possible shelters for organic molecules (Cairns-Smith et al., 1992; Kuhn and Waser, 1994b). Minerals serving as host structures were suggested by T. Lee et al. (1993) and Arrhenius et al. (1997).

Inorganic membranes

> The membrane is the message.
>
> M. J. Russell, R. M. Daniel, and A. J. Hall, "On the emergence of life via catalytic iron-sulfide membranes"

This section is relevant to the discussion about prebiotic membranes. An inorganic membrane as a compartmentation mechanism is a central feature in the scenario suggested by Russell and his collaborators (Russell et al., 1993; MacLeod et al., 1994; and Russell et al., 1994). Discussing the chemistry of the seawater and hydrothermal solutions of the Hadean era (the earliest era of Earth's history), they suggested that certain supersaturated solutions, especially iron sulfide, would precipitate colloidal gels when mixed with ocean water under certain conditions. According to their theory, life emerged from bubbles containing a highly reduced hydrothermal solution capable of budding. The gelatinous iron-sulfide membrane thus formed has catalytic properties and could develop osmotically driven reactions. These inorganic vesicles thus served as a first step in the origin-of-life process.

The role of minerals in Kuhn and Waser's theory on molecular self-organization, genesis of life, and the genetic code Kuhn and Waser's theory (1994a, 1994b, and references therein) is anchored in a fluctuating environment of hydration-dehydration cycles, like Kuhn's earlier works (Kuhn, 1976). Amino acids are assumed to be supplied by the environment. RNA-like strands formed by the environment (by an unspecified mechanism) are named "primary strands"; their formation is a rare event. Once such a strand is formed, however, it can template-replicate by the use of available activated building blocks. The scenario takes place in a porous system that provides compartmentation to the growing strands, so that their diffusion outside is hampered (Kuhn and Waser, 1994b).

The suggested evolution of the genetic code is an attempt to bridge the RNA world with the "amino acids world." Mechanistically, it is based on an "assembler strand," a predecessor of mRNA "along which RNA hairpin molecules are lined up, forming a picket-fence-like aggregate" (Kuhn and Waser, 1994a). The hairpin strands are the predecessors of tRNA molecules, each carrying its specific amino acid; when lined up along the assembler, these amino acids are linked to form a peptide. Lateral weak interactions between the hairpins, together with their interaction with the assembler, enhance their alignment and catalyze peptide-bond formation (see Wood, 1991).

The fluctuating, porous environment supplied specific sites that drove the replication of strands and aggregates. Amino acids and activated nucleotides G, C, A, and U were continuously synthesized and supplied to those specific sites. In the initial stages G and C were the most important nucleotides because of their strong base pairs with three hydrogen bonds. The dominant amino acids were glycine and alanine, where the code represents the relationships between the sequence of the nucleotides along the assembler strand and the amino acids. The order in which amino acids are incorporated into the genetic code code reflects their availability, polarity, and structural features. The authors suggest six evolutionary stages in which the code evolves.

It should be noted that the function of the first oligopeptides synthesized according to this hypothesis is not catalytic: Their function is agglutination, thus assisting in the restraining of the porous medium on which the whole assembly rests. Therefore, the sequence of building blocks in the peptides is not important. The chirality of nucleotides is solved by an old suggestion that Kuhn and Waser call a "frozen accident" (following Crick; see chapter 10). Systems containing either D- or L-ribose were present at a certain time; the winning system is the one selected for further evolution. The detailed scenario includes the chirality problem, peptide-bond formation, amino acid attachment to the 3′-end of RNA, the primordial ribosome, and replication. Computer model building of various molecules supports the suggested chemical mechanisms of the proposed template-directed reactions.

Prebiotic rhythms

A biological rhythm is also related to regulating agents and mechanisms. Is it conceivable to relate the rhythmic attribute of extant living forms to the rhythm of the primordial environment, or to a prebiotic regulation agent? At present, such wild speculations should be presented as stimulating agents and thought-provoking appetizers rather than concrete hypotheses. This topic is further discussed in chapter 22 in relation to the effects of the prebiotic environment.

Synchronizing messengers?

The diversified and ubiquitous effects of nitric oxide (NO) in many biological systems have stimulated novel suggestions regarding its possible role in molecular evolution (Anbar, 1995). This is of special interest in view of the attribution of the sole rhythmical role in prebiotic chemistry so far, to the environment (Lahav, 1994). According to Anbar (1995), "Nitric oxide must have been around in small concentrations when life first evolved in a partially oxidizing environment, long before the evolvement of atmospheric O_2. NO was probably produced photochemically from CO_2 and N_2 in the upper atmosphere, and by cavitation from N_2 and water in the oceans." Moreover, NO could have been involved in a variety of reactions: "Due to its binding to iron and to aromatic π electrons, NO must have exerted some effect on many different aspects of early biochemistry." Presumably, according to Anbar, NO could have served as a regulatory agent, synchronizing various reactions and processes.

The presence of NO in the primordial atmosphere was discussed earlier by Mancinelli and McKay (1988) in the context of the evolution of the nitrogen cycling, but its synchronizing effects were not known at that time. The possible involvement of this free radical gas in prebiotic reactions has yet to be studied.

Fluctuating environments

Except for deep hydrothermal vents on the ocean floor, all the prebiotic environments suggested so far as possible sites for the molecular evolution process fluctuate according to the rhythm of Earth's movement in the solar system. Fluctuating environments are dynamic systems, characterized by cyclic changes of various parameters, which reflect the rhythmical variations of our planet's surface. This rhythm was suggested (Kuhn, 1976; Kuhn and Waser, 1994b; Lahav and Chang, 1976; Lahav, 1994; Popa, 1997) as a powerful mechanism for various reactions and processes related to the origin of life.

Affected processes during a dehydration process include photochemical reactions, increasing concentrations of soluble species, shifting the direction of exchange reactions, precipitation, and the buildup of surface acidity on the surfaces of various minerals. In addition, owing to temperature rise, hydrogen bonding becomes less effective, and enhanced thermal decomposition of various organic molecules may take place. Moreover, catalytic sites on mineral surfaces become active, activation energy of adsorbed molecular species may be reached, and reactions such as condensation can take place. Upon rehydration, the temperature drops, dissolution as well as dissociation of various compounds takes place, and exchange reactions between soluble and adsorbed species again respond to the new situation. Also, freezing effects are similar to those of dehydration. For instance, freezing- or dehydration-induced surface acidity of clay minerals can bring about similar changes in organic molecules attached to these surfaces. Therefore, fluctuations between, for instance, wet and dry states bring about new environmental conditions; during the very transition between the two hydration states, interesting chemical and physicochemical reactions may

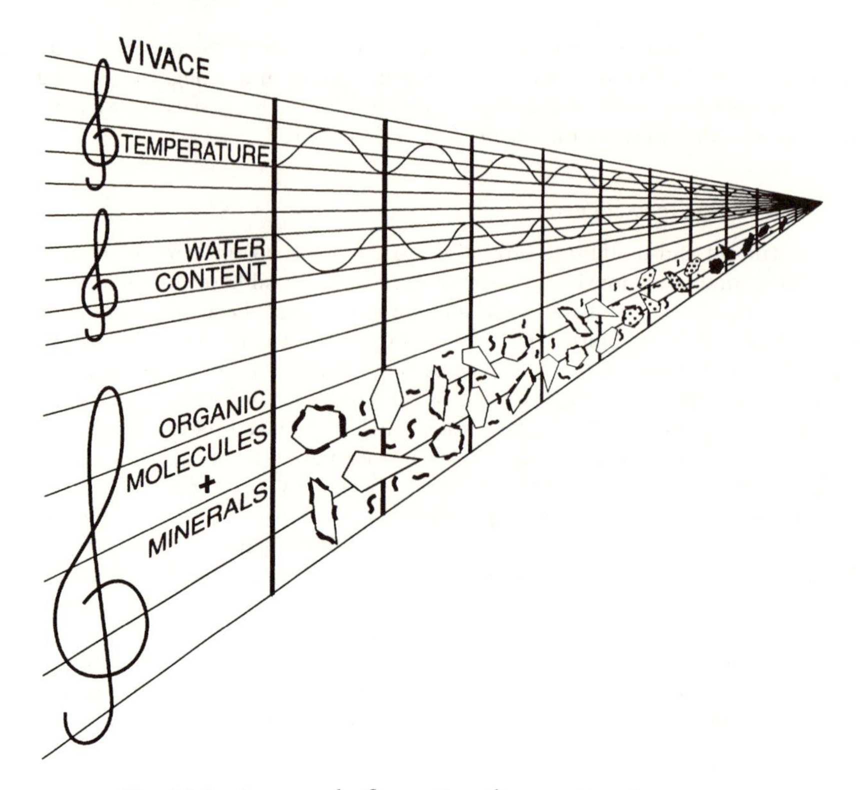

Fig. 20.1. A score of a fluctuating clay-water environment.

take place. Moreover, the rhythm of fluctuating environments gives rise to sequences of reaction cycles (fig. 20.1). Photochemical rhythmicity at sea surface (Morowitz, 1992; chapter 19) is another example of a fluctuating environment.

Was Darwin's "warm little pond" a cycling environment fluctuating between wet-cool and dry-warm conditions? (See also Deamer et al., 1994.)

Condensation reactions

Several researchers have focused on the condensation reaction of amino acids to form peptides in clay–amino acid systems under dehydration conditions and elevated temperatures of less than 100°C (Lahav et al., 1978; Chang, 1993; Zamaraev et al., 1997). In experiments of this kind it was found that amino acids can be condensed to form peptides on certain clays. White and Erickson (1980) showed that the dipeptide histidyl-histdine enhances the condensation rate of glycine on the clay mineral kaolinite. In another study it was found that polyribonucleotides also enhance peptide-bond formation (White and Erickson, 1981). Thus, it seems reasonable to assume that peptides could have formed in prebiotic fluctuating environments containing kaolinite. A prebiotic process of this kind may go on and on, one cycle after the other, where the products are non-template-directed peptides. It is noted, however, that due to hydrolytic processes, such peptides are limited in size.

Based on these works on condensation reactions of amino acids catalyzed by clay minerals in fluctuating environments, and the effect of Cu-montmorillonite on such condensation reactions (Lawless and Levi, 1979), Hartman (1995b) proposed that Cu-montmorillonite could have functioned as a primordial aminoacyl-tRNA synthetase. The biogeochemical relevance of Cu-montmorillonite is presently unknown. Similarly, the possible importance of the recent report (D. H. Lee et al., 1996) on self-replication of peptides in molecular evolution has yet to be explored.

Most important, the sequence of amino acids in these peptides is not instructed by a preformed pattern. In other words, there is no evidence that the sequence of amino acids of these peptides is directed by a clay template. And even though they are not random peptides, because of both the chemical properties of the amino acids and the varying reactivities of different sites on the clay surface, they seem to be of limited importance in the evolution of synthesis of template-directed peptides.

The adsorbed-template model

This model (Gibbs et al., 1980; Lahav and White, 1980) deals with a possible mechanism for prebiotic information transfer, where the model organic molecules are nucleotides and polynucleotides. In view of the difficulties in synthesizing nucleotides under prebiotic conditions (chapter 15), the model molecules are better referred to as "nucleotidelike," or "nucleotide analogues," thus encompassing compounds capable of serving as templates. The main assumptions in this model are:

1. The environmental system is a solid-liquid prebiotic fluctuating environment in which polynucleotides are the informational molecules; they are characterized by negatively charged moieties (such as phosphate groups) under the assumed prevailing conditions. Moreover, these molecules are synthesized in the form of activated mononucleotides, which can undergo random condensation to form random oligonucleotides. The aqueous solution also contains soluble salts, with cations such as Na^+ and Mg^{2+}.
2. Under the prevailing conditions of this model system, the minerals possess either positively charged crystals or domains of positive charges when suspended in aqueous solution. One candidate for such a mineral is apatite ($Ca_5(PO_4)_3X$, where X is generally either OH^- or F^-), which is positively charged above a pH of around 6. Another candidate is kaolinite clay mineral ($Al_4Si_4O_{10}(OH)_8$), the crystallites of which carry positive charges at their edges. Much less studied groups of positively charged minerals include ferric or ferro-ferric oxides, which were formed extensively during the prebiotic era and precipitated in the form of banded iron formations (chapter 13).

The negatively charged groups of the nucleotides are the phosphate moieties. Thus, mono- and oligo-nucleotides tend to be adsorbed on the positively charged mineral surfaces by means of their phosphate groups, whereas the bases are exposed to further interactions (fig. 20.2). Because of their size, oligomers would be attached stronger than monomers to the mineral surfaces (Lazard et al., 1987, 1988), thus forming an "adsorbed template." Presumably, these adsorbed templates can interact with solution nucleotides, where their base moieties can hydrogen-bond with the complementary bases of the solute nucleotides. If neighboring mononucleotides on an adsorbed template undergo condensation reaction, a complementary strand is formed on this template. Under certain conditions, such as elevated temperatures, the complementary domain is detached from its "mother template," capable of undergoing a similar cycle of reactions where it serves now as the adsorbed template.

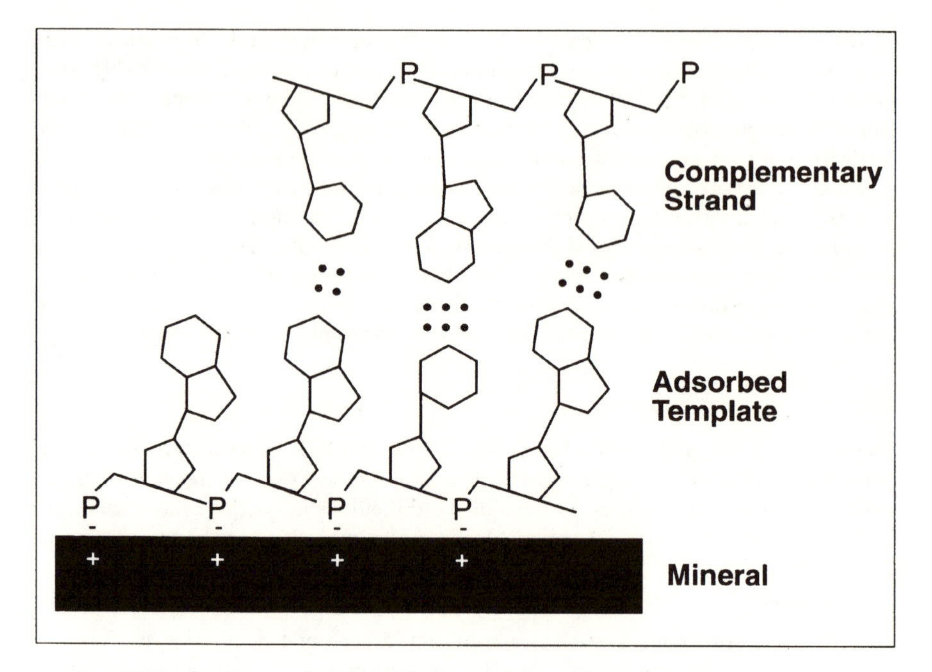

Fig. 20.2. A scheme of an adsorbed template. (Adapted from Lahav, 1994)

The adsorbed-template model system has been partially supported by experimental work in which the minerals were apatite ($Ca_{10}(PO_4)_6(OH)_2$; Gibbs et al., 1980) or gypsum ($CaSO_4.2H_2O$; Chan et al., 1987; Lazard et al., 1987, 1988). These two minerals represent calcium-based minerals, capable of adsorbing the negatively charged nucleotides, which are characterized by different solubilities: Apatite is a sparingly soluble mineral, whereas gypsum is a rather soluble mineral (Lahav and Chang, 1982). These two model minerals are relevant to chemical evolution in a fluctuating system. The cycling environment of such a system supplies a sequence of conditions under which different reactions take place. For instance, with apatite as a scaffold, the hydrogen bonding between the adsorbed template—the polynucleotide—and the mononucleotides is efficient at relatively low temperatures, whereas at high temperatures complementary domains are likely to dissociate. With gypsum as a scaffold, the dissociation of the adsorbed domains, as well as their template, from the solid surface takes place upon dissolution, during the appropriate stage of the fluctuating environment. The condensation reaction in this model will be discussed later.

An interesting indication regarding the validity of this model system is the finding (Lazard et al., 1987) that the binding curves of purine nucleotides to gypsum were sigmoidal rather than hyperbolic, indicating cooperativity in the binding. The latter observation supports the plausibility of base-stacking interactions on the template. The adsorbed template model was recently studied experimentally by Winter and Zubay (1995).

How about the biological continuity principle? The adsorbed-template model was suggested for the simplification of both self-replication of oligonucleotides and directed peptide-synthesis processes (Lahav, 1991, 1993, 1994; chapter 23). The fluctuating environment was also suggested to manipulate other prebiotic reactions, such as RNA hairpin structures by denaturation and renaturation (Di Giulio, 1994a). In these hypotheses, the association of complementary oligonucleotide strands, the primitive tRNA-mRNA relationships, and the charging of amino acids onto the primordial tRNAs are all viewed as simplified features that are continuous to the basic template-directed processes of extant cells.

The adsorbed-template reaction in a fluctuating environment: Could it have served a primordial self-replication mechanism? Two basic conditions should be fulfilled in order to render the adsorbed-template model a viable mechanism in chemical evolution:

1. Condensation of the mononucleotides that are hydrogen-bonded to their complementary sites on the template. Substitutes for the extant reaction mechanisms have to be devised, therefore, by students of the origin of self-replication.
2. Repeated cycles in a system fluctuating within adequate ranges of environmental parameters such as temperature, hydration, or pH, which are not damaging to the organic molecules under study.

When these two conditions are met, then it is conceivable that this system would be able to support the evolution of random populations of oligonucleotides into more ordered populations. Indeed, the adsorbed-template-fluctuating-environment under study may serve as the beginning of an information-transfer process; it is hypothesized as being characterized by a buildup in population of oligonucleotides that are interrelated in their synthesis process: some of them are complementary to other ones, and by the same token, some of them are copies of others. In other words, in such a system copying of domains of strands of oligonucleotides is attainable, at least in principle.

Minerals and the origin of homochirality

The origin of chirality is still a controversial issue. In a recent meeting (see Cohen, 1995), Bonner argued in reference to the big gap between the origin of homochirality and the origin of life that homochirality must have preceded life: "I happen to think that you have to understand the origin of homochirality before you can bridge that gap. Stepwise, one has to deal with the origin of homochirality first, and then how do you get to living organisms." Moreover, said Bonner, "I spent 25 years looking for terrestrial mechanisms for homochirality and trying to experimentally investigate them and didn't find any supporting evidence." This failure has led him to suggest an extraterrestrial source for the homochiral molecules. According to Greenberg (1995), one such extraterrestrial source was the organic molecules of comets.

In contrast to Bonner, Miller and Bada argued that rather than a precondition for life, homochirality is an "artefact of life."

The importance of the microenvironment

Several authors have noted that the symmetry breaking of biomonomers should be looked for at the molecular level. Thus, it is not necessary to achieve chiral purity (or excess of one chiral molecule) on a large scale in order to bring about chiral purity of the chemical entities of the molecular evolution era. Local excess of chiral molecules have been shown experimentally by Brack (1993b) in peptides, and by the school of Meir Lahav and Leslie Leizerowitz (see Addadi et al., 1982; Weissbuch et al., 1994) with minerals and adsorbed organic molecules. This subject is discussed later in relation to the possible role of minerals in asymmetry breaking.

Adsorption and symmetry breaking

The involvement of asymmetric crystal lattices in the adsorption and catalytical processes that brought about the enantiomeric homogeneity of living creatures was suggested already in 1934 by Schwab as noted above. More recently it was suggested by MacDermott (1993), based on theoretical considerations, that "symmetry-breaking is altogether easier if a surface is involved."

One obvious mineral candidate for such a function was quartz, which is found in nature in well-defined enantiomorphic crystals. Various experimenters used quartz as an adsorbent in an attempt to find a relation between the adsorption of chiral organic molecules and the chirality of the crystals. The experimental procedure in these experiments, which was aimed at the asymmetric adsorption of the chiral organic molecules onto D- or L- quartz crystals, involved grinding the quartz crystals in order to increase their specific surface area. However, in the many experiments that have been carried out in this manner, the adsorption of the homochiral molecules was not significantly affected by the chirality of the quartz crystals: "The chirality of the samples was essentially random" (Bonner, 1991). This conclusion cannot be accepted without criticism, however, because of the following reasons.

Preparation of the minerals

The preparation of the quartz samples was made by mechanical grinding, which must have influenced the adsorption of the organic molecules (homochiral amino acids) under study. The grinding of the quartz crystals brings about the formation of a "non-quartz" layer. Similar grinding effects were shown in kaolinite clay mineral by IR measurements (Yariv, 1975, and references therein). Moreover, grinding of quartz, kaolinite, and many other minerals is characterized by photon emission (triboluminescence) (Lahav et al., 1982), which indicates that the crystal surfaces undergo changes. Thus, in the cited adsorption experiments on ground D- and L-quartz crystals, asymmetric adsorption would probably not take place.

The microscopic reaction, not the batch reaction, is relevant

The differences between enantiomorphic crystals with regard to adsorption of different homochiral molecules are expected to take place on the different faces of these crystals (Addadi et al., 1982). Thus, batch adsorption experiments not only should avoid grinding of the crystals but may also be insensitive to enantiomorphic asymmetry when the whole crystal surfaces are measured. The systematic studies of Meir Lahav and Leslie Leizerowitz's school resulted in a new understanding of the potential importance of minerals in symmetry breaking and surface recognition. In their words, "the search for minerals which may interact with chiral molecules should not be limited to chiral crystals but should also encompass nonchiral minerals which express chiral surfaces, such as gypsum, kaolinite, talc, etc." (Weissbuch et al., 1994).

Thus, if a selective adsorption of chiral amino acids or sugar entities, or their derivatives, on certain crystal faces, were observed, then the problem of biological homochirality would be possible to comprehend: The differential adsorption creates homochiral microenvironments in which further prebiotic processes may take place.

Clay minerals as symmetry breakers

Clay minerals have, for a long time, been favorable minerals for researchers of the origin of life. No wonder that several attempts were made with regard to their possible involvement in prebiotic symmetry breaking. Indeed, in spite of the old observation that crystal structure of clays has no known chirality, several researchers claimed to have found asymmetric effects in clays. These claims could not be supported experimentally (see Bonner, 1991). However, in view of this postulate (Weissbuch et al., 1994), the possible role of clay minerals in symmetry breaking should be explored by an entirely different methodology.

Pyrite crystals as symmetry breakers

This idea is part of Wächtershäuser's scenario for the pyrite-pulled chemoautotrophic origin of life and will be discussed in the framework of this hypothesis (chapter 21).

Clay minerals as participants in information-transfer reactions with organic molecules: Genetic takeover ("clay life")

Clay minerals, according to several researchers, could have been involved in the chemical evolution process not only by means of adsorption and catalysis but also by participating in information-transfer processes. According to Cairns-Smith (1966, 1982), minerals were the first genetic-material organisms on Earth, characterized by the central attributes of life. The transition into organic life is described as a "genetic takeover." According to Dyson (1985, p. 34), this is a "double-origin" theory.

The minerals suggested by Cairns-Smith as most likely candidates for his proposition are clay minerals of the montmorillonite group. These clays are characterized by information (their isomorphic substitution and crystal lattice defects) and can replicate, thus transferring their information content from one "generation" to the other. The evolutionary potential of these "living clays" or "replicating clays," as they have been called, is postulated to have been in their influence on prebiotic organic molecules. The latter thus evolved into more and more complex molecules and aggregates, with the help of the "clay organisms." Eventually the "genetic takeover" was completed and the more sophisticated organic life discarded their predecessors, embarking on their own trail of evolution and life.

So far, no experimental evidence has been published to support Cairns-Smith's proposition, nor have molecular fossils related to Cairns-Smith's hypothesis been found in living organisms. It is interesting, however, to analyze it according to central attributes of life, as discussed in chapter 11. In table 20.1 a comparison between "clay life" and life as we know it is given with regard to several features that seem to be directly comparable. Lack of factual support makes it difficult to include more attributes in this comparison. Theoretical analysis of the informational aspects of the structure of both minerals and polynucleotides (see Lapides and Lustenberg, 1990) may be helpful in this problem.

Table 20.1 Various fundamental characteristics of living organisms compared to corresponding attributes of Cairns-Smith's "clay organisms"

Feature	Living organisms	"Clay organisms"
Information storage	One-dimensional sequence of polynucleotides	Three-dimensional array of crystal lattice defects*
Recognition mechanism	Hydrogen bonding between purines or pyrimidines moieties of nucleotides	Mainly electrostatic forces that affect crystal lattice growth and formation.**
Stages in self-replication process	Information is read twice in self-replication processes by formation of acomplementary strand	Information is read once, by direct transfer of crystal lattice information from a "mother" lamella to its "daughter" lamella**
Compartmentation	Lipid membrane with embedded proteins	Clay lamellae and their adsorption characteristics**

Source: Adapted from Lahav, 1994.

*Established experimentally. The stability of these defects and their exact pattern are known only partially.

**Implied. No experimental data.

"Adsorption compartmentation" has been suggested earlier and seems to be a possible stage in the transition from nonliving to living entities (Lahav, 1994). However, the speculated informational attributes of "living clays" differ fundamentally from the known attributes of living creatures. The critical analysis should focus on information transfer, since it is hard to see a continuity in the transition from mineral information transfer to molecular-biological (genetic) information transfer. The essence of the mechanism of genetic information transfer is template-directed reaction, by which the sequence of building blocks of a biopolymer is transferred into a corresponding sequence of building blocks of another biopolymer. So far no such process has been found in minerals. In view of the characteristics of information storage in biopolymers, on the one hand, and in minerals, on the other, it seems unlikely that information transfer of the biological kind between the two chemical entities is possible.

It is noted that minerals can, in principle, exert specific effects on both inorganic (Van Bladeren et al., 1997) and organic molecules (Weissbuch et al., 1994). For instance, certain adsorption sites on a clay crystallite may favor certain organic monomers. In the presence of a condensing agent, and under appropriate conditions, several such adsorbed organic monomers may condense into a small oligomer. A similar system of this kind was recently described (Bar-Nun et al., 1994), in which amino acids were condensed to form dipeptides on another organic molecule that served as a template. However, a template-directed genetic system is characterized by a sequence of sites onto which the building blocks of the polymer under consideration would be attached and then condense. As far as I know, no such experimental system, in which information was transferred between a mineral template and an organic polymer or oligomer, has ever been reported and verified. At the most one may hope to observe effects of minerals on properties of organic-molecular populations such as size distribution or intensity of certain constituents. Even if such a process is discovered, its relevance to transfer of biological information between minerals and organic molecules would be uncertain.

Thus, it is necessary to differentiate between the concept of "clay life," on the one hand, and the claim that it could have given rise to biological life, on the other. The first concept is intellectually stimulating, may be partially tested, and is arbitrary with regard to the definition of life. The second seems to violate the biological continuity principle.

Therefore, the belief in "clay life" does not imply a "genetic takeover." The interactions between the prebiotic organic molecules and the clays are chemical and physicochemical by nature, regardless of whether the clays are considered "alive" or inorganic entities. Moreover, by parsimonial considerations it seems unnecessary to invoke "mineral life" in dealing with the hypothetical prebiotic evolution of organic molecules; the role of clay surfaces with their catalytic properties may preferentially be described in terms of the immediate environment of the organic molecules under study.

It should be noted that "mineral life" may have another connotation: life based on chemistry or organizational principles unknown to us. This consideration is beyond the scope of the present discussion but will be mentioned again in chapter 23.

Summary

The involvement of minerals in life processes has been one of the oldest beliefs in human civilizations. Mineral involvement in central processes of the origin of life was also among the most popular scientific ideas in the search for the origin of life, varying from mere adsorption to "clay life." The list of candidates for such a role includes a large variety of minerals varying in their composition, solubility, specific surface area, adsorption characteristic, electrical charge, isomorphic substitutions and defects, and catalytic properties. The most popular minerals in the context of the study of the origin of life have been a group of aluminosilicates called clay minerals. The hypotheses suggested for the involvement of minerals in the origin-of-life reactions, processes, and scenarios include adsorption, compartmentation, symmetry breaking, information transfer, and condensation and adsorbed-template reactions.

According to the "clay life" theory, particles of montmorillonite clay mineral could have served as the first "living" entities on Earth, taken over by organic life in a process called "genetic takeover." Although the arbitrary inclusion of these particles in the realm of "life" is possible, so far, this speculation has not been supported by experimental evidence.

21

Bottom-Up Biogeochemical Reconstruction

Mineral Involvement in Energy Production and Transfer

Ladies and gentlemen, throughout my lecture I have presented to you nothing but speculations.

G. Wächtershäuser, "Order out of order"

The role of minerals in the emergence of de Duve's "thioester world"

Thioesters are energy-rich metabolites. Hydrolysis of thioesters yields a carboxylate anion and a thiol (sulfhydryl)—SH. Equation 21.1 shows the hydrolysis reaction of acetyl coenzyme A to coenzyme A (CoA) and acetate, where the standard free energy of the reaction, $\Delta G^{0'}_{hydrolysis}$, is -7.5 kcal mol^{-1}:

$$\text{acetyl-CoA} + H_2O \rightleftharpoons \text{acetate} + \text{CoA} + H^+ \quad (21.1)$$

It is noted that the standard free energy of the hydrolysis of acetyl-CoA molecule is about that of ATP. ATP and Coenzyme A are considered to be among the primary carriers of activated groups in metabolic reactions.

The thioester hypothesis (the "thioester world") was suggested recently by de Duve (1991, 1992, 1995a; see chapter 19) as an important stage in the main reconstructed stages of the origin of life; according to this theory, the emergence of life involved four main successive stages, or "worlds": the prebiotic world, the thioester world, the RNA world, and the DNA world.

Essentially, the thioester world is a back-extrapolation from biology, based on the chemical properties and functions of thioesters in extant living organisms. However, more than in the case of the RNA world, it is also a part of the prebiotic environment that supplies the needed building blocks for both the thioesters themselves and the presumed oligomers that they help to synthesize. Moreover, thioesters also served as an integral part of the electron-transfer system, which connects the thioesters to the prebiotic iron cycle (de Duve, 1991).

De Duve's scenario adopts the RNA world and the DNA world that succeeded it but suggests the thioester world as an essential link between the primeval prebiotic world and the RNA world. De Duve's solution to the unlikely emergence of the RNA world by itself is the introduction of a stage of protometabolism that paved the road to the RNA world. Central to this bridge between the prebiotic world and the RNA world is the Haldane-Oparin primeval soup, seemingly supported by the Miller-Urey experiment (chapters 5, 15). The two needed functions in the synthesis of novel compounds in Miller's list are catalysis and energy manipulation. Thus, metabolism should have evolved from the very beginning of the processes that later brought about the emergence of the first living entities.

A major assumption in de Duve's scenario is the presence in the prebiotic environment of protoenzymes, which are primitive catalysts with little efficiency and specificity. Thus, the principle of biological continuity is applicable to the evolution of metabolism: protometabolism must be continuous with extant metabolism. The source of energy for the synthesis of a large variety of organic compounds, including protoenzymes, was the thioester bond. The central role of iron in biological electron transport and the linkage between thioesters and iron in ancient Fe—S proteins prompted de Duve (1991) to suggest that the thioester world may be better named the thioester-iron world.

The role of minerals in the origin of photosynthesis and metabolism according to Granick and Hartman

Solar photons are considered by most researchers to be the only source of energy sufficient for the continuing evolution of life. The photons cause reactive molecules through excited states to undergo redox chemical reactions and form reactive products (e.g., aldehydes), the free energy of which is greater than that of the reactants. By application of the principle of biological continuity, it has been proposed that photosynthesis is a very ancient feature of primordial life. However, the origin of photosynthesis is not a recipe for the origin of life and is still an unresolved problem (Mauzerall, 1992).

In searching for the beginning of photosynthesis, it was Granick (1957) who suggested that the origin of photosynthesis be connected with some common minerals. These minerals contained metal ions and served as both catalysts and coordinating templates. The inorganic catalysts served as centers to which organic molecules gradually associated, thus increasing the complexity of these systems. Biosynthetic chains gradually emerged where metal catalysts were transformed into metalloenzymes, thus increasing their efficiency.

According to Hartman, the origin and evolution of photosynthesis involved self-replicating iron-rich clays. His recent scenario (Hartman, 1992b; see also Mauzerall, 1992) involves four stages representing the evolutionary phases of the relevant chromophores and metabolic functions, including the incorporation of nitrogen and phosphorus. Furthermore, his evolutionary scheme also represents the transition of the wavelength of the light that drives the photosynthetic reactions according to the chromophore in each phase. The metabolic compounds and functions include ferredoxin, a small protein that contains an iron-sulfur group, which fulfills many metabolic functions. It is a very ancient protein and, among other functions, serves as the first soluble electron acceptor of photosynthetic electron transport. The four phases of Hartman's scenario are as follows:

1. Reduction of carbon dioxide Carbon dioxide is reduced to oxalate and formate by ferrous ions driven by UV light, where the chromophore and electron donor is ferrous ion.
2. Entry of sulfur The entry of sulfur results in the formation of thioesters, as well as disulfides and Fe_2S_2 and Fe_4S_4 cores, which are ferredoxin analogs, and the first organic chromophores, quinones, which are driven by blue light. Quinones are lipid-soluble substances involved in electron transport. They serve as hydrogen-atom acceptors and electron donors. The main photochemical reactions are now the formation of thioesters and the reduction of the ferredoxin cores. Thus, rather than the ferrous ion acting as a sole chromophore, it is now incorporated into the metalo-organic chromophore; the iron found in the ferredoxin analogs is now involved in electron transport, where the central electron donors are ferrous and sulfide ions (Hartman, 1992b). More about the role of ferredoxin and its preparation can be found in Buchanan, 1992, and Bonomi et al., 1985, respectively, and Österberg, 1974, 1977.
3. Introduction of nitrogen The introduction of nitrogen involves the entry of a molybdenum ferredoxin analog that can reduce nitrogen to ammonia. The molybdenum ferredoxin analog is driven by photochemically reduced ferredoxin. The introduction of nitrogen into the evolving system led to the formation of predecessors of extant chromophores, as well as the chromophores themselves, including chlorophyll. Biosynthesis of amino acids, based on the citric acid cycle, also became possible (see also Morowitz, 1992, pp. 167–170); so did the formation of peptide membranes (see Brack and Orgel, 1975; Zhang et al., 1993).
4. Introduction of phosphate The entry of phosphate heralds the synthesis of phospholipids and nucleotides.

Hartman labels these four phases the iron world, the sulfur world, the nitrogen world, and the phosphate world, respectively. The first photosynthetic process began in absorption in the UV range; it evolved through blue and yellow to absorption in the red wavelengths, which characterizes chlorophyll. The concomitant chromophores evolved from ferrous ion through quinones and other compounds to chlorophyll. The electron transport chain evolved from Fe^{++} through the Fe_2S_2 and Fe_4S_4 clusters to the hemes (Hartman, 1992b). This is an interesting speculation that in fact encompasses the beginning of photosynthesis and metabolism. This scenario demonstrates the transition from the inorganic to the organic, from energetic to less energetic photons, and from the simple to the complex.

Hartman's scenario poses some new questions; among these, the introduction of template-directed reactions into the evolutionary scheme is instructive. Thus, should the peptides functioning as the primordial ferredoxin be template-directed? If the introduction of a ferredoxin analog of phase 2 of the scenario should indeed be based on template-directed peptides, then the question is, "What was the primordial template in this scenario?"

Another problem that should be addressed in the development of Hartman's scenario is the compartmentation mechanism of his primordial evolving entities, before the advent of lipid-peptide membranes. As to the sites in which the origin of photosynthesis could have occurred, the answer would be that these sites must have been exposed to solar radiation. This concept will be further discussed in chapter 22.

Is photosynthesis an accidental by-product of near-IR phototaxis?

In view of the difficulties in understanding the intermediate stages in the evolution of photosynthesis, Nisbet et al. (1995) suggested that the beginning of photosynthesis is related to the infrared (IR) radiation of hydrothermal vents. In earlier publications these authors supported the suggestion that the origin of life was associated with these deep vents. Thus, the present biogeochemical hypothesis does not involve the direct effect of minerals; it is discussed here in the context of the beginning of photosynthesis.

According to this scenario for the early organisms that developed photosynthesis, this was an accidental by-product of phototaxis based on near-IR radiation. Photosynthesis arose from bacteriochlorophyll-like molecules used by these bacteria for phototaxis. In contrast with Hartman's scenario, in which the first reaction is driven by UV light absorbed by ferrous ion, Nisbet and his associates start their scenario with a chromophore at the IR range. In the debate that followed this suggestion, the availability of oxygen (Allen, 1995) and the relevant electron transport reactions (Björn, 1995) were also implicated. It is interesting to note that according to Nisbet and his associates the evolution of photosynthesis is a relatively late event, which was preceded by the appearance of those early organisms that developed phototaxis. According to Hartman (1992b) and Morowitz (1992), the evolution of photosynthesis shows a transition from the inorganic to the organic and is thus primordial.

The role of iron-sulfur minerals in the evolution of universal ancestral metabolic complex according to Edwards

A combination of Wong's coevolution theory, a kind of pyrite world (Wächtershäuser, 1988), and metabolite channeling was proposed by Edwards (1996) as the central attributes of an autotrophic origin of life, including the emergence of the genetic code. In Edward's scenario the metabolite channeling concept is extended to the origin of life, where "the ancestral structure giving rise to the protocell was mineral-based complex." Pyrite mineral served as the surface area, equipped with catalytic sites based on metal-sulfur clusters. Biomolecules were synthesized on specific sites of the pyrite surfaces either photo- or chemoautotrophically. The metabolic complexes thus formed could have evolved without the recognition mechanisms characterizing extant enzyme-substrate interactions, similar to the first metabolic cycles suggested by Wächtershäuser in his pyrite world.

The evolution of the genetic code and its machinery are related to metabolite channeling. Biosynthetically related molecules could arise in close proximity, resulting in the establishment of the codon structure of the genetic code in a way similar to Wong's (1975, 1976, 1980) coevolution scenario. Unlike the pyrite world suggeted by Wächtershäuser, which takes place in the depth of the ocean and is a chemoautotrophic process, Edwards (1996) suggests a photoautotrophic scenario that must be exposed to solar energy and is thus located at the upper layer of a hydrous system, such as calm coastal waters. In his scenario the pyrite functions also as a semiconductor. The high quantum yield of pyrites is attributed to the presence of surface iron-sulfur clusters that may be considered, as in Granick's (1957) theory, analogous to the iron-sulfur cores of ferredoxins. In a cycling process with a redox system based on iron-sulfur clusters, exergonic reactions analogous to those suggested earlier by Wächtershäuser would also bring about the reduction of cluster ligands such as CO_2, CO, and NO_2^- to organic species. The next cycle would start when the clusters in the oxidized state were again reduced photochemically. Iron may be replaced by nickel (as in Wächtershäuser, 1996) or by molybdenum.

The chemical entity that gave rise to a protocell was thus a mineral-based complex, termed the "universal ancestral metabolic complex" (UAMC). It has also been suggested that this system was involved in protein synthesis, in the context of the RNA-world theory, and that it explains the specific loading of tRNA with amino acids during the evolution of the genetic code, by invoking the importance of the relative positions of tRNA and amino acids in the UAMC complex. No detailed mechanisms were given for the emergence of this process.

The iron-sulfur world according to Wächtershäuser

The most comprehensive scenario for the origin of life was proposed by Wächtershäuser (1988, 1990, 1992a, 1992b, 1993, 1994a, 1994b, 1996, 1997). It has also been one of the most controversial propositions in the origin-of-life field. Moreover, although controversial, it has been recognized recently by many origin-of-life researchers as a viable hypothesis with an outstanding explanatory power for some of the most important molecular-evolution reactions, chemical entities, and processes that comprise the presumed emergence of life, eons ago. In his speculative proposition, Wächtershäuser has made a most imaginative utilization of the principle of continuity and its implied principle of simplification, thus endowing biochemistry with a novel, comprehensive historic-phylogenetic dimension. Thus, whether true or not, it is a most interesting theory, encompassing many central aspects of life processes. The detailed discussion of this theory is didactic.

Wächtershäuser uses the term "retrodiction" in his back-extrapolation from extant biology toward the ancient processes that gave rise to the emergence of life. The scenario is based on a number of assumptions and on many scientific speculations. Various central aspects of this theory were recently reviewed by Lahav (1994) Maden (1995), and Wächtershäuser (1997).

In all his contributions, Wächtershäuser stresses the scientific methodology of the philosopher Karl Popper. This reference to Popper's methodology should not be misunderstood as an attempt to get "support of validity" from a higher authority.

According to Wächtershäuser's scenario, the molecular evolution process starts with fixation of CO_2 on the surfaces of pyrite (FeS_2) crystals, where the energy source for this carbon-reduction reaction is the formation reaction of the pyrite. The carbon dioxide fixation is an autocatalytic process, and it is the first step in the establishment of a network of reaction cycles of organic molecules on the surfaces of the pyrite crystals; this is the beginning of metabolism. The primordial reaction networks are called "surface metabolism," since the molecules under consideration are adsorbed to the pyrite surfaces.

The idea of minerals as a source of energy for organic reactions is not entirely new. Minerals have been suggested as a source of energetic photons emitted during luminescence reactions. Thus, photons have been shown to be absorbed by amino acids adsorbed to the mineral surface and were hypothesized to be potentially useful in certain prebiotic photochemical processes (Lahav et al., 1982, 1985; Coyne, 1985).

However, in Wächtershäuser's scenario the energy source is a mineral-formation reaction. Moreover, the scenario is not based on an isolated set of organic reaction connected vaguely to a largely unknown process; it is the beginning of a detailed scenario designed as a bridge between the known biochemistry, on the one hand, and the presumed primordial reactions on the surface of the pyrite, on the other. The methodology is thus an interpolation between these two points of reference, while exploring the origin of the pathways of metabolism. By using this methodology, Wächtershäuser not only obeys the principle of continuity but also delineates the pathways along which the primordial biochemistry had evolved.

Furthermore, rather than assuming the involvement of minerals in information-transfer processes between organic molecules and clay minerals, as Cairns-Smith (1982) does in his "genetic takeover" theory, Wächtershäuser bases his scenario on energy flow between pyrite and organic molecules. The pyrite is considered the cradle of the general chemical evolution process, so this scenario is also sometimes called the "pyrite world." The central role of pyrite stems from its exergonic formation, its positively charged surfaces, and its presumed ubiquity in certain prebiotic environments.

But Wächtershäuser's theory encompasses much more than that, claiming to be able to delineate the web of reaction pathways through which biochemistry and biology have emerged. These include the origin of the central biomolecules and metabolic cycles, as well as cell formation and division and perhaps biological homochirality. Moreover, the choice of pyrite as the cradle of life implies a geochemical set of boundary conditions and guidelines with which one can identify the sites of these processes, on the primordial earth. The central tenets of the iron-sulfur world (also called by Wächtershäuser, the "iron-nickel-sulfur world") are the subject of the next section.

The iron-sulfur-world's main thesis

At the present stage of the discussion, the main thesis of this theory can be phrased as follows: The first living entities were chemoautotrophic organisms attached to pyrite surfaces, which obtained their energy from the formation reaction of this mineral (Wächtershäuser, 1992a).

This is a biogeochemical point of view; it stresses the source of energy in the transition from geochemistry to biology. Energy is a commodity that comes from outside (i.e., from the environment) in all chemical evolution scenarios; in the present case it comes from the growth of pyrite crystals. The involvement of a mineral in energy-flow reactions does not contradict the biological continuity principle; the mineral surface serves as an immediate environment with which the organic molecules interact, not as an intrinsic part of these molecules.

Pyrite formation

Pyrite is the most stable iron mineral under anaerobic conditions and was presumably present in certain sites of Earth's surface, such as the hydrothermal vents at the bottom of the primordial ocean. Wächtershäuser's scenario starts with the oxidative formation of pyrite under the presumed reducing conditions of the specific prebiotic environment (discussed later). The reaction he had suggested originally, where a variety of chemical entities can serve as electron acceptors, was:

$$FeS + H_2S \rightarrow FeS_2 + 2H^+ + 2e^- \quad (21.2)$$

This suggestion conflicted with the standard doctrine of geochemical pyrite formation, and a collaboration was needed between Wächtershäuser, the theorist, and Stetter and his associates, the experimentalists, in order to resolve the disagreement. In a laboratory experiment, the two researchers reacted iron sulfide with hydrogen sulfide, at 100°C and under strict anaerobic conditions (Drobner et al., 1990). The products of the reaction were pyrite and molecular hydrogen, according to this reaction:

$$FeS + H_2S \rightarrow FeS_2 + H_2 \qquad (21.3)$$

The reaction shows that protons can serve as the electron acceptors. The reaction is exergonic, as seen from the negative standard free energy of the reaction: At 25°C and pH0 (actually, this value is fictitious because FeS cannot exist under these conditions), $\Delta G^0 = -38.4$ kJ/mol. This, then, suggests that it could serve as "a thermodynamic sink for the reducing power of aqueous FeS/H_2S" systems (Blöchl et al., 1992).

The electrons are also available for the reduction of carbon, where protons and carbon dioxide compete as terminal electron acceptors. Wächtershäuser suggests the formation of the reductive carbon dioxide cycle, as the primordial carbon fixation reaction; this is the archaic autocatalytic version of the reductive citric acid cycle (chapter 9). In other words, since the reaction is exergonic, the energy released can be used for the establishment of various organic reaction cycles, where the beginning of life is (according to Wächtershäuser) *the establishment of the first primordial metabolic cycles.* Accordingly, the inorganic nutrients are Fe^{2+}, H_2S, and CO_2. Note that by associating life with these primordial metabolic cycles, Wächtershäuser circumvents the prebiotic heterotrophy/autotrophy relationships discussed in chapter 19.

Importance of the reaction rate

The reaction rate of H^+ reduction is low. Should the reaction be fast, the useful energy should be rapidly depleted (Wächtershäuser, 1992a, p. 92). Instead of such a rapid depletion, this theory postulates the buildup of a high chemical potential that can be tapped by the first hypothetical organisms (p. 92). This first energy source was thus continuously available, like an "endless battery."

Importance of pH and temperature

The reaction becomes less exergonic above pH8; below pH6.5, the reaction becomes more exergonic with increasing temperature. This is interpreted as an indication that the pyrite-pulled metabolism (discussed later) would be thermoacidophilic. The phylogenetic implications of this possibility are discussed later in relation to the hypothesized thermophilicity of the earliest living creatures.

Surface metabolism and surface metabolists

The most fundamental organic reaction on the pyrite surfaces is CO_2 fixation according to the following equation, where $\Delta G^0 = -420$ kJ/mol.

$$4CO_2 + 7FeS + 7H_2S \rightarrow (CH_2COOH)_2 + FeS_2 + 4H_2O \qquad (21.4)$$

Wächtershäuser's scenario is essentially a scenario for the carbon-fixation reaction and the evolution of the resulting primordial metabolic pathways. The idea of interacting "metabolic" cycles was already suggested as a general principle for the origin of life by Yčas (1955). This time, however, it is a much more specific set of metabolic cycles.

The fixation reaction leads to the production of the carboxylate anion ($—COO^-$), which in turn is adsorbed on the positively charged surface of the pyrite crystals in its nascent state. Additional organic anions are also formed, based on the pyrite-pulled reaction cycles that take place on the crystallites surfaces; these include HS^-, $—S^-$, $—COO^-$, $—COS^-$, and $—O—PO_3^{2-}$, which also bind to the pyrite surfaces. The strength of this binding is a function of the charge and size of the organic anions, as well as on other factors, such as temperature. The relative amounts of each species in the adsorbed state depend, among other factors, on the concentrations of the other species, including inorganic species such as HCO_3^-. The adsorbed anions can react with each other, thus establishing reaction cycles of various kinds. Organic molecules that are not adsorbed onto the pyrite surfaces diffuse away and disappear in the environment. The adsorption of the constituents of this metabolism functions as a compartmentation agent in the same way that was conceived by Lahav and White (1980) and H. D. White (1980) (chapters 17, 20).

The physicochemical attributes of the adsorbed ions

The difference between the present system and the suggested involvement of minerals in prebiotic chemistry according to "conventional wisdom" is instructive (Wächtershäuser, 1992a, p. 95): An ideal catalyst is not blocked or poisoned by the reaction products. This means that the free energy of the surface bonding, as well as the activation energy of surface detachment of the products, are low. The desired product-surface relationships for Wächtershäuser's surface metabolism is characterized by a high bonding energy, which for normal catalytic processes is considered undesirable (poisoning). Moreover, the surface bonding and the surface reactions are characterized as a system *in statu nascendi*. This characterization of the reactivity of the pyrite surface is essential for the surface metabolism.

Surface metabolists The "surface metabolist" is the primordial composite structure, attached to the surface, in which surface metabolism takes place. It is defined as a system containing all the surface-bonded organic molecules and their pyrite surfaces (Wächtershäuser, 1992a, p. 96). The environment of these chemical entities is defined as that system encompassing the non-surface-bonded molecules in the liquid phase. These molecules serve as a source of inorganic "nutrients"; the liquid phase serves as a sink for detached organic products of the surface metabolism.

Chemical symbiosis The concept of "chemical symbiosis," developed earlier by King (1982; see Wächtershäuser, 1992a, p. 113), fits into the surface metabolism scenario. It has been restricted by Wächtershäuser to the fusion of entire autocatalytic entities, not for the associations between molecules in biomolecular reactions.

Mechanisms of selection The system under study provides a simple selection mechanism, "selective detachment," for the molecules involved, based on their adsorption. In short, the organic molecules are formed in situ on the pyrite surfaces; the surrounding aqueous solution thus serves not as a source of organic nutrients (as in the case of Bernal's adsorption reaction) but, rather, as a sink for desorbed molecules. Considering the assumed low concentrations of organic molecules in this system, he views this sink as a very clean "sewer" (the purest sewer in the entire history of life; see Wächtershäuser, 1992a, p. 95). If we apply the principle of biochemical continuity, this suggestion corroborates with the observation that all constituents of the central metabolic pathways of extant organisms, including coenzymes, are polyanions.

It is interesting to note that Stilwell (1980), following the prebiotic-soup paradigm, suggested another selection mechanism for "amino acids, sugars, nucleotides, cations and protons from the primordial oceans into a lipid vesicle type of protocell." Lipid membranes were considered also as "chemical shelters" against the destructive effects of hydrated electrons; moreover, organic anions are much less vulnerable than neutral and cationic molecules, with regard to decomposition by hydrated electrons, thus being able to accumulate in the primordial soup (Scott, 1981).

Migration on the mineral surfaces The activation energy for surface migration is on the order of 10–20% of the activation energy for surface detachment (Wächtershäuser, 1992a, p. 96); thus, there is a temperature range in which adsorbed species can diffuse and migrate along the surface, while remaining attached to it. Wächtershäuser even suggests that bonded polyanionic polymers are able to migrate from one "inhabited" crystal onto a neighboring, "uninhabited" crystal in direct contact with it. In this way, even nonpyrite crystal surfaces may become occupied by the metabolists. Thus, surface metabolism is characterized by the lack of *lateral isolation.*

The surface reaction system suggested by Wächtershäuser has not been studied experimentally. Its principles are derived, however, from surface chemistry and solution chemistry.

Thermodynamic feasibility The thermodynamic suitability of the primordial energy source of pyrite formation for organic reduction reactions (Wächtershäuser, 1992a, p. 91) shows that an archaic metabolism can be established on the pyrite surfaces, with a linear and direct electron flow from H_2S to CO_2, without the extant energy-coupling reaction. Moreover, the source of energy under study is powerful enough for the formation of the basic metabolic cycles.

Trophic method The origin of life, according to Wächtershäuser, is chemoautotrophic, without need of photochemistry. The attribution of chemoautotrophy to this scenario is consistent with the attribution of "life" to the first metabolic cycles on the pyrite surface. If, however, the organic molecules performing the first metabolic cycles are not considered living entities, then the trophic method is a combination of autotrophy and heterotrophy, as discussed in chapter 19.

Continuity of evolution of bioenergetics According to Wächtershäuser, all the prebiotic-soup hypotheses imply discontinuity in the evolution of energy flow. No such discontinuity is implied in Wächtershäuser's hypothesis, where the evolving surface metabolism is based on the same kind of chemical free energy from the beginning.

Geochemical site and environmental continuity The pyrite-pulled chemo-auto origin evolved in anaerobic hydrothermal systems or volcanic sites, which are characterized by elevated temperatures and H_2S, CO_2, N_2, NH_3, and Fe^{++} (or FeS). Recall that pyrite, which is ubiquitous in these sites, is the most stable iron mineral under anaerobic conditions.

Following several geochemists, Wächtershäuser assumes that these hydrothermal systems, with their anaerobic aqueous microenvironments, existed in the prebiotic era, as they exist today. Furthermore, during that time, environmental continuity (or, in Wächtershäuser's vocabulary, "uniformitarianism"), the continued existence of these hydrothermal systems, was maintained during the time period needed for the evolution of these biochemical systems. Moreover, it is not necessary to assume the global existence of this system: The auto origin is limited to microenvironments rather than being spread over large zones (Wächtershäuser, 1992a, p. 93). Due to the vast volume of the primordial sea, compared with the small domains of hydrothermal systems, the required local conditions of neutrality (or acidity), high temperatures, and anoxicity could have been maintained simultaneously with global conditions of an ocean with alkaline pH, lower temperatures, and not extremely low concentrations of atmospheric oxygen. Similar considerations may be applied in the case of thaw-freeze cycles suggested by Bada et al. (1994), where sites of high temperature may coexist with a cold upper layer of the ocean (chapter 15).

Environmental continuity considerations should also deal with the transition from the environment of the surface metabolists to other zones on Earth. It is conceivable that mechanisms such as upwelling of the water of the primordial ocean would help in transmitting organic molecules, including surface metabolists, to other zones, thus enabling these entities to embark on their evolutionary pathways. The chemical problems involved in such transitions have not been worked out.

It is interesting to note Wächtershäuser's use of the term "uniformitarianism" regarding the continuous existence of hydrothermal vents, with their dissolve species. This was the term used by Lyell's school, including Darwin, in the description of the geological past of planet Earth (chapter 4).

"Iron-sulfur world" versus the "prebiotic soup"

Assuming the pyrite formation as a source of energy for prebiotic reactions, Wächtershäuser is placing this reaction at the very beginning of the origin of life. His basic assumption is thus the emergence of surface metabolists characterized by chemoautotrophy (Wächtershäuser, 1992a, p. 88). This is a most fundamental strategical choice in the reconstruction of life from scratch, which contrasts with the prebiotic-soup (Oparin-Haldane-Miller-Urey) theory. The latter assumes that the first chemical entities in the processes that initiated the beginning of life were heterotrophic, feeding on preformed organic molecules. Wächtershäuser proposes that the beginning of life is associated with the emergence of autotrophic entities in the very site where the pyrite-pulled reactions take place. The two propositions are, according to Wächtershäuser, mutually exclusive. As noted earlier in this chapter and in chapter 19, there may be a combination of autotrophy and heterotrophy in certain cases.

The importance of Wächtershäuser's proposition is not due just to his slogan "Autotrophs are producers. . . . Heterotrophs are consumers," which implies that the origin of life *must* have been autotrophic. One can imagine a pyrite-pulled carbon fixation reaction where some of the products were dispersed in their microenvironment, thus forming their own "prebiotic soup." This is not the case with Wächtershäuser's theory. The importance of his theory rests on the central attribute of an autotrophic scenario formed at the very site of the pyrite-pulled reduction reaction, where all chemical reactions take place in the same small site (Wächtershäuser, 1994b).

Requirements for the establishment of surface metabolism

It is now necessary to describe in more detail the central tenets of the surface metabolism theory. This description should start with the energy flow needed from the immediate environment of the first primordial carbon-fixation reactions and chemical entities that eventually gave rise to living creatures. Wächtershäuser (1992a, pp. 89–90) lists the conditions an adequate energy source should satisfy, as follows:

1. It must be a source of reducing power.
2. The reducing potential must be sufficient for all reductions in the metabolism.
3. The electron flow must proceed directly and linearly from the reducing agent to carbon dioxide.
4. The energy flow must be somewhat inhibited in the absence of a metabolism. Otherwise, a high chemical potential would not build up for being tapped by a metabolism.
5. The energy source must be operative within the organism, because the reducing agent is required at several steps along the metabolic pathways. Therefore, it must be mild and selective. This excludes, for example, ultraviolet light.
6. The energy source must be geochemically plausible.

Some biochemical continuity postulates

Since extant organisms are sources of redox energy, the energy flow in the present scenario satisfies the biological continuity principle. This is exemplified by extant biomolecules that are considered evolutionary successors of sulfur-iron predecessors. According to this analysis, cysteine and CoA are evolutionary successors of H_2S, extant S-dependent energy sources are evolutionary successors of H_2S + FeS, and Fe—S clusters are evolutionary successors of pyrite (Wächtershäuser, 1992b).

The reductive citric acid cycle: The first autocatalytic carbon-fixation cycle

In extant metabolism there are two carbon-fixation cycles, where the CO_2 acceptor molecule increases by two carbon atoms each cycle, thus functioning as an autocatalytic cycle. These are the reductive pentose-phosphate cycle (RPC; the Calvin cycle) and the reductive citrate cycle (RCC). As early as the 1970s, it was suggested that the latter cycle preceded the RPC.

It was suggested that the RCC was primordial in the evolution of metabolism and as a precursor of the extant oxidative citric acid cycle (Wächtershäuser, 1990). Earlier, this cycle was also suggested by Cairns-Smith (1982, p. 358) in his back-extrapolation toward the primordial metabolic cycles. Adopting this theory, Wächtershäuser was able to reconstruct an archaic autocatalytic cycle of carbon fixation (Kalapos, 1997). Moreover, the primordial version of this enzyme-free cycle, where the sole reducing agent was a FeS/H_2S system, coincides with life. Obviously, this archaic cycle underwent some changes in the course of evolution.

Autocatalysis and reconstruction of the first cycle The RCC is a most central cycle, providing the starting point for most biosynthetic pathways—lipids, sugars, amino acids, the bases, and all the coenzymes. Moreover, it is an autocatalytic carbon-fixation process, in which the number of CO_2 acceptors doubles with every turn of the cycle (Wächtershäuser, 1994a). The overall 16-step RCC turn can be summarized as follows:

$$\text{oxaloacetate} + 4\ CO_2 \rightarrow 2\ \text{oxaloacetate} \qquad (21.5)$$

The first archaic cycle that was established on the pyrite surface was reconstructed from the extant RCC by replacing all reducing agents and thioester-activation and carbonyl groups by their corresponding primordial inorganic predecessors (Wächtershäuser, 1994b).

Autocatalytic cycles, feedback loops and evolution Rather than replication errors as the only source of inheritable variations (the established view among researchers), Wächtershäuser introduced a more general principle of inheritable variation. The mechanism of variation in the autocatalytic reactions under consideration is side reactions, which are common in organic chemistry. In the surface-bonded reaction cycles, side reactions are expected to be detrimental in two ways: by decreasing the efficiency of the autocatalytic production cycles and by poisoning the pyrite surface.

A simple catalyst with indirect positive feedback If, however, an occasional product of one such side reaction exhibits a catalytic effect on a slow step that is a rate-limiting step in the autocatalytic cycle, the detrimental effect will end up being helpful for this catalytic cycle. The result of the activity of this simple catalyst would be a positive feedback, rather than the negative feedback that characterizes other side reactions, the products of which have no such catalytic effects. As a result, a positive feedback loop is "grafted" to the primordial cycle thus improving its survival in new areas (Wächtershäuser, 1994b).

A catalyst with indirect and direct positive feedbacks In its new state, the system is very sensitive to environmental changes that affect the production of the side-products. Such a precarious state can be avoided if a side product can catalyze a whole class of reactions, including a slow step in the branch pathway to itself. Having been synthesized once de novo, it may form a feedback loop into its own reaction pathway (Wächtershäuser, 1994b). The autocatalytic cycle thus changes the system profoundly, bringing about a transition from a chemical system directed by the actual environment to a chemical system that came about as a result of the history of this system. In other words, the system remembers a past event of the first effective formation of the first catalyst. This memory effect is the physical basis for the origin of life. These catalysts, characterized by an indirect positive feedback and a direct positive feedback, Wächtershäuser calls "vitalyzers."

A catalyst with direct positive feedback A third form of catalyst entities is characterized by lack of positive feedback into their own production cycle (Wächtershäuser, 1994b). These catalysts promote a process of decay, thus establishing a destructive parasitic loop. Based on these features, they are called "virulysts."

Heterotrophs, vitamins, and viruses These opportunistic entities use organic molecules (food) produced by others. Obligate heterotrophs are dependent on external vitalizers, called "vitamins." External virulysts may also be taken in; they are called "viruses." Wächtershäuser thus postulates that both autotrophs, heterotrophs, and viruses can be traced back to the primordial surface-metabolism era. These three catalytic constituents—simple catalysts, vitalysts, and virulysts are products of the production network. Their complexity is minimal, where their structure is characterized by two functional requirements: (1) a polyanionic moiety for anchoring onto the positively charged mineral surfaces, and (2) a catalytic moiety.

Coenzymes evolved from the earliest vitalizers Extant coenzymes are involved in metabolism and catalyze a very large class of reactions. According to Wächtershäuser, coenzymes are vitalysts; their long evolutionary history is reflected by functional improvement as well as evolution of their anchoring function. The evolution of enzymes marked a transition from the essentially ionic anchoring to covalent anchoring of these new entities.

The genetic machinery Nucleic acids are vitalizers, following the twofold evolutionary pathway of coenzymes—namely, catalytic functions, on the one hand, and an anchoring role, on the other. In the present scenario, however, they are not primordial.

Predecessors of nucleic acids It is accepted by many students of the origin of life that the extant RNA molecules were preceded by RNA-like molecules that may have been able to function as templates. According to Wächtershäuser (1988, 1994b), the proposed molecule is made not of a phosphodiester backbone but, rather, of a hemiacetal backbone to which triosephosphate groups (the anchoring agents) and bases (serving as catalytic groups) are attached. This hypothetical precursor (de Duve and Miller, 1991) has been termed "tribonucleic acid" (TNA; fig. 21.1).

The advent of nucleic acids The genetic machine, based on its predecessor TNAs, evolved stepwise. For instance, the earliest TNAs are suggested to have been acid-base catalysts not yet involved in base pairing. The latter feature evolved at a later stage. The earliest TNA molecules could not act as templates, but they gave rise to nucleic acids: At a certain stage, purines were formed by modifications of the bases of the TNA, and with them, base pairing developed. It is not necessary to go into the

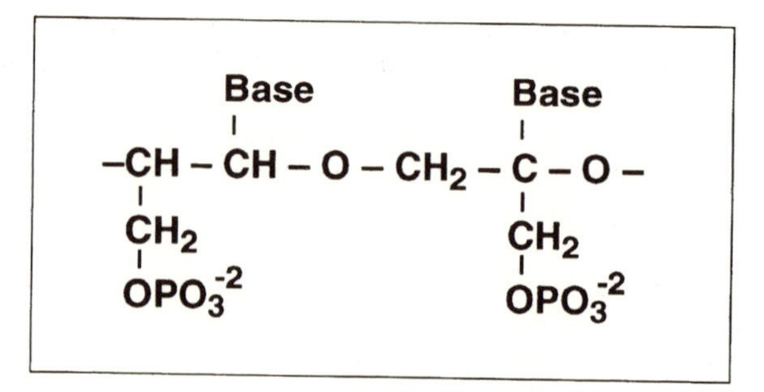

Fig. 21.1. The backbone of Wächtershäuser's tribonucleic acid (TNA) according to de Duve and Miller (1991). See also Wächtershäuser, 1988.

details of the formation of the nucleic acids in order to see the dramatic changes that their formation would bring about: Helix formation by means of base pairing and folding would result in the detachment of the molecule from the adsorbing mineral surface. This is the point of departure of the nucleic acids' pathway from surface metabolism. In Wächtershäuser's scenario, the first nucleic acid folding process is virulytic, which becomes a burden on the metabolism. Furthermore, nucleic acids are the ultimate vitalizers (Wächtershäuser, 1994b).

Template-replication syntheses Admittedly, the evolution of the translation, transcription, and replication machineries based on temple-directed reactions poses a problem that has not yet been solved. It is noted that molecules that were formed by the reductive citrate cycle did not have to wait for the evolution of template-directed synthesis. This latter feature came after the establishment of the central primordial metabolic cycles.

Information Wächtershäuser objects to the use of the terms "information" and "memory" in chemistry. According to his chemical viewpoint, the process of reproduction is an autocatalytic production cycle of synthetic chain reaction. Thus, "the mechanism of evolution may be given a very simple formulation: the appearance of branch products with a dual catalytic feedback—a feedback into the production cycle and a feedback into their own branch pathways. . . . This is evolution by autocatalytic expansion loops" (Wächtershäuser, 1994a, p. 4285). He also objects to the popular concept of genetic machinery and the idea that the origin of life is directly connected to polymer sequences. Thus, he makes a distinction between analog information, which represents the beginning of life, and digital information, which represents a later evolutionary stage of life.

Additional environmental effects

The main environmental changes are those related to the pyrite surface, the ambient solution, temperature, and pressure.

Biochemical homochirality Pyrite is known to have a cubic crystal structure; thus, it has neither optical anisotropy nor optical activity. However, it was suggested recently (Wächtershäuser, 1992b) that the structure of pyrite crystals depends on the temperature under which they were formed. According to this claim, pyrite crystals grown at high temperatures of magmatic or metamorphic origin are cubic. On the other hand, pyrite crystals grown at lower temperatures of sedimentary or hydrothermal origin are not cubic. The interpretation of the possible structures of the latter pyrite crystals is that they show optical activity and could have been involved in symmetry breaking. If this is true, then the origin of biological homochirality coincides with the origin of life (Wächtershäuser, 1992a, p. 120). The next stage in this scenario would be a gradual transition, by means of adaptation, from the pyrite surface metabolism to the enzymatic metabolism.

Thus, rather than the relatively inert (compared to pyrite) homochiral quartz crystals studied first by Schwab and later by Bonner and his associates (chapter 20), the pyrite crystals suggested by Wächtershäuser are synthetically, energetically, and electrostatically involved in the functioning of the organic molecules. So far, Wächtershäuser's proposition of symmetry breaking has had no experimental support. Moreover, in view of the possible effect of nonchiral mineral surfaces on the adsorption of chiral organic molecules (Weissbuch et al., 1994; chapter 20), the needed symmetry breaking may also be obtained on regular pyrite crystals.

Thermophilicity of surface metabolism

Kinetic considerations led Wächtershäuser to conclude that surface reactions require elevated temperatures; in other words, surface metabolists are thermophiles. Moreover, thermodynamic calculations indicate that under the presumed condition of the system under study, temperature increase makes the reaction more exergonic. This is a very interesting observation, since, based on biochemical and molecular-biological considerations, it has been suggested that the most ancient creatures known to us are certain bacteria living in hot environments (thermophilic bacteria). Moreover, the last common ancestor has been postulated to have been a thermophilic creature (Pace, 1991; Stetter, 1992, 1994; chapter 10). Thus, back-extrapolation from these ancient thermophilic bacteria toward the hypothesized common ancestor may suggest that the environmental conditions in which the first living entities evolved were characterized by high temperatures. Recall that this issue is still controversial (chapter 10).

Barophilicity of surface metabolists

It was argued by Wächtershäuser (1992a, p. 103) that pressure would enhance the autotrophic metabolism and, thus, the chemo-auto origin of the pyrite world occurred under high pressure. This corroborates with the proposed hydrothermal system origin.

Environmental fluctuations Temporal fluctuations affect the surface metabolists, including acquisition of inheritable novelties. Evolution is a spatiotemporal affair, and it is expected that surface metabolists would evolve into a number of different varieties.

Stevens and McKinley (1995) recently reported evidence for an active, anaerobic subsurface lithoautotrophic microbial ecosystem in deep basaltic aquifers. The energy source of these microorganisms appears to be derived from geochemically produced hydrogen, and they can persist independently of photosynthetic products. It remains to be seen whether these microbial systems have anything to do with Wäshtershäuser's hypothetical creatures.

Cellularization

The idea that "life" can be pronounced within a less-than-cellular organizational level is an arbitrary issue, to a large extent. Thus, surface metabolists are not cellular forms of life. The chemoautotrophic proposition, however, explains the appearance of cells. The process of carbon dioxide fixation brings about both the lowering of the oxidation level of the carbon atom (from the level of CO_2 toward the level of CH_4), and the formation of C—C bonds. Thus, a progressive formation of —(CH_2)— units takes place, and these units are added to the organic entities bonded (by means of anionic groups) to the pyrite surfaces. In other words, the surface-bonded organic molecules tend to grow spontaneously in their surface-bonded state. Gradually, the pyrite surfaces, which are initially hydrophilic, become more hydrophobic (more lipophilic), pushing the water molecules away from the surface; the surrounding of the surface metabolist becomes more and more lipophilic. One of the hypothesized results of this process has to do with the enhancement of the formation of polypeptides and nucleic acids. This, however, is not directly connected to the cellularization process and will not be discussed here. More information can be found in Wächtershäuser, 1992a (p. 104).

Now a complex process starts that culminates in the formation of cells with amphiphilic membrane (Wächtershäuser, 1992a, pp. 104–109). Without going into the details of this process, here are some of the most important steps and stages of this process:

1. Self-lipophilization of the pyrite surface and the establishment of a close environment—"ambience," a reaction milieu near the pyrite surface. The accumulation of lipids protects the constituents of the surface metabolism against detachment.
2. A two-dimensional phase separation, which takes place at a certain point as the surface concentration of polar lipids increases. Compact lipid domains in the form of membranes are established, with a polar boundary of hydrophilic heads facing the aqueous surroundings and a lipophilic interior facing the anionic entities adsorbed on the pyrite surface.
3. The establishment of hydrophilic, metabolically active domains within a continuous membrane domain, on the pyrite surface.
4. The growth of Pyrite in the form of crystals, simultaneously with the accumulation of lipid molecules on the pyrite surface. The decrease in the surface/volume ratio with the increasing crystal size, results in the formation of a closed membrane enveloping the crystal.
5. The gradual separation of the membrane from the pyrite surfaces at certain domains, and the formation of semicellular structures.
6. The loosening of the connection between the envelope and the mineral, and their detachment, where the lipidic envelope surrounds the pyrite crystal.
7. The emergence of cytosol metabolism and heterotrophy. These processes include the emergence of catabolism, activated groups (including ATP), enzymatic energy coupling, and heterotrophy.

The direction of evolution from hyperthermophilicity to mesophily

This postulate is discussed in chapter 16.

Are iron-sulfur clusters the most ancient biological record?

Iron-sulfur proteins are proteins in which the iron is at least partially coordinated by sulfur. In most cases, the iron is bound either to the sulfur of cystein residues of the protein or to inorganic sulfurs in a prostetic group known as iron-sulfur clusters (fig. 21.2). They are formed spontaneously when sufficient amounts of reduced and soluble sulfur and iron are available, and their chemical versatility encompasses involvement in nitrogen fixation, photosynthesis, and electron transport (Rouault and Klausner, 1996).

The Fe—S clusters in many ancient proteins of extant cells and their contemporary functions may be considered a "molecular fossils" of the pyrite world, and evolutionary precursors to the iron-sulfur enzymes. Pyrite formation also preceded the thioester world (de Duve, 1991), in which thioesters preceded phosphate-bond ener-

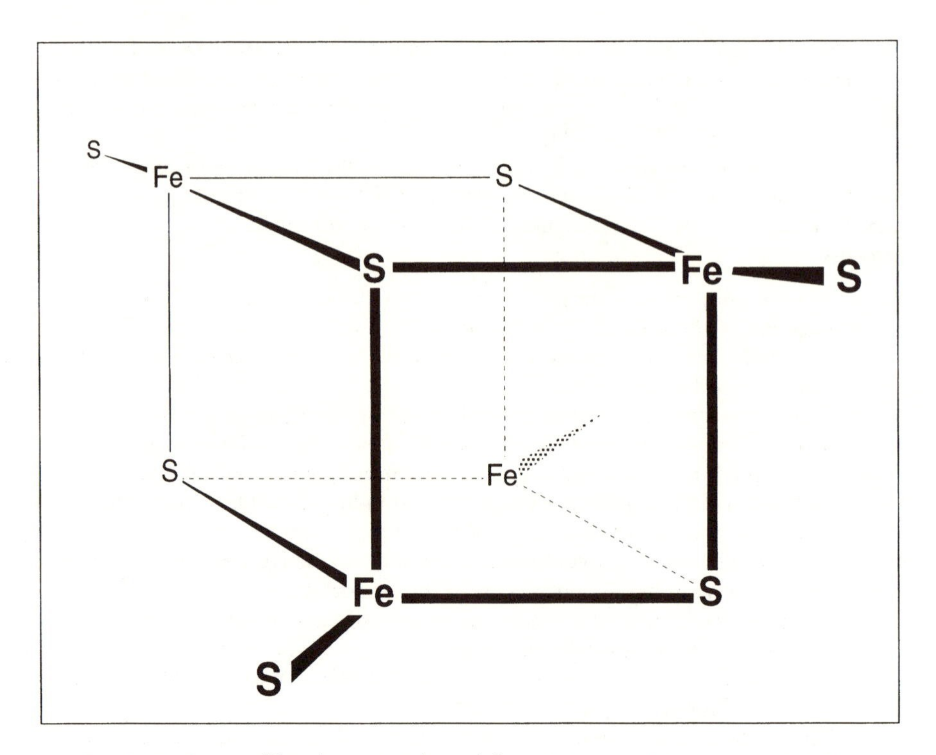

Fig. 21.2. Iron-sulfur cluster. (Adapted from Prince and Grossman, 1993)

getics. Moreover, nitrogen fixation, with its iron-sulfur-vanadium or iron-sulfur-molybdenum centers, was also originally driven by pyrite formation.

The recent identification of Ni/Fe/S in the active site of the enzymes hydrogenases (Happe et al., 1997), which are among the most ancient enzymes, seems also to corroborate with the iron-nickel-sulfur world.

The process of evolution as a process of liberation

It is adequate to end this section by citing the last paragraph of Wächtershäuser, 1996, as follows: "Metaphorically speaking, the process of evolution is a process of liberation—of liberation of the narrower chemical confines of an iron-sulphur world and from a two-dimensional existence on pyrite surfaces. This process of liberation has now been going on for some four billion years. And it is still going on. But only at a price; at the price of unfathomable complications and ever more sophisticated controls."

A few comments

Definition of "minimal life" The metabolists are characterized by metabolism and compartmentation (by means of adsorption) but are not template-directed replicators. The definition of the emergence of life in the present case is arbitrary.

The establishment of the primordial metabolic cycles Whereas pyrite formation can serve as the source of energy for organic reactions, the driving force for the establishment of the primordial cycles should be worked out. The present model contrasts with the "fluctuating environment" models, since it cannot tolerate a dehydration process that stops metabolic activities (Wächtershäuser, personal communication).

The metabolists as producers of organic compounds to be transmitted elsewhere Leaking of organic compounds to the bulk sea occurs all the time. It may be enhanced by temperature rise in an environment characterized by thermal fluctuations. Higher temperatures may bring about the desorption of some of the adsorbed anions, since desorption is enhanced at elevated temperatures. Upwelling of ocean water, a mechanism by which water cycling takes place, may transport those desorbed molecules to the bulk water and to the sea surface (see Polzin et al., 1997). The expected concentrations of these organic compounds in the bulk ocean water are very low and, according to Wächtershäuser, insignificant. However, some organic molecules, especially amphiphilic molecules, that happen to reach the sea surface may undergo concentrating processes (Morowitz, 1992). More generally, dilution based on the total volume of the prebiotic sea may not be achieved. Can such a dilute "prebiotic sewage" serve as a basis for heterotrophic scenarios of the Oparin-Haldane-Miller-Urey school?

Even more intriguing is the ability of the surface metabolists to reach additional prebiotic niches. Is it conceivable that an occasional effective upwelling event would transmit surface metabolists from their origination site to the prebiotic sea surface? As I noted earlier, at a certain stage of their evolution, some of the surface metabolists were probably transferred to other environments, such as that of the sea surface. Does it mean that the two apparently mutually exclusive hypotheses can represent two stages of one evolutionary pathway?

The pyrite-world theory is experimentally testable Experiments designed to test various mechanisms and assumptions of the pyrite-world theory have already begun. Indeed, the details of many propositions have yet to be worked out in order to be experimentally tested. For instance, the cellularization process and the formation of lipophilic layers on the pyrite surfaces may interfere with the surface metabolism reactions. Another example is the symmetry breaking of organic molecules adsorbed on pyrite crystals.

Laboratory experiments on prebiotic synthesis In a recent report (Keefe, Miller, McDonald, and Bada 1995) it was found that "no amino acids, purines, or pyrimidines are produced from carbon dioxide with the ferrous sulfide and hydrogen sulfide system." The conclusion reached by the researchers was that "the proposed autotrophic theory . . . lacks the robustness needed to be a geological process and is, therefore, unlikely to have played a role in the origin of metabolism or the origin of life." This issue has not been settled so far (Hafenbradl et al., 1995), since these authors did not show pyrite formation (which is at the heart of the pyrite-world theory) in their test tubes.

Synthesis of amides by carbon fixation on nickel-iron sulfide Insisting that his theory is experimentally testable, Wächtershäuser has embarked on various laboratory experiments designed to test some of the consequences of his postulates. In one such experiment, Ni—Fe—S clusters were used as a model for a primordial catalyst. It is recalled that Ni is almost always associated with iron in natural environments; these clusters are therefore a reasonable model for a prebiotic catalyst in Wächtershäuser's scenario. Working under conditions simulating volcanic or hydrothermal settings, several prebiotically relevant reactions were studied. For instance, it was observed (Huber and Wächtershäuser, 1997; see Crabtree, 1997) that "a mixture of nickel sulfide and iron sulfide is catalytic for converting methyl mercaptane and carbon monoxide into acetic acid." These experiments "lend support for a hyperthermophilic, chemo-autotrophic origin of life in an iron-nickel-sulfur world." The researchers point out, "It may strike us as ironic that nickel, one of the last biocatalytic metals to be recognized in biology . . . may well turn out to be among the very first in the history of life" (Huber and Wächtershäuser, 1997).

FeS/FeS_2 redox systems Guided by Wächtershäuser's theory, Kaschke et al. (1994) conducted laboratory experiments on the redox system Fes/FeS_2 using cyclohexanone as a model compound for carbonyl groups. Their conclusion was that "the [Fes/FeS_2] redox system is able to reduce the carbonyl group of the model compound cyclohexanone," thus providing evidence in support of the possible importance of the pyrite-world theory.

A panoramic view The theory of chemoautotrophic origin of life is characterized by a panoramic scenario dealing with multireaction processes. In spite of being highly speculative, it encompasses processes that seem both to corroborate with central biochemical metabolic cycles and to be consistent with geochemical and physicochemical considerations.

Wächtershäuser's theory is not directly comparable to other origin-of-life theories. One reason for this is that no other theory suggested so far deals with such a wide scope of the origin of life saga. Thus, scenarios dealing with more specific topics, such as prebiotic synthesis or the emergence of self-replication, cover only a limited number of aspects of the panoramic view dealt with by Wächtershäuser. Another reason is that the chemoautotrophic theory differs dramatically from all other theories suggested so far.

A cosmological outlook From the chemical point of view, Wächtershäuser considers the universe as a monotonous world with the same 92 or so stable elements anywhere. They are formed by the same few nuclear processes, and their proportions would be similar everywhere. And since iron is always associated geochemically with nickel, it is expected that nickel sulfide would also be able to be involved in reactions similar to those of pyrite, namely, serving as a source of energy for the origin-of-life process. More about this cosmic outlook can be found in Wächtershäuser, 1996.

Historical outlook: Why so late? For more than five decades, the heterotrophic paradigm of Oparin-Haldane-Urey-Miller has been the established school of thought of students of the origin of life. A few investigators rejected this paradigm (see, for instance, Woese, 1979, 1980a) but could not come up with an attractive alternative suggestion. Wächtershäuser (1992b) provides two reasons for this: (1) the way the "prebiotic soup" theory circumvented the problem, resorting to the vast ocean, with its many zones apparently available and adequate for chemical evolution; (2) the restrictive feature of the chemoautotrophic theory, according to which all chemical reactions must take place in the same locale. The problem has thus been to suggest the kinds of chemical reactions that can corroborate with these restrictive conditions.

Summary

Mineral involvement in the chemical evolution process via energy production and transfer was described by several hypothetical approaches differing by their scope. The oldest one (Granick, 1957; Hartman, 1992) focuses on the relatively limited aspect of the origin and evolution of photosynthesis. It deals with several aspects of this evolutionary transition, such as the main molecules involved and the energy source of each of them. According to Hartman, the origin of life and photosynthesis took place in hot springs, rich in ferrous ion as well as other transition metals and cations, and gases such as CO_2, N_2, and H_2S, on the surface of the primordial earth. The iron-rich clay minerals produced under such conditions were involved in the evolution of photosynthesis, which includes CO_2 fixation, nitrogen fixation, the entry of sulfur, and the incorporation of phosphate into the evolving system.

The second one, the thioester world (de Duve, 1991), serves as a bridge between the prebiotic world and the RNA-world theory. The third hypothesis (Edwards, 1996) is a combination of Wong's coevolution theory, a kind of pyrite world, and the cellular metabolite channeling adjusted to the problem of the origin of life. Wächtershäuser's controversial scenario covers a large portion of the spectrum of the origin of biochemistry and life. The origin of life was chemoautotrophic, and took place in or near hydrothermal vents at the bottom of the primordial oceans. Its energy source is the oxidative formation of pyrite from hydrogen sulfide and ferrous ions. This energy source is large enough to pull an autocatalytic carbon dioxide–fixation cycle and the first ensuing metabolic cycles. The products of the autocatalytic CO_2 fixation reaction are organic anions, which are adsorbed onto the positively charged pyrite surfaces. The adsorption-induced compartmentation is the most primitive mechanism for retaining the organic anions thus produced in close vicinity. Chemical reactions between the adsorbed organic anions result in the establishment of the first metabolic cycle, the archaic form of the reductive citrate cycle (RCC). This pyrite-pulled autocatalytic cycle can be reconstructed from the extant RCC. Inheritable variations occur by branch products with dual catalytic feedback into both the reproduction cycles and their own branch pathways. Nucleic acids, the genetic apparatus, and template-directed syntheses appeared at a later stage. Cellularization is initiated by the formation of lipophilic zones on the pyrite surfaces, followed by the expansion of the lipophilic layer to produce a cell.

22

Possible Sites for Molecular Evolution Scenarios and Their Rhythms

> The primitive organic synthesis may have taken place (1) in the ocean, (2) in some body of fresh water, or (3) on the land, or, more specifically, (4) in the soil.
>
> T. C. Chamberlin and R. T. Chamberlin,
> "Early terrestrial conditions that may have favored organic synthesis"

With the background given up to this point, it is possible now to better evaluate the many environments that have been suggested as possible sites for the molecular evolution reactions, processes, and scenarios. These environments, together with their concomitant (and sparse) scenarios, will be now reviewed briefly. Some of these have been discussed already and therefore will only be mentioned in order to present a complete picture of the present state of our knowledge. In view of the importance of minerals and their ubiquity in the hypothesized prebiotic environments, it may be convenient to divide the following presentation accordingly. Such a division would be arbitrary and should be supplemented by additional features, such as the implied trophic method, or the rhythm characterizing each scenario.

A large variety of environments were suggested by many researchers, ranging from the sea bottom to water droplets of the ocean waves and from systems fluctuating between wet and dry states on land to water droplets in the atmosphere (fig. 22.1). The involvement of minerals in each of these scenarios is either inherent or implied.

Bulk sea water and its boundaries

The "soup theories" are based on the primordial sea. However, the popularity of the bulk water as a likely site for the origin of life has decreased among researchers. An obvious drawback of this large body of water is the dilute solution it has. Therefore, certain specific sites that are directly connected to the sea—the sea bottom, sea surface, and tide pools bordering the sea—are considered more interesting candidates for the origin of the first living entities than the bulk sea water because of their potential to accumulate organic molecules.

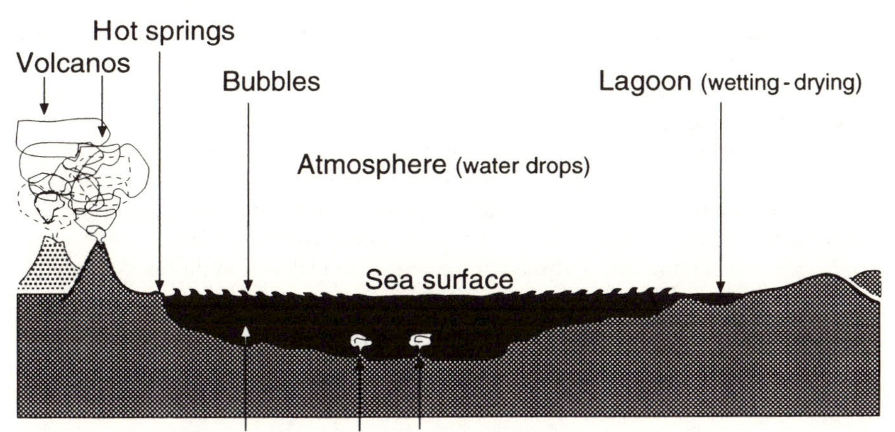

Fig. 22.1. A scheme of the main environments suggested for origin-of-life scenarios.

Sea surface and formation of protocells

According to this scenario (Morowitz, 1992; chapter 19), the surface of the primordial sea served as the site of accumulation of certain organic compounds, and the formation of the first protocells, which became the primordial living entities.

Bubble-aerosol-droplet cycles at the ocean-atmosphere interface

This little-explored interface was first suggested by Lerman (1986, 1993; see Chang, 1993). The main features of Lerman's hypothesis are as follows:

1. Materials from all sources in the atmosphere reach the upper layer of the ocean and mix with the organic compounds synthesized in situ. These include surface active compounds capable of forming bilayer membranes, as well as other organic molecules.
2. Bubble formation takes place continuously in the upper ocean water layer as a result of wind and wave action. Bubbles bursting at the surface tend to eject aerosols and particulates into the overlying atmosphere. Presumably, aerosol droplets being exposed to various energy sources can undergo a dehydration process; rehydration then occurs as a result of water condensation around appropriate nuclei (see also Negrón-Mendoza and Albarrán, 1993). Thus, a hydration-dehydration process similar to the fluctuating systems suggested earlier for certain bodies of water may take place in these microenvironments, where the hypothesized molecular evolution processes are very fast. Moreover, as a result of aerosol return to the upper water layer, chemical enrichment of this layer is expected.
3. Cavitation and sonochemistry, as well as photochemical processes taking place in the system under study, may lead to processes of prebiotic syntheses. In addition, scavenging of sonochemically produced hydroxyl radicals and other oxidants by Fe^{2+} may reduce side reactions and favor organic synthesis.

This system thus consists of "a network of microenvironments which is almost certain to have existed on the early earth. It provided consecutive cycles of selective chemical concentration, catalysis, and organization of increasingly complex organic molecules" (Lerman, 1993).

Minerals are not an intrinsic part of the bubble-aerosol formation. However, they are assumed to have been in the environment, and they may be easily introduced into this model system. They could have served as adsorbents, catalysts, and photochemical screens against destructive UV irradiation. The adsorbed-template model (chapter 20) is of special interest in this environment. Also included in this model are lipid molecules that can form liposomes.

Hot volcanic springs on the surface of the early earth

Abundance of amino acids and formation of proteinoids

As noted in chapters 19, the proteinoids are not considered relevant to the origin of life by most researchers, even though the geochemical setting—volcanic rocks near a body of water—is plausible. The involvement of rocks has to do with their high temperature, not with specific physicochemical reactions such as adsorption or catalysis. The controversy among scientists on this issue is still rather heated, as reflected by the following rebuttal by S. Fox (1995) directed to early and recent criticism:

> Since 1959 . . . the only defensible retracement model for the protocell to be published is the proteinoid microsphere. . . . In summary, retracing by evolutionary experiments to and beyond the first cellular life has generated the view that a Proteinoid World preceded the Protocell World that preceded the RNA World, and the DNA World followed the RNA World.

Abundance of reduced iron and sulfur and clays

According to Hartman (1992b), the origin of life and photosynthesis would take place in hot springs rich in ferrous ion, other transition state metals (e.g., Mo, Cu, and Zn), Mg, Al, silicate ions and gases such as CO_2, N_2, and H_2S. Under appropriate conditions, iron-rich clays would form, and carbon fixation would take place on these clays. The oxalic acid and other organic acids thus formed would catalyze the formation of the clays. As explained in chapter 17, the chain of reactions in this system would establish the first metabolic functions and the origin of life.

Sea bottom

This system does not seem to be an efficient arena for molecular evolution if it is in a state near chemical equilibrium. Another problem that should be addressed by any specific scenario is that the sea bottom is bound to be buried by fresh sediments that render it unproductive to evolutionary processes (Lahav and Chang, 1976).

Hydrothermal systems at the sea bottom

Several researchers suggested by the late 1970s that if hydrothermal systems were present already in the molecular evolution era, then they could have served as adequate sites for the origin of life. In the absence of any scenario at that time, however, these tentative suggestions were based mainly on circumstantial evidence. These arguments, plus new ones by more recent works, provide some additional historical and biogeochemical background to Wächtershäuser's pyrite-world scenario, and will now be reviewed briefly.

Deep-sea hydrothermal vents have been considered by several researchers as adequate sites for organic synthesis, and also for the origin of life, because of the following reasons (see Corliss, 1990; Corliss et al., 1981; Holm, 1992; Holm and Hennet, 1992; Cairns-Smith et al., 1992; Lowe, 1994; MacLeod et al., 1994; Russell et al., 1994; Shock, 1992; Stetter, 1994; Henley, 1996):

1. The most primitive organisms found in modern environments are thermophiles; therefore, the last common ancestor is also hypothesized to have been a thermophilic organism (chapters 16, 19).
2. Deep-sea hydrothermal vents are characterized by reducing conditions and elevated temperatures, where various minerals are formed. It is believed that these characteristics have not been changed since the prebiotic era. It was hypothesized by Shock (1990, 1992) that organic synthesis from CO_2 and N_2 at high temperatures with mineral assemblages of these systems can take place on the ocean floor.
3. Deep-sea hydrothermal systems were the least affected by impact meteorites, asteroids, and comets; therefore, if initiations of life did start there, they would be likely to survive impacts strong enough to destroy any other molecular evolution system in different locations.
4. Hydrothermal environments are dynamic systems far from equilibrium. This is, it is recalled, one of the prerequisites for life's emergence.
5. Hydrothermal systems are considered sites of primary formation of various chemicals, mainly hydrogen cyanide, which can serve as a precursor for abiogenic amino acids and nucleotides. Energy sources, metal ions, catalysts, and mechanisms for organic syntheses have been proposed by many authors (see Chang, 1993; Wächtershäuser, 1988; Shock et al., 1995). The hypothetical chemical entities that presumably evolved in this environment have been chemolithoautotrophic, that is, entities able to "live" by the utilization of available inorganic (rock) chemical energy.

Minerals considered important in hydrothermal systems are sulfides and clay minerals. Porous systems formed by minerals in hydrothermal vents are of special interest, since they seem to be characterized by favorable conditions regarding molecular evolution processes; these include a complex porous system, as well as varying pH, redox, and chemical composition (Cairns-Smith et al., 1992).

Hydration-dehydration fluctuating environments

These include lagoons, tidal pools, lakes, and puddles (Kuhn, 1976; Lahav and Chang, 1976; Miller and Orgel, 1974; see Lahav, 1994, for a review). The minerals in these systems are either sparingly soluble or soluble (Lahav and Chang, 1982). A biogeochemical scenario based on a fluctuating environment will be presented in chapter 23. A possible role of fluctuating environments in the origin of biological chirality was suggested by Popa (1997).

Atmosphere

Atmospheric water droplets

According to Woese (1980a), who suggested this scenario, life on Earth might have originated during the cooling process of the initially hot surface. In the global atmospheric reflux column thus established, water droplets formed at the cool upper atmosphere would evaporate upon descending to the hotter lower region and be carried upward again. These droplets could have been sites of prebiotic evolution.

Oberbeck et al. (1991) expanded this model system to include also organic compounds supplied by or synthesized during injection of extraterrestrial material. Thus, prebiotic reactants were supplied by interplanetary dust, comets, and meteorites, or synthesized in the atmosphere from simple compounds by various energy sources. Condensation was enhanced by nuclei such as clays and salts. The cycles of condensation-evaporation facilitated the rate of prebiotic reactions similar to the wetting-drying and bubble-aerosol-droplet cycles. Polymerization processes could have also taken place in the atmosphere; life originated either in the droplets (Woese, 1979, 1980a) or in the ocean (Oberbeck et al., 1991). Like in the bubble-aerosol model, minerals are not an inherent feature of the reflux systems under study. But, since they were presumably present in the atmosphere, they could be involved in the hypothetical molecular evolution processes in the water droplets.

Volcanic ash-gas clouds

Volcanic eruptions temporarily enrich the atmosphere with gases and solid particles. Their possible prebiotic roles were reviewed by Basiuk and Navarro-González (1996).

Was there a rhythmic continuity along the history of life?

In a recent review on the rhythm of biological processes, Paolo Sassone-Corsi (1994) noted: "One parameter common to all living organisms is time. Biological processes such as the cell cycle, cellular differentiation, and embryonical development involve the fine temporal tuning of complex molecular mechanisms. . . . Indeed, circadian and seasonal rhythms are central to most biological systems."

Is the rhythm of extant life based on evolutionary processes that still reflect rhythmic features of the primordial environment of life?

The importance of the environment in biology during its entire history, starting from the environments in which life had emerged, is well established. The environmental parameters, with their varying intensities and dynamics, range from temperature and irradiation to the presence of water and solid surfaces, from solar radiation and solar wind to interplanetary magnetic field (Breus et al., 1996), and from amphiphilic vesicles or nitric oxide to organic monomers and polymers. In the following discussion it is assumed that cyclic effects connected to phenomena such as the solar wind, gravitational field, or magnetic field are irrelevant to the present discussion, because of their very low frequency.

Incorporation of and adaptation to environmental information has been stressed recently by Elitzur (1995) as a fundamental attribute of living systems. Is it conceivable to search for a biological record of an elusive attribute such as an environmental rhythm? To the best of my knowledge, rhythmic continuity along the history of biology has never been explored in the context of molecular fossils from the primordial stages of the origin of life.

The first suggestion that a rhythmic continuity along the history of life may have been initiated simultaneously with the origin of life in fluctuating environments was proposed in 1976 (Lahav, unpublished). It seems that in view of the enormous increase of our present knowledge compared with that of 1976, such a proposition should be revisited.

Rhythms of fluctuating environments

Diurnal environmental rhythm is a fundamental feature of all environments exposed to solar radiation. It has been suggested that hydration-dehydration cycles bring about processes of consecutive reactions of adsorption-desorption, condensation-hydrolysis, or dissolution-crystallization of organic compounds and minerals in various natural systems (Kuhn, 1976 [see Kuhn and Waser, 1994a, 1994b]; Lahav and Chang, 1976, 1982; Lahav et al., 1978; Lahav, 1994). Several general scenarios based on such fluctuating environments have been proposed. A detailed scenario of one such scenario is discussed in chapter 23.

It is tempting to suggest that the circadian rhythm of living entities hypothesized to have evolved in fluctuating environments may have originated and then internalized (see Elitzur, 1995) during the very beginning of the molecular evolution era.

Hypothetical living entities without externally imposed rhythm

The most likely candidate for an environment in which diurnal cyclic effects are negligible is deep hydrothermal vents on the ocean floor. The relevant scenario is Wächtershäuser's pyrite world (chapter 21). Apparently, the complex systems of metabolic cycles, according to this scenario, are not driven by the earth's diurnal rhythm. Alternative driving forces are chemical potentials (and their concomitant diffusion processes) and vent effluent. Unless there are unknown rhythms in the activity of the hydrothermal vents, their role in origin of life scenarios is independent of the earth's diurnal rhythm. Thus, if one adopts the pyrite-world scenario for the beginning of life, the circadian rhythm of extant living organisms can be explained by a late adaptation and internalization, upon radiation of the primordial living entities to environmental niches characterized by the diurnal rhythm.

Was it possible to avoid the influence of environmental rhythms?

Any molecular evolution scenario should address either the absence or the existence of biological rhythms; these include the initiation, internalization, and maintenance of such rhythms. Sassone-Corsi (1994) has phrased one aspect of this intriguing problem as follows: "Were most cells sensitive to light/dark cycles some millions of years ago? Does what we study today as the cycle represent a vestigial circadian rhythm?"

From the point of view of the origin of life, any scenario must address the rhythm of the environment, either from the very beginning or from a later stage.

Summary

It was noted by Jim Ferris (cited by Simoneit, 1995) that those who work according to the paradigm symbolized by the spark discharge experiment (Miller-Urey experiment) may be called Arcists, whereas those who explore the possible origin of life in hydrothermal vents may be referred to as Ventists. By the same token, one may speak about Bubblists and Atmospherists. Obviously, a proposed location for the origin of life does not say much about the mechanisms and chemical pathways through which life could have evolved. Thus, the sea surface has been invoked in the origin of life by rather different mechanisms and chemical routes—amphiphilic vesicles (Morowitz, 1992) and bubble bursting (Lerman, 1986). Similarly, hot springs were suggested for proteinoid formation (S. Fox, 1988), on the one hand, and for the beginning of photosynthesis (Hartman, 1992b), on the other.

A survey of the sites proposed for the origin of life may be used as an indication of the ingenuity of the researchers who suggested them. It may also be used as a warning that with the present level of understanding it is difficult—in fact, impossible—to indicate with certainty which site (and its concomitant scenario) should be preferred over the others as the primordial cradle of life. Thus, we are left with our own bias in exploring the site and scenario we like best.

23

Artificial Life and Computer Modeling of Biogeochemical Scenarios

> There cannot be a disembodied theory of life, and the work of simulation can only be a help not a re-constitution.
>
> Francisco Varela, "On defining life"

> Artificial life is the philosophical heir of artificial intelligence, which preceded it by several decades. Whereas artificial-intelligence researchers seek to understand the mind by mimicking it on a computer, proponents of artificial life hope to gain insights into a broad range of biological phenomena.
>
> John Horgan, "From complexity to perplexity"

> Discovering how to make . . . self-reproducing patterns more robust so that they can evolve to increasingly more complex states is probably the central problem in the study of artificial life.
>
> Doyne J. Farmer and Aletta d'A Belin, cited in John R. Koza, "Artificial life"

> All of the above self-reproducing entities are very brittle in that they do not correctly function if there is almost any perturbation. . . . In contrast, living organisms exhibit considerable ability to continue to function under perturbation.
>
> John R. Koza, "Artificial life"

Kinds of artificial life

Though the main context of the present use of the term "artificial life" has to do with "computer life," it is not the only imaginable form of artificial life in the broad sense of the words. In a broad sense, every biochemical pathway or molecule that has not been found in nature and can be synthesized in the laboratory may be considered artificial. By the same token, hypothesized reactions, processes, and scenarios of the origin of life may include unknown parts that are "artificial" in the sense that they are invented in order to bridge over gaps in our understanding. Thus, in the context of the search into the origin of life one can speak about *artificial bio-*

chemistry, which is considered biology-based artificial biochemistry (see also Chyba and McDonald, 1995). Extending these considerations to a more general framework, artificial life may be divided, somewhat arbitrarily, into the following categories:

1. Artificial reconstruction of the beginning of biochemistry Artificial reconstruction of the beginning of biochemistry includes biochemical hypotheses and experimental works designed to simulate partially understood, natural biochemical reactions, processes, and scenarios. The most appropriate example of this group is the origin of life; it includes, in fact, all the scientific work related to the biochemical life we know.
2. Artificially expanded biochemistry These are chemical hypotheses and experimental works designed to synthesize molecules, reactions, processes, and scenarios that are not known in biology. In the words of Eschenmoser (1994, p. 394), who focused on nucleic acids: "The problem of the origin of the nucleic acids' structure can be approached experimentally by systematically studying the chemistry of structural alternatives, molecular structures which—according to chemical reasoning—could have been, but were not, chosen by Nature to become (or to survive as) biomolecules." A notable example of this group is the attempt to expand the genetic lexicon by the incorporation of amino acids that are not normally encoded by natural genes (Benner, 1994).
3. Artificial, nonbiological chemistry of "living" entities Chemical hypotheses and experimental works designed to synthesize chemical entities capable of replication and selection have started recently by Rebek and others (see Orgel, 1995). Speculative "living" entities based on such chemical entities are still imaginary; they include nanobiology (Maron et al., 1993; Nussinov and Maron, 1993) and hypothetical "living" entities based on another chemistry.
4. Artificial reconstruction of alien, nonbiological living entities Chemical hypotheses and experimental works designed to simulate partially understood, alien forms of life. This group is still in the domain of science fiction. It will be revived into life if and when extraterrestrial life is discovered somewhere in our universe.
5. Artificial, abstract, computer-based life (Artificial Life) This category includes the design and synthesis of abstract patterns of life, the constraints and logic of which are based on and inspired by either biological or any other form of life.

Of the above five categories, number 1 is the subject matter of the present book; items 2, 3, and 4 are outside the scope of the present book. Number 5 is of great interest and will be briefly discussed now.

Computer-based artificial life

> One of the most attractive features of the manifesto of Artificial Life research is that it legitimizes the study of any life-like phenomenon that can reproduce in a computer.
>
> Murray Shanahan, "Evolutionary automata"

A new field of exploration has started recently under the title "Artificial life" (for recent reviews see Emmeche, 1991, 1994; Koza, 1994; Knudsen et al., 1991; Langton, 1989; Langton et al., 1992). Artificial life (or AL, or A-life) was defined by Langton (1989, p.1) as follows:

> It complements the traditional biological sciences concerned with the analysis of living organisms by attempting to synthesize life-like behaviors within computers and other artificial media. By extending the empirical foundation upon which biology is based beyond the carbon-chain life that has evolved on Earth, Artificial Life can contribute to theoretical biology by locating *life-as-we-know-it* within the larger picture of *life-as-it-could-be.*

The first step in a discussion of A-life has to do with its characterization. A central assumption of A-life is that living organisms may be viewed as a combination of material form and a logical content, which may be separated (Langton, 1989, p. 11). According to Ray (cited by Emmeche, 1994, p. 4), life is a process characterized by rhythmic pattern of matter and energy, where the important feature is "the pattern, the process, the *form.*"

According to Emmeche (1994), the central attributes of A-life comprise four interrelated ideas concerning what A-life is and three principles concerning the way in which it must be constructed. The first four ideas have been summarized by Emmeche (1994, pp. 17–20) and are given here briefly as follows:

1. A-life is the biology of any possible form of life.
2. A-life attempts to synthesize lifelike processes or behavioral patterns in computers .
3. The artificial aspect of A-life has to do with the components of which it consists—silicon chips, computational rules, formulas, and the like. Whereas the latter are designed by human beings, their behavior is produced by A-life itself.
4. The essence of life is a form of a process, not the matter. Thus, life is independent of the medium.

The three additional attributes are as follows:

5. The synthesis of A-life is carried out by bottom-up construction. According to this principle, the construction of computer-based information starts at the level of a few simple rules, and many small units.
6. Information processing is carried out in parallel, in analogy to information processing in the brain.
7. A-life is characterized by emergence, which is different from the predesigned feature of robot manufacturing.

But even if this biologically inspired A-life research would be helpful in understanding our own biology (Morowitz, 1994), can it be of any help in improving one's understanding of the origin of life?

Our point of departure

Our interest, it is recalled, is the emergence of the animate from the inanimate or, in more pragmatic terms, the emergence of the central attributes of life in the prebiotic world. Has this been the goal of researchers of A-life so far?

Emmeche's fifth item in the list of ideas and principles speaks about the bottom-up-construction, which he also calls "simulated self-organization." His starting point is different from ours, however. His bottom-up approach (Emmeche, 1994, p. 19) is exemplified by DNA-coded proteins, where the starting point is extant biology. Another example is self-replication, which is a starting point for researchers of A-life.

Thus, whereas extant coded protein synthesis and self-replication are the starting points of present A-life research, these features are, in fact, the end of our voyage in the search for the origin of life. Therefore, it is not surprising that in spite of postulates regarding the improvement of one's understanding of molecular evolution and the origin-of-life processes with the help of A-life research, very little has been gained so far in this regard. Our fundamental question has been and still is: How did life emerge from inanimate matter? The involvement of A-life research in the problems of the origin of life is still largely an unexplored area.

In view of the lack of an accepted model for the origin of biological life, any biologically inspired model of the emergence of A-life may be too far-fetched at present. Therefore, rather than dealing with the emergence of life, it is much more practical to focus first on just a few central features of life. Self-replication and coded protein synthesis are two cornerstones among the attributes of a minimal living system (Adami, 1995; Emmeche, 1994). Both these attributes belong to a more general group of template-directed reactions. It is thus appropriate to explore the use of computer modeling in the study of the emergence of template-directed reactions under prebiotic conditions. Such an exploration may bring one to the twilight zone between the origin of life and artificial life.

In what follows I briefly review two biogeochemical models dealing with the emergence of central attributes of life in a plausible geochemical environment. These systems served as the basis for some computerized calculations. In spite of the preliminary stages of these research programs, they may serve to inspire A-life researchers in their effort to be involved in the most fundamental aspect of biology, namely, the search into the origin of life.

More information on A-life research and its relation with biology can be found in Adami, 1995; Rasmussen et al., 1992; Schuster, 1995; and Moya et al., 1995.

"From its origin," Moran and his colleagues point out, "A-Life has been oriented in two main directions: applying the *biologically inspired* solutions to the development of new techniques and methods; and the use of artificial media (particularly computation) to model, replicate, and investigate life processes" (1995).

The autogen according to White

The autogen (White, 1980; White and Raab, 1982) is a theoretical construct suggested by David White based on Lahav and White's (1980) biogeochemical scenario involving a fluctuating environment, minerals, and adsorbed-template mechanism (chapter 20). It consists of two short oligonucleotide sequences coding for two simple catalytic peptides. The autogen is capable of self-reproduction, where both replication and translation are involved. The starting point of the theory is random oligomerization processes in a fluctuating environment, which lead eventually to a rapid autocatalytic growth. The main requirements and assumptions of the autogen theory can be summarized as follows (White, 1980):

1. Availability of free energy and monomers (amino acids and nucleotides).
2. Spontaneous synthesis of oligopeptides and oligonucleotides.

3. Kinetic stability of oligomers.
4. Localization of oligomers by means of a compartmentation of a kind. One possible mechanism is adsorption.
5. Primitive translation by unspecified mechanism.
6. Crude oligonucleotide replication.
7. Two short peptides or families of peptide sequences that catalyze replication and translation without selectivity toward any particular oligonucleotide sequence.

The autogen model is a simplification of the hypercycle (chapter 17), thus rendering that theoretical construct an adequate model system for the very beginning of the molecular evolution process (fig. 23.1). It lacks a mechanism for coded peptide synthesis, but with appropriate (simplified and arbitrary to a large extent) assumptions and kinetic parameters, it shows the central features of a hypercycle. Thus, the emergence of the autogen culminates in the emergence of a feedback system that leads, via several stages, to the autogen, itself an autocatalytic entity. The theoretical development of the emergence of this entity is controversial (Joyce, 1983).

In principle the model under study can be used for the evaluation of the time needed for the emergence of the autogen, under certain simplifying assumptions, including the kinetics of the reactions involved. White and Raab (1982) carried out such estimates and came to the conclusion that this emergence time is very short—

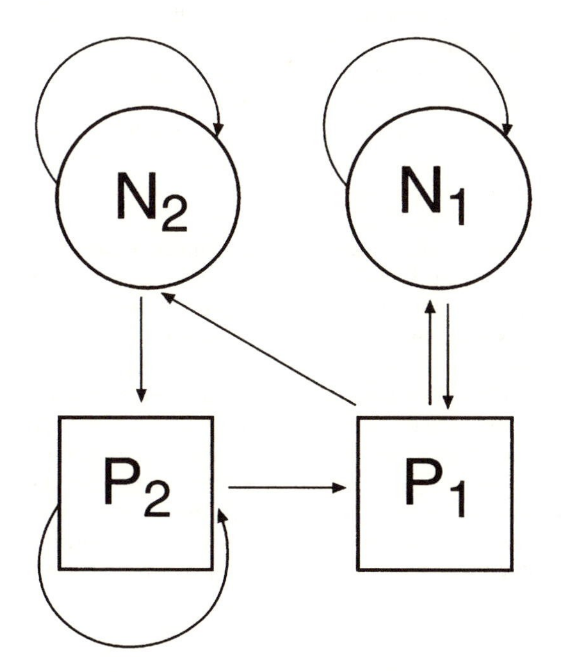

Fig. 23.1. Hypercyclic relationships and functional organization of the autogen. N_1 is the protogene that codes, by an unspecified mechanism, for P_1, the replication protoenzyme. N_2 is the protogene that codes, by an unspecified mechanism, for P_2, the translation protoenzyme. The translation reaction produces both P_1 and P_2. Arrows represent the involvement of the oligomer under consideration in the process under consideration. (Adapted from White and Raab, 1982)

on the order of human-life time. However, in view of theoretical difficulties (not to be discussed here) and lack of experimental data regarding the kinetics of the reactions under study, this model was not accepted by most scientists. Despite this lack of acceptance, though, various features of the autogen model offer an interesting insight into the problem of its emergence time. Thus, because of the autocatalytic nature of the emergence of the autogen, this process is fast: "The autogen would most likely nucleate and grow to dominance either rapidly (10–100 cycles of 1 day each) or not at all" (White and Raab, 1982).

Computerized coevolution model for the emergence of template-and-sequence-directed syntheses according to Nir and Lahav

> Theories are like toothbrushes: Everyone uses only his own.
> Popular saying

> All we want is to be wrong in an interesting way.
> Günter Wächtershäuser

The biogeochemical scenario

It is instructive to describe now in some detail the background and central assumptions of a more recent computer model dealing with an attempt to simulate a prebiotic system in which the emergence of template-directed (TD) and template-and-sequence-directed (TSD) reactions (Lahav, 1991; Lahav and Nir, 1997; Nir and Lahav, 1997) take place. This scenario is similar to that suggested in the autogen model but has a more detailed molecular mechanism for the TSD reactions. Its rather detailed discussion exemplifies the problematics of central attributes of the origin-of-life process.

Consider a prebiotic system with a constant influx of preformed homochiral organic monomers. The latter building blocks are of two kinds, namely, a "family" of organic catalysts (i.e., amino acids) and a "family" of organic templates (i.e., mononucleotides). It is assumed that these two central molecular species can be activated and condensed into oligomers and polymers with the help of catalysts. They can also interact with each other, as well as with other components of their environment, such as minerals, metal ions, organic monomers, amphiphilic molecules, and energy-rich molecules. Assuming an evolutionary process based on these two molecular species and their polymers, as well as a fluctuating environment (chapter 20), the question is, "What is the minimal system with regard to the number of constituents, system complexity and dynamics, and functions, capable of supporting the emergence of evolvable ensembles of organic entities characterized by TSD syntheses?"

Constituents

In addition to the amino acids and mononucleotides, the constituents include a proto-ATP (which may be a pyrophosphate; see chapter 19), catalytic minerals, and various metal cations. For simplicity, two amino acids and two protoribonucleotides are assumed; the oligomers formed from these building blocks are peptides and proto-RNAs, respectively. A small fraction of the latter oligomers can function as proto-tRNAs and proto-mRNAs, as will be seen shortly.

Catalytic peptides

One central assumption is the role of small catalytic peptides in the reactions under study; it is assumed that these reactions can be catalyzed by peptides of various associating moieties and sizes, where the efficiency of catalysts with the same active sites normally increases with their size. Indications regarding catalytic activity of small peptides were published recently (Walse et al., 1996; Atassi and Manshouri, 1993; Matthews et al., 1994).

The scenario thus starts with the spontaneous formation of a population ("library") of peptides, with the help of specific sites on mineral surfaces capable of catalyzing peptide-bond formation (chapter 20). Some of these peptides are able to catalyze the organic reactions under study; in the minimal system under study, four catalytic peptides are needed for the catalysis of the formation of the following chemical bonds: peptide bonds, phosphodiester bonds, and two amino acid–sugar bonds, one for each amino acid–proto-tRNA molecular pair. The model is flexible and can accommodate the organic precursors of the catalytic peptides. For instance, in the loading reaction of an amino acid onto a proto-tRNA it is possible to introduce CoA as a predecessor of proto-tRNA (Di Giulio, 1996a; chapters 15, 17).

It is helpful to differentiate the peptide molecules according to their active sites; peptides with the same catalytic site are able to catalyze the same reaction, albeit with different efficiencies. In order to explain this, consider the peptide $x_1—x_2—x_3—\mathbf{x_4}—\mathbf{x_5}—\mathbf{x_6}—x_7$, which is capable of catalyzing a certain reaction. Here x_1–x_7 are any one of the amino acids under consideration, and bold characters designate the catalytic site.

Catalytic condensation of mononucleotides

Based on the recent findings of Ferris and his associates (chapter 20) it is assumed that non-TD proto-RNA strands can be formed under plausible prebiotic conditions on the surfaces of catalytic minerals. It is further assumed that some of these primordial peptides can catalyze phosphodiester-bond formation between any pair of protonucleotides, resulting in the formation of proto-RNAs. This includes phosphodiester-bond formation between protonucleotides hydrogen-bonded to a proto-RNA template.

Formation and loading of proto-tRNAs

A small fraction of the proto-RNAs thus formed can adopt the secondary structure of a proto-tRNA, which is similar to the microhelices studied by Schimmel and his coworkers (chapter 17). In the presumed fluctuating environment, proto-tRNAs are formed during the cold period, where the low temperatures enhance the formation of the hydrogen bonds needed for the formation of such secondary structure. In the presence of an adequate catalytic peptide and proto-ATP, a proto-tRNA molecule is assumed to be loaded by an amino acid.

Catalytic takeover by peptides

Extending Cairns-Smith's (1982) semantics, the present model system allows for two stages of catalytic takeover of the catalytic functions under study. The primordial mineral catalysts are taken over by random peptides, which are eventually taken over by TSD peptides, according to the following scheme:

$$\text{minerals} \rightarrow \text{random peptides} \rightarrow \text{TSD peptides} \qquad (23.1)$$

Environment

A fluctuating environment undergoing hydrating (cool)-dehydrating (warm) cycles is assumed (chapter 20), where possible candidates range from Lerman's bubble-aerosol-droplets (chapter 22) to diurnal cycles. Temperature is assumed to be below the range of thermal decomposition of the molecules under study. The fluctuating environment is characterized by a sequence of consecutive reactions (table 23.1), which constitutes the driving force of the assumed reaction cycle. Minerals and amphiphilic vesicles are considered as possible compartmentation mechanisms, where the first one is probably the primordial one (chapter 21). The presumed homochirality implies homochiral microenvironments.

Emergence of template-and-sequence-directed syntheses

The emergence of the molecular populations, with their concomitant reactions in the model system under study, is a continuous process. Its presentation becomes easier, however, if we divide it into (arbitrary) stages, each characterized by specific functions as follows:

Stage 1. Emergence of a non-template-directed population of primordial catalytic peptides, formed with the help of catalytic minerals.
Stage 2. Emergence of a non-template-directed, secondary population of catalytic peptides, formed with the help of the appropriate catalytic peptide that takes over the catalytic functions in the next stages.
Stage 3. Emergence of proto-RNA and the first, sporadic TD reactions of com-

Table 23.1 Main reactions in one fluctuating-environment cycle according to the coevolution model.

Cool, hydrated
1. Formation of secondary proto-tRNA configuration with its recognition sites.
2. Formation of complementary strands on adsorbed templates of proto-RNA, i.e., template-directed phosphodiester-bond formation and complementary strand elongation.
3. Peptide-bond formation catalyzed by catalytic peptides (activation requires the presence of proto-ATP).
Warm, dehydrated
1. Phosphodiester-bond formation and non-template-directed elongation of proto-RNA strands, catalyzed by peptides.
2. Dissociation (melting) of double-stranded structures of complementary strands and domains of proto-RNA molecules, including proto-tRNAs, thus leading to the formation of separate strands.
3. Peptide-bond formation catalyzed by minerals.
4. Peptide-bond formation catalyzed by catalytic peptides.

Source: Adapted from Lahav and Nir, 1997.

plementary strand formation, based on the adsorbed-template model and catalyzed by non-template-directed catalytic peptides.

Stage 4. Emergence of proto-tRNA, proto-mRNA, and the first TD reactions of peptides, concomitant with functions such as amino acid loading on proto-tRNAs, formation of proto-mRNAs, and template-directed (but not yet TSD) peptides.

Stage 5. Emergence of cycles of TSD syntheses, autocatalysis, and feedback loops. The correspondence between recognition domains of the proto-mRNAs and the sequence of amino acids in the peptides may be considered the beginning of an ad hoc genetic code of the system under study.

Stage 6. Takeoff. The system under study is now equipped with a "memory," an initiation of a genetic code, and the potential to "discover" new functions in the evolutionary pathway on which it would embark.

The computer simulation model: Nir and Lahav's "toy model"

The design of the computer model follows, with various simplifying assumptions, the biogeochemical model I just described. It demonstrates transition from monomers toward populations of non-template-directed oligomers, from which template-directed oligomers would eventually emerge.

The central algorithm of the model deals with the emergence of a population of proto-RNA strands. The algorithm describing the growth of these oligomers and their population dynamics is based on Nir et al., 1983; see also Nir, Klappe, and Hoekstra, 1986; Nir, Stegmann, and Wilschut, 1986). The proto-RNA strands are assumed to consist of complementary units, C and G (the monomers, or building blocks); for example, C—C—G—C—G. In addition to the activated building blocks C and G, a population of random peptides is also assumed. Some of these peptides can catalyze certain reactions; for instance, peptides characterized by the same active site and designated P_2 can catalyze phosphodiester-bond formation, resulting in a population of oligomers, the proto-RNAs. The mathematical characterization of the strands is given by means of the three-dimensional matrices that define the length and sequence of the proto-RNA strands. In the current program the growth process of the chain involves, at each step, an addition or deletion of one unit, C or G, at the growing end of the chain, as in this equation, where C_{GC} and D_{CG} are the association elongation and dissociation constants, respectively:

$$\mathrm{C{-}G{-}G{-}C{-}G{-}G + C} \underset{D_{CG}}{\overset{C_{GC}}{\rightleftharpoons}} \mathrm{C{-}G{-}G{-}C{-}G{-}G{-}C} \qquad (23.2)$$

At time $t = 0$ the initial concentration of the catalytic peptide (P_2) is zero or extremely small, which may imply a very slow growth process, depending on the temperature. The program calculates the concentrations of the strands as a function of time, which is given in units of day and night (table 23.1). Simplification is necessary regarding the strand length: At the present stage of the program development the maximal strand length is 10 monomers. With two kinds of monomers—C and G—the number of all possible 10-mer strands equals 2^{10} (1,024).

Proto-tRNA strands, which are characterized by specific recognition domains (fig. 23.2), consist only a small fraction of the proto-RNA population; in a population of

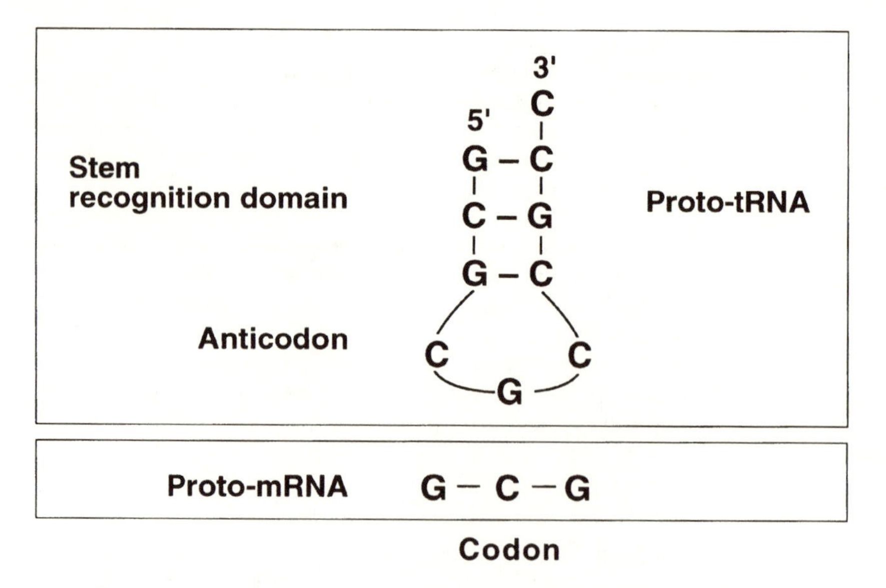

Fig. 23.2. A scheme of a simplified 10-mer proto-tRNA molecule.

1,024 proto-RNAs, only eight strands are proto-tRNAs. In its present stage, the model does not use the antisense reading (chapter 17), but this feature can easily be introduced.

Once a 10-mer proto-RNA strand is formed, it is assumed to be able to serve as a template in complementary-strand formation catalyzed by a peptide catalyst. Thus, the population of 10-mer proto-RNAs starts to increase, according to the sequence of reactions of the fluctuating environment. This population includes proto-tRNAs, which can be loaded by an appropriate amino acid, helped by other peptide catalysts. Strands that can accommodate three proto-tRNAs loaded by amino acids serve as proto-mRNAs, according to Lahav's (1991) coevolution model. The system thus starts to template-direct the synthesis of peptides, some of which are the catalytic peptides that catalyze the reactions under study. The latter then enhance their own formation and that of proto-RNAs. The process is characterized by the establishment of a feedback system and selective enhancement of the TSD syntheses (TSD-syntheses takeover). Thus, the exponential growth of the proto-tRNAs is followed by an exponential growth of the catalytic peptides, as exemplified in fig. 23.3.

Typically, the synthesis of proto-tRNA (and the concomitant catalytic peptide molecules) shows a lag period followed by a very fast increase, which is the autocatalytic reaction.

Involved molecules

The model can accommodate for additional features of prebiotic molecules, such as a peptide covalently bonded to a proto-RNA; PNA templates; ribozymes and proto-tRNAs larger than the ones assumed above; and molecules and ions involved in metabolic cycles.

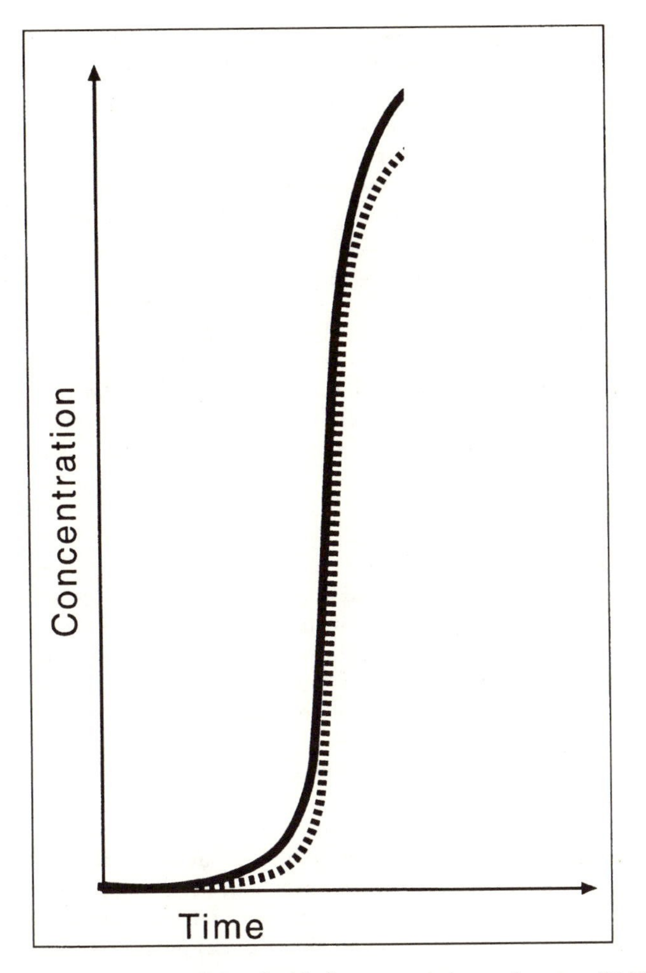

Fig. 23.3 A scheme of an increase of added concentrations of proto-tRNA (line) and a template-directed peptide (dotted line) as a function of time.

Size of molecular populations and vesicles

It is instructive to compare the minimal system capable of initiating and maintaining the TSD processes under study, with other hypothetical entities, namely the protocell, minimal cell, and chemoton (chapter 17). In contrast to the latter constructs, which are characterized by a relatively small number of full-fledged and template-directed primordial systems, the present model treats non-template-directed molecular populations that are enriched gradually with TSD molecules. The enrichment process starts from negligible concentrations of TSD molecules of peptides and proto-RNAs. These consist, at first, of only a minute fraction of the entire populations of their respective non-template-directed molecules; gradually the proportion of these molecules increases.

Compartmentalized entities thus must contain large numbers of the relevant chemical species in order for the small percentage of TSD molecules to be formed and their proportion in the entire molecular population to increase. Therefore, the takeover process, with its concomitant TSD molecules, can emerge only when relatively large populations of molecules of these kinds are present in close proximity, i.e., in the same compartment. In other words, in spite of the involvement of only a tiny fraction of the molecular populations of non-template-directed peptides and proto-RNAs in the first stage of the establishment of TSD systems and takeover modulus, the latter processes can emerge only above a certain population size. This may be considered a "size threshold" of either the relevant population or the amphiphilic vesicle in which they are encapsulated. Though this threshold is not known at present and is model-dependent, it can be determined experimentally, at least in principle. Moreover, in spite of the presumed humble beginning of the TSD system establishment in these compartments, the expected size of such vesicles is likely to be relatively large, perhaps in the range of the theoretical size of a protocell.

Potential developments of the model

The TSD model system can be used in the study of evolutionary processes and mechanisms and can serve both as a tool and as a guide for the reconstruction of the origin of life. Such a reconstruction will probably always remain compatible rather than an exact simulation of the processes of the origin of life. The computer experimenter will thus be able to invent novel kinds of "biology," explore their properties, and devise the methods of their synthesis in the laboratory.

Evolutionary features The development of the present model focuses on the establishment of the central mechanistic aspects of the TSD system; it opens up the road for the introduction of evolutionary features into this system. This development should be connected with compartmentation, without which the present system is rather inefficient and wasteful. At a certain stage of the development of this model, evolutionary selection may be operated by means of random compartmentation. For instance, only compartments that include the essential molecules of the TSD system would be able to develop efficiently, whereas others would die out. One likely candidate mechanism of prebiotic encapsulation is that suggested by Deamer et al. (1994), which functions in a fluctuating environment (see chapter 21).

In addition to central features of the model—that is, corroboration with the principle of continuity and the gradual transition from the inorganic to the organic world—it is possible to incorporate it into other models that focus on additional aspects of the origin of life, such as the origin of metabolism. In this case, a gradual transition from preformed building blocks such as amino acids and protomononucleotides would take place, by which a metabolic system based, for instance, on CO_2 fixation is added to the existing system; the present TSD system is capable of synthesizing the peptide catalysts needed for such a development. Thus, the TSD model under study can be the central feature of more complicated systems.

Primordial exon-intron relationships? Additional features of extant cells that may be related to much simpler features of the chemical entities of the system under study are exon-intron relationships. Indeed, it is tempting to suggest that the primordial exon-intron features of Lahav's coevolution scenario could have served as the predecessor of extant exon-intron relationships (chapter 17 and Lahav, 1989).

Obviously, it is not known whether extant exon-intron relations are connected to the primordial features of complementary domains suggested here.

Primordial gene duplication Mistakes in the copying process of proto-RNAs were presumably much more common in the prebiotic era than in extant cells. It is suggested that mutation mechanisms in that era included not only small steps of one building block at a time, but also entire strands of proto-RNAs. Thus, a primordial "gene duplication" could greatly facilitate the evolutionary process of the present TSD system.

Estimates of the time needed for the emergence of TSD syntheses First estimates of the time needed for the emergence of the TSD system under study can be made by employing the rate constants of the following reactions: proto-RNA replication, amino acid loading on proto-tRNA, and peptide-bond formation.

The size of the system under study and its relative simplicity in comparison with the smallest known extant cell, or the theoretical minimal cell, indicates that their emergence time is likely to be relatively short. In laboratory work the experimenter can manipulate experimental parameters, such as fluctuating rate and temperature regime, in order to facilitate the emergence time of TSD systems.

Genetic code Given the relationships between the major components of the model, the development and evolution of the genetic code is inevitable. This model thus serves as a mechanistic model for the emergence of genetic code

Directed evolution, ad hoc genetic codes, and synthetic life

If indeed the emergence time of TSD systems is relatively short, future experimental work along the guidelines of the coevolution theory may be possible. It is expected to bring about the synthesis of chemical entities characterized by central features of extant life. This would signify the initiation of a new scientific discipline that may be called "synthetic life." The concomitant evolutionary processes of such systems may be directed by manipulating their composition and concentrations, as well as their environment. Moreover, any such experimental system is expected to be characterized by its own ad hoc genetic code.

Summary

So far, the new discipline of computer-based artificial life (AL) has not dealt with the very transition from inanimate to animate. This transition, however, may serve as a novel point of view for AL researchers. It is reasonable to expect that both the inspiration and first models for such an undertaking would come from the discipline of molecular evolution. Moreover, it should not be surprising if a new school, "Origin of artificial life," would emerge from such an effort. Furthermore, even before such a discipline is established, computer modeling of molecular evolution processes and scenarios may serve an invaluable role in the search of the origin of life.

The little use of computer modeling in molecular evolution scenarios stems, essentially, from the scarcity of such scenarios and their speculative and controversial characteristics. The great potential of computer models in molecular evolution makes them an invaluable substitute for the vast number of laboratory experiments that should be carried out in order to select plausible pathways for the emergence of life.

The two models presented here demonstrate the research potential of computer modeling when connected to a plausible biogeochemical model. The coevolution scenario is of particular interest because it seems to be the only tool now available to explore the mechanism of the emergence of central attributes such as autocatalysis, feedback systems, and the genetic code.

The unique position of the TSD system in the transition from inanimate to animate is the result of its ability to "memorize" every evolutionary "innovation." This is accomplished by the formation of feedback loops by which beneficial catalytic peptides, as well as their corresponding proto-mRNAs and proto-tRNAs, are selected and encoded in the system's memory, rendering the system with autocatalytic attributes. Moreover, it seems that the encoding mechanism suggested for the establishment of the TSD system under study is applicable to other central features, that is, metabolic cycles and homochirality.

Epilogue

Having made a long and tortuous journey in search of the origin of life, some readers may feel disappointed: The alarming number of speculations, models, theories, and controversies regarding every aspect of the origin of life seem to indicate that this scientific discipline is almost in a hopeless situation. Still others may find comfort in the significant progress already made and the knowledge accumulated in this interdisciplinary discipline. A common denominator to the latter group of researchers is the working hypothesis that the transition from inanimate to animate in the origin-of-life process can be studied scientifically without being hampered by "the nature of knowledge itself" (Pattee, 1995). However, none of the theories advanced so far encompasses all the aspects of the emergence of the central functions of extant cells, thus "bridging the gap between life and inanimate matter" (Arrhenius et al., 1997). Furthermore, because some of these theories differ so widely from each other, bridging the gap between them seems difficult, perhaps even impossible, at present. Can we observe the initiation of promising directions that might lead us into the beginning of a new era in the study of the origin of life?

Obviously, every researcher focuses on the potential of his or her own school to bring about a significant progress or even a breakthrough of a kind in the understanding of the origin of life. It is difficult, however, and probably impossible, to point out which of the present schools of thought or their combination have the potential to serve as the basis of the next paradigm. Therefore, rather than searching for an unambiguous pathway leading from one of the existing schools to a vague pathway, it may be helpful to find a common denominator among various points of views and examine to what extent changes in this common denominator would affect our understanding of the origin of life and its research strategies.

One such common denominator is the estimate of the time needed for the emergence of a chemical system characterized by the most central attribute of life. The history of these estimates has been characterized by dramatic changes during the last few decades, from more than 2 billion years according to Bernal in 1967, to several tens of years according to the controversial theory of White and Raab in 1982. Should one of these estimates be supported by novel observations or theories and thus be considered prebiotically plausible, then it is likely to bring about a paradigmatic change in the thinking of origin-of-life researchers. The affected topics are expected to range from life on other planets in our galaxy, including our solar system, to the synthesis of "living entities" in the laboratory.

But even if a paradigmatic change in the study of the origin of life took place and "living" chemical entities were synthesized out of inorganic compounds in our laboratories, we should recognize our own limitations in deciphering the transition from inanimate to animate. It was noted by Haldane ("in a rare moment of modesty," according to Haynes, 1987, p. 15) that "the universe is not only queerer than we suppose, it is queerer than we can suppose" (cited by Haynes, 1987, p. 15). In spite of the apparent brilliance of Haldane's expression, it may well be wrong with regard to the origin of life. Hopefully, our brains are capable of understanding the essence of the processes of the origin of life, but we shall probably never be able to know all the details and intricacies of this process. And since the road to a better understanding of nature has always been characterized by errors and mistakes, origin-of-life researchers are bound to make numerous oversights in their attempts to improve our understanding of the transition from inanimate to animate. In this spirit, I think, it was commented by Sigmund Freud (told and cited by Gould, 1994) that great revolutions in science have a common (and ironic) denominator: "They knock human arrogance off one pedestal after another of our conviction about our previous conviction about our own self-importance."

Thus, whether a better understanding of the origin of life would be obtained by focusing on the time of emergence of living entities, or by an unexpected discovery of another kind, humbleness should be central to our emerging world-picture.

References

Abel, E. L. (1973) Ancient views on the origin of life. Rutherford, New Jersey: Fairleigh-Dickinson University Press.

Adami, C. (1995) On modeling Life. In: R. A. Brooks and P. Maes (eds.) Artificial life IV proceedings of the fourth international workshop on the synthesis and simulation of living systems. Cambridge: MIT Press. pp. 268–274.

Addadi, L., Z. Berkovitch-Yellin, N. Domb, E. Gati, M. Lahav, and L. Leizerowitz (1982) Resolution of conglomerates by stereoselective habit modifications. Nature **296:** 21–26.

Alberts, B., D. Bray, J. Lewis, M. Raff, K. Roberts, and J. D. Watson (1994) Molecular biology of the cell. 3rd ed. New York: Garland Publishing, Inc.

Alexander, J. (1948) Life, its nature and origin. New York: Reinhold Publishing Company.

Allan, D. J. (1970) The philosophy of Aristotle. London: Oxford University Press.

Allen, J. F. (1995) Origins of photosynthesis. Nature **376:** 28.

Ambler, R. P., and M. Daniel (1991) Proteins and molecular paleontology. Phil. Trans. R. Soc. Lond. B **333:** 381–389.

Amirnovin, R. (1997) An analysis of the metabolic theory of the genetic code. J. Mol. Evol. **44:** 473–476.

Anbar, M. (1995) Nitric oxide a synchronizing chemical messenger. Experimentia **51:** 545–550.

Anders, E. (1989) Prebiotic organic matter from comets and asteroids. Nature **342:** 255–257.

Anderson, S., H. L. Anderson, and J. K. M. Sanders (1993) Expanding roles for templates in synthesis. Acc. Chem. Res. **26:** 469–475.

Angert, E. R., K. D. Clements, and N. R. Pace (1993) The largest bacterium. Nature **362:** 239–241.

Argyle, E. (1977) Chance and the origin of life. Origins of Life **8:** 287–298.

Arnez, J. G., and D. Moras (1997) Structural and functional considerations of the amino-acylation reaction. TiBS **22:** 211–216.

Arrhenius, T., G. Arrhenius, and W. Paplawsky (1994) Archean geochemistry of formaldehyde and cyanide and the oligomerization of cyanohydrin. Origins of Life Evol. Biosphere **24:** 1–17.

Arrhenius, G., B. Sales, S. Mojzsis, and T. Lee (1997) Entropy and charge in molecular evolution—the case of phosphate. J. Theor. Biol. **187:** 503–522.
Ash, R. D., S. F. Knott, and G. Turner (1996) A 4-Gyr shock age for a Martian meteorite and implications for the cratering history of Mars. Nature **380:** 57–59.
Atassi, M. Z., and T. Manshouri (1993) Design of peptide enzymes (pepzymes): Surface-simulation synthetic peptides that mimic the chymotrypsin and trypsin active sites exhibit the activity and specificity of the respective enzyme. Proc. Natl. Acad. Sci. USA **90:** 8282–8286.
Atkins, P. W. (1981) The creation. Oxford: W. H. Freeman.
——— (1994) The second law. New York: Scientific American Library.
Awramik, S. M. (1992) The oldest records of photosynthesis. Minireview. Photosynthesis Research **33:** 75–89.
Bacon, F. (1909) Of studies. In: C. W. Eliot (ed.) The Harvard classics, Volume 3. New York: P. F. Collier and Sons. p. 128.
Bada, J. L. (1991) Amino acid cosmogeochemistry. Phil. Trans. R. Soc. Lond. B. **331:** 349–358.
——— (1997) Extraterrestrial handness? Science **275:** 942–943.
Bada, J. L., C. Bigham, and S. L. Miller (1994) Impact melting of frozen oceans on the early Earth: Implication for the origin of life. Proc. Natl. Acad. Sci. U.S.A. **91:** 1248–1250.
Bada, J. L., S. L. Miller, and M. Zhao (1995) The stability of amino acids at submarine hydrothermal vent temperatures. Origins of Life Evol. Biosphere **25:** 111–118.
Badash, L. (1989) The age-of-the-Earth debate. Scientific American. August: 78–83.
Baldauf, S. L., J. D. Palmer, and W. D. Doolittle (1996) The root of the universal tree and the origin of eukaryotes based on elongation factor phylogeny. Proc. Natl. Acad. Sci. U.S.A. **93:** 7749–7754.
Baltscheffsky, H. (1993) Chemical origin and early evolution of biological energy conversion. In: C. Ponnamperuma, and J. Chela-Flores (eds.). Chemical evolution. Hampton, Virginia: A. Deepak Publishing.
——— (1997) Major "anastrophes" in the origin and early evolution of biological energy conversion. J. Theor. Biol. **187:** 495–501.
Baltscheffsky, H., and M. Baltscheffsky (1994) Molecular origin and evolution of early biological energy conversion. In: S. Bengtson (ed.) Early life on Earth. Nobel Symposium **84.** New York: Columbia University Press. p. 81–90.
Banin, A., and J. Navrot (1975) Origin of life: Clues from the relations between chemical composition of living organisms and natural environments. Science **189:** 550–551.
Bar-Nun, A. E. Kochavi, and S. Bar-Nun (1994) Assemblies of free amino acids as possible prebiotic catalysts. J. Mol. Evol. **39:** 116–122.
Barinaga, M. (1994) Archaea and eukaryotes grow closer. Science **264:** 1251.
Basiuk, V. A., and R. Navarro-González (1996) Possible role of volcanic ash-gas in the Earth prebiotic chemistry. Origins of Life Evol. Biosphere **26:** 173–194.
Bastian, H. C. (1872) Quoted in J. Farley, 1977, p. 124.
——— (1911) The origin of life. New York: G. P. Putnam's Sons, Knickerbocker Press.
Beale, L. S. (1871) Quoted in J. Farley, 1977, p. 89.
Beaudry, A. A., and G. F. Joyce (1992) Directed evolution of an RNA enzyme. Science **257:** 635–641.
Béland, P., and T. F. H. Allen (1994) The origin and evolution of the genetic code. J. Theor. Biol. **170:** 359–365.
Benner, S. A. (1994) Expanding the genetic lexicon: Incorporating non-standard amino acids into proteins by ribosome-based synthesis. Tibtech **12:** 158–163.
Benner, S. A., M. A. Cohen, G. H. Gonnet, D. B. Berkowitz, and K. P. Johnsson (1993) Reading the palimpsest: Contemporary biochemical data and the RNA world. In: R. F. Gesteland and J. F. Atkins (eds.) The RNA world. New York: Cold Spring Harbor Laboratory Press. pp. 27–70.

Benner, S. A., A. D. Ellington, and A. Tauer (1989) Modern metabolism as a palimpsest of the RNA world. Proc. Natl. Acad. Sci. U.S.A. **86:** 7054–7058.

Bernal, J. D. (1951) The physical basis of life. London: Routledge and Kegan Paul.

——— (1965) Molecular structure, biochemical function, and evolution. In: T. H. Waterman and H. J. Morowitz (eds.) Theoretical and mathematical biology. New York: Blaisdel.

——— (1967) The origin of life. London: Weidenfeld and Nicolson.

Bernard, C. (1878a) Lectures on the phenomena of life. Translated by H. E. Hoff, R. Guillemin, and L. Guillemin (1974). Springfield, Illinois: Charles Thomas.

——— (1878b) An introduction to the study of medicine. Translated by H. C. Greene (1927). New York: Macmillan.

Berry, S. (1995) Entropy, irreversibility, and evolution. J. Theor. Biol. **175:** 197–202.

Bertalanffy, L. Von (1933) Modern theories of development: An introduction to theoretical biology. English translation. New York: Harper Torchbooks, the Science Library, Harper and Brothers.

Bertolaet, B. L., H. M. Seidel, and J. R. Knowles (1995) Introns and the origin of protein-coding genes. Science **268:** 1367.

Bishop, J. C., S. D. Cross, and T. G. Waddell (1997) Prebiotic transamination. Origins of Life Evol. Biosphere **27:** 319–327.

Björn, L. A. (1995) Origins of photosynthesis. Nature **376:** 25–26.

Blalock, J. E. (1990) Complementarity of peptides specified by "sense" and "antisense" strands of DNA. Tibtech **8:** 140–144.

Blalock, J. E., and E. M. Smith (1984) Hydropathic anti-complementarity of amino acids based on the genetic code. Biochemical and Biophysical Communication **121:** 203–207.

Blankenship, R. E. (1992) Origin and early evolution of photosynthesis. Minireview. Photosynthesis Research **33:** 91–111.

Bloch, D. P., B. McArthur, R. Widdowson, D. Spector, R. E. Guimaraes, and J. Smith (1983) tRNA-rRNA sequence homologies: Evidence for a common evolutionary origin? J. Mol. Evol. **19:** 420–428.

Blöchl, E., M. Keller, G. Wächtershäuser, and K. O. Stetter (1992) Reactions depending on iron sulfide and linking geochemistry and biochemistry. Proc. Natl. Acad. Sci. U.S.A. **89:** 8117–8120.

Blomberg, C. (1997) On the appearance of function and organization in the origin of life. J. Theor. Biol. **187:** 541–554.

Böhler, C., A. R. Hill, and L. E. Orgel (1996) Catalysis of the oligomerization of O-phospho-serine, aspartic acid, or glutamic acid by cationic micelles. Origins of Life Evol. Biosphere **26:** 1–5.

Böhler, C., P. E. Nielsen, and L. E. Orgel (1995) Template switching between PNA and RNA oligonucleotides. Nature **376:** 578–581.

Bohr, N. (1933) Light and life. Nature **131:** 457–459.

Bonner, W. A. (1991) The origin and amplification of biomolecular chirality. Origins of Life Evol. Biosphere **21:** 59–111.

——— (1995) Chirality and life. Origins of Life Evol. Biosphere **25:** 175–190.

Bonomi, F., M. T. Werth, and D. M. Kurtz Jr. (1985) Assembly of $[Fe_nS_n(SR)_4]^{2-}$ (n = 2, 4) in aqueous media from iron salts, thiols, and sulfur, sulfide, or thiosulfate plus Rhodanese. Inorganic Chem. **24:** 4331–4335.

Brack, A. (1993a) Chiralty and the origin of life. In: J. M. Greenberg, C. X. Mendoza-Gómez, and V. Pirronello (eds.) The chemistry of life origin. Dordrecht: Kluwer Academic Press. pp. 345–355.

——— (1993b) Early proteins. In: J. M. Greenberg, C. X. Mendoza-Gómez, and V. Pirronello (eds.) The chemistry of life's origins. Dordrecht: Kluwer Academic Press. pp. 357–388.

——— (1993c) Liquid water and the origin of life. Origins of Life Evol. Biosphere **23:** 3–10.

——— (1994) Are peptides possible support for self-amplification of sequence information? In: G. R. Fleischaker, S. Colonna, and P. L. Luisi (eds.) Self-production of supramolecular structures. Dordrecht: Kluwer Academic Press. pp. 115–124.

Brack, A., and L. E. Orgel (1975) β-structures of alternating polypeptides and their possible prebiotic significance. Nature **256:** 383–387.

Braterman, P. S., A. G. Cairns-Smith, and R. W. Sloper (1983) Photo-oxidation of hydrated Fe^{2+}—significant for banded iron formations. Nature **303:** 163–164.

Breus, T. K., K. Y. Pimenov, E. V. Syutkina, F. Halberg, G. Cornelissen, Y. I. Gurfinkel, S. M. Chibisov, and V. A. Frolov (1996) Biological effects of solar activity. ISSOL '96, 8th ISSOL meeting, 11th International Conference on the Origin of Life. Orleans, France. July. Poster Abstract 133.

Brock, D. T., and M. T. Madigan (1991) Biology of microorganisms. Englewood Cliffs, New Jersey: Prentice-Hall.

Browning, J. (1869) Quoted in J. Farley, 1977, p. 75.

Buchanan, B. B. (1992) Minireview. Carbon dioxide assimilation in oxygenic and anoxygenic photosynthesis. Photosynthesis Research **33:** 147–162.

Buchner, L. (1855) Quoted in J. Farley, 1977, p. 72.

Buechter, D. D., and P. Schimmel (1993) Aminoacylation of RNA minihelices: Implications for tRNA synthetase structural design and evolution. Critical Reviews in Biochemistry and Molecular Biology **28** (4): 309–322.

Bujdak, J., K. Faybikove, A. Eder, Y. Yongyai, and B. M. Rode (1995) Peptide chain elongation: A possible role of montmorillonite in prebiotic synthesis of protein precursors. Origins of Life Evol. Biosphere **25:** 431–441.

Bujdak, J., and B. M. Rode (1995) Clays and their possible role in prebiotic peptide synthesis. Geologica Carpathica—Series Clays **4:** 37–48.

——— (1996) The effect of smectite composition on the catalysis of peptide bond formation. J. Mol. Evol. **43:** 326–333.

Burchfield, J. D. (1975) Lord Kelvin and the age of Earth. New York: Science History Publications.

Cairns-Smith, A. G. (1966) The origin of life and the nature of the primitive gene. J. Theor. Biol. **10:** 53–88.

——— (1982) Genetic takeover and the mineral origins of life. Cambridge: Cambridge University Press.

Cairns-Smith, A. G., A. J. Hall, and M. J. Russell (1992) Mineral theories of the origin of life and an iron sulfide example. Origins of Life Evol. Biosphere **22:** 161–180.

Calnan, B. J., B. Tidor, S. Biancalana, D. Hudson, and A. D. Frankel (1991) Arginine-mediated RNA recognition: The arginine fork. Science **252:** 1167–1171.

Calvin, M. (1951) Reduction of carbon dioxide in aqueous solutions by ionizing radiation. Science **114:** 416–418.

Cassirer, E. (1950) The problem of knowledge. New Haven: Yale University Press.

Casti, J. L. (1992) That's life—yes, no, maybe. In: J. and K. Tran Thanh Van, J. C. Mounolou, J. Schneider, and C. McKay (eds.) Frontiers of life. Singapore: Fong and Sons Printers. pp. 41–60.

Castresana, J., and M. Saraste (1995) Evolution of energetic metabolism: The respiration-early hypothesis. TiBS **20:** 443–448.

Cech, T. R. (1993a) The efficiency and versatility of catalytic RNA: Implications for an RNA world. Gene **135:** 33–36.

——— (1993b) Structure and mechanism of the large catalytic RNAs: Group I and group II introns and ribonuclease P. In: R. F. Gesteland and J. F. Atkins (eds.) The RNA world. New York: Cold Spring Harbor Laboratory Press. pp. 219–237.

Cech, T. R., and O. C. Uhlenbeck (1994) Hammerhead nailed down. Nature **372:** 39–40.

Cedergren, R., and P. Miramontes (1996) The puzzling origin of the genetic code. TiBS **21:** 199–200.

——— (1997) Reply. TiBS **22:** 50.

Chakrabarti, A. C., R. R. Breaker, G. F. Joyce, and D. W. Deamer (1994) Production of

RNA by a polymerase protein encapsulated within phospholipid vesicles. J. Mol. Evol. **39:** 555–559.

Chakrabarti, A. C., and D. W. Deamer (1994) Permeation of membranes by the neutral form of amino acids and peptides: Relevance to the origin of peptide translocation. J. Mol. Evol. **39:** 1–5.

Chamberlin, T. C., and R. T. Chamberlin (1908) Early terrestrial conditions that may have favored organic synthesis. Science **28:** 897–911.

Chan, S., J. Orenberg, and N. Lahav (1987) Soluble minerals in chemical evolution II. Characterization of the adsorption of 5′-AMP and 5′-CMP on a variety of soluble mineral salts. Origins of Life Evol. Biosphere **17:** 121–134.

Chan, Wing-Tsit (1963) A source book in Chinese philosophy. Princeton: Princeton University Press.

Chang, S. (1993) Prebiotic synthesis in planetary environments. In: J. M. Greenberg, C. X. Mendoza-Gómez, and V. Pirronello (eds.) The chemistry of life's origins. Dordrecht: Kluwer Academic Press. p. 259–299.

——— (1994) The planetary setting of prebiotic evolution. In: S. Bengtson (ed.) Early life on Earth. Nobel Symposium 84. New York: Columbia University Press. pp. 10–23.

Chang, S., D. DesMarais, R. Mack, S. L. Miller, and G. E. Strathearn (1983) Prebiotic organic syntheses and the origin of life. In: J. W. Schopf (ed.) Earth's earliest biosphere: its origin and evolution. Princeton: Princeton University Press. pp. 53–92.

Chela-Flores, J. (1994) Are viroids molecular fossils of the RNA world? J. Theor. Biol. **166:** 163–166.

Chyba, C. F. (1997) A left-handed solar system? Nature **389:** 234–235.

Chyba, C. F., and Gene D. McDonald (1995) The origin of life in the solar system: Current issues. Annu. Rev. Earth Planet. Sci. **23:** 215–249.

Chyba, C., and C. Sagan (1992) Endogenous production, exogeneous delivery, and impact-shock synthesis of organic molecules: An inventory for the origins of life. Nature **355:** 125–132.

——— (1997) Comets as a source of prebiotic organic molecules for the early earth. In: P. J. Thomas, C. F. Chyba, and C. P. McKay (eds.) Comets and the origin and evolution of life. New York: Springer-Verlag. pp. 147–173.

Clark, B. C. (1988) Primeval procreative comet pond. Origins of Life Evol. Biosphere **18:** 209–238.

Cohen, J. (1995) Getting all turned around over the origin of life on earth. Science **267:** 1265–1266.

Cohen, P. T. W. (1997) Novel protein serine/threonine sulfates: Variety is the spice of life. TiBS 22: 245–250.

Collins, J. R., G. H. Loew, B. T. Luke, and D. H. White (1988) Theoretical investigation of the role of clay edges in prebiotic peptide bond formation. Origins of Life Evol. Biosphere **18:** 107–119.

Connell, G. J., and E. L. Christian (1993) Utilization of co-factors expand metabolism in a new RNA world. Origins of Life Evol. Biosphere **23:** 291–297.

Cooper, G. W., M.R. Onwo, and J. R. Cronin (1992) Alkyl phosphonic acids and sulfonic acids in the murchison meteorite. Geochim. Cosmochim. Acta 56: 4109–4115.

Corliss, J. B. (1990) Hot springs and the origin of life. Nature **347:** 624.

Corliss, J. B., J. A. Baross, and S. E. Hoffman (1981) An hypothesis concerning the relationship between submarine hot springs and the origin of life on Earth. Oceanol. Acta Proc. 26th Int. Geol. Congr., Geology of the Ocean symp., Paris. pp. 59–69.

Cowan, D. A. (1995) Protein stability at high temperatures. In: D. K. Apps and K. F. Tipton (eds.) Essays in biochemistry **29:** 193–207. London: Portland Press.

Coyne, L. M. (1985) A possible energetic role of mineral surfaces in chemical evolution. Origins of Life Evol. Biosphere **15:** 162–206.

Crabtree, R. H. (1997) Where smokers rule. Science **276:** 222.

Crick, F. H. C. (1968) The origin of the genetic code. J. Mol. Biol. **38:** 367–379.

——— (1981) Life itself. New York: Simon and Schuster.

——— (1993) Foreword. In R. F. Gesteland and J. F. Atkins (eds.) The RNA world. New York: Cold Spring Harbor Laboratory Press. pp. xi–xiv.
Crick, F. H. C., S. Brenner, A. Klug, and G. Pieczenik (1976) A speculation on the origin of protein synthesis. Origins of Life Evol. Biosphere **7**: 389–397.
Crick, F. H. C., and L. E. Orgel (1973) Directed Panspermia. Icarus **19**: 341–346.
Cronin, J. R., and S. Chang (1993) Organic matter in meteorites: Molecular and isotopic analysis of the Murchison meteorite. In: J. M. Greenberg, C. X. Mendoza-Gómez, and V. Pirronello (eds.) The chemistry of life's origins. Dordrecht: Kluwer Academic Press. pp. 209–258.
Cronin, J. R., and S. Pizzarello (1997) Enantiomeric excess in meteoritic amino acids. Science **275**: 951–955.
Croswell, K. (1994) Vermin of the skies. New Scientist, 27 August. pp. 26–29.
Csanyi, V., and G. Kampis (1985) Autogenesis: The evolution of replicative systems. J. Theor. Biol. **114**: 303–321.
Danchin, A. (1992) From stars and minerals to life: Is the paradigm changing? In: J. and K. Tran Than Van, J. C. Monolou, J. Schneider, and C. McKay (eds.) Frontiers of Life. Singapore: Fong & Sons Printers. pp. 399–414.
Darwin, C. (1859) The origin of species. A Mentor Book. New York: New American Library.
Davis, P. C. W. (1996) The transfer of viable microorganisms between planets. In: Evolution of hydrothermal ecosystems on Earth (and Mars?). Chichester: John Wiley & Sons; Ciba Foundation Symposium 202. pp. 304–317.
Davis, Wanda L., and C. P. McKay (1996) Origins of life: A comparison of theories and application to Mars. Origins of Life Evol. Biosphere **26**: 61–73.
Dawkins, R. (1986) The blind watchmaker. London: Longmans.
——— (1989) The selfish gene. Oxford: Oxford University Press.
de Duve, C. (1991) Blueprint for a cell: The nature and origin of life. Burlington, NC: Neil Patterson Publishers.
——— (1992) The thioester world. In: J. and K. Tran Thanh Van, J. C. Mounolou, J. Schneider, and C. McKay (eds.) Frontiers of Life. Singapore: Fong & Sons Printers. pp. 1–20.
——— (1993) Co-chairman remarks: The RNA world: Before and after? Gene **135**: 29–31.
——— (1995a) The beginning of life on earth. American Scientist **83**: 428–437.
——— (1995b) Vital dust: life as a cosmic imperative. New York: Basic Books.
——— (1996) The constraints of chance. Scientific American, January: p. 112.
de Duve, C., and S. L. Miller (1991) Two-dimensional life? Proc. Natl. Acad. Sci. U.S.A. **88**: 10014–10017.
De Loof, A. (1993) Schrödinger 50 years ago: "What is Life?" "The ability to communicate," a plausible reply? Int. J. Biochem. **25**: 1715–1721.
Deamer, D. W. (1994) Sources and syntheses of prebiotic amphiphiles. In: G. R. Fleischaker, S. Colonna, and P. L. Luisi (eds.) Self-production of supramolecular structures. Dordrecht: Kluwer Academic Press. pp. 217–229.
——— (1997) The first living systems: A bioenergetic perspective. Microbiology and Molecular Biology Reviews **61**: 239–261.
Deamer, D. W., and G. R. Fleischaker (eds.) (1994) Origins of life: The central concepts. Boston: Jones and Bartlet Publishers.
Deamer, D. W., and E. Harang (1990) Light-dependent pH gradients are generated in liposomes containing ferrocyanide. BioSystems **24**: 1–4.
Deamer, D. W., E. Harang-Mahon, and G. Bosco (1994) Self-assembly and function of primitive membrane structures. In: S. Bengtson (ed.) Early life on Earth. Nobel Symposium 84. New York: Columbia University Press. pp. 107–123.
Deamer, D. W., and J. Oro (1980) Role of lipids in prebiotic structures. BioSystems **12**: 167–175.
Deamer, D. W., and A. G. Volkov (1996) Oil/water interfaces and the origin of life. In:

A. G. Volkov and D. W. Deamer (eds.) Liquid–liquid interfaces: Theory and methods. Boca Raton, Florida: CRC Press Inc. pp. 363–374.

Décout, J.-L., and M.-C. Maurel (1993) N6-substituted adenine derivatives and RNA primitive catalysts. Origins of Life Evol. Biosphere **23:** 299–306.

Décout, J.-L., J. Vergne, and M.-C. Maurel (1995) Synthesis and catalytic activity of adenine containing polyamines. Macromol. Chem. Phys. **196:** 2615–2624.

Degani, C., and M. Halman (1967) Chemical evolution of carbohydrate metabolism. Nature **216:** 1207.

Delarue, M. (1995) Partition of aminoacyl-tRNA synthetases in two different structural classes dating back to early metabolism: Implications for the origin of the genetic code and the nature of protein sequences. J. Mol. Evol. **41:** 703–711.

DeLong, E. F., K. Y. Wu, B. B. Prézelin, and R. V. M. Jovine (1994) High abundance of archaea in antarctic marine picoplankton. Nature **371:** 695–697.

Delsemme, A. (1997) The origin of the atmosphere and of the oceans. In: P. J. Thomas, C. F. Chyba, and C. P. McKay (eds.) Comets and the origin and evolution of life. New York; Springer-Verlag. pp. 29–67.

Dennett, D. C. (1996) Darwin's dangerous idea. A Touchstone Book. New York: Simon & Schuster.

Di Giulio, M. (1994a) On the origin of protein synthesis: A speculative model based on hairpin RNA structures. J. Theor. Biol. **171:** 303–308.

——— (1994b) The phylogeny of tRNA molecules and the origin of the genetic code. Origins of Life Evol. Biosphere **24:** 425–434.

——— (1995) The phylogeny of tRNAs seems to confirm the predictions of the coevolution theory of the origin of the genetic code. Origins of Life Evol. Biosphere **25:** 549–564.

——— (1996a) The origin of protein synthesis: On some molecular fossils identified through comparison of protein sequences. BioSystems **39:** 159–169.

Di Giulio, M. (1996b) The β-sheet of proteins, the biosynthetic relationships between amino acids, and the origin of the genetic code. Origins of Life Evol. Biosphere **26:** 589–609.

——— (1997) On the origin of the genetic code. J. Theor. Biol. **187:** 573–581.

Dick, T. P., and W. A. Schamel (1995) Molecular evolution of transfer RNA from two precursor hairpins: Implications for the origin of protein synthesis. J. Mol. Evol. **41:** 1–9.

Diener, T. O. (1989) Circular RNAs: Relics of precellular evolution? Proc. Natl. Acad. Sci. U.S.A. **86:** 9370–9374.

Ding, P. D., K. Kawamura, and J. P. Ferris (1996) Oligomerization of uridine phosphoimidazolides on montmorillonite: A model for the prebiotic synthesis of RNA on minerals. Origins of Life Evol. Biosphere **26:** 151–171.

Dobzhansky, T. (1973) Nothing in biology makes sense except in the light of evolution. American Biology Teacher **35:** 125–129.

Doolittle, R. F. (1995) Of archae and eo: What's in a name? Proc. Natl. Acad. Sci. U.S.A. **92:** 2421–2423.

Doolittle, R. F., Da-Fey Feng, S. Tsang, G. Cho, and E. Little (1996) Determining divergence times of the major kingdoms of living organisms with a protein clock. Science **271:** 470–477.

Doolittle, W. F. (1994) Evolutionary creativity and complex adaptations: A molecular biologist's perspective. In: J. H. Campbell and J. W. Schopf (eds.) Creative evolution?! Boston: Jones and Bartlet Publishers. pp. 47–73.

Doolittle, W. F., W. Lam, and L. Schalkwyk (1991) Evolution and basic features of gene and genome transfer. In: S. Mohan, C. Dow, and J. A. Coles (eds.) Prokaryotic structure and function: A new perspective. Cambridge: Cambridge University Press.

Drobner, E., H. Huber, G. Wächtershäuser, D. Rose, and K. O. Stetter (1990) Pyrite formation linked with hydrogen evolution under anaerobic conditions. Nature **346:** 742–744.

Dyson, F. (1985) Origins of life. Cambridge: Cambridge University Press.

Edwards, M. R. (1996) Metabolite channeling in the origin of life. J. Theor. Biol. **179:** 313–322.

Eigen, M. (1971) Self-organization of matter and the evolution of biological macromolecules. Naturwissenschaften **58:** 465–523.

——— (1992) Steps towards life. Oxford: Oxford University Press.

Eigen, M., B. F. Lindemann, M. Tietze, R. Winkler-Oswatitsch, A. Dress, and A. von Haeseler (1989) How old is the genetic code?: Statistical geometry of tRNA provides an answer. Science **244:** 673–679.

Eigen, M., and P. Schuster (1977) The hypercycle: A principle of natural self-organization. Part A: emergence of the hypercycle. Naturwissenschaften **64:** 541–565.

Eigen, M., and R. Winkler-Oswatitsch (1981a) Transfer-RNA: The early adaptor. Naturwissenschaften **68:** 217–228.

——— (1981b) Transfer-RNA, an early gene? Naturwissenschaften **68:** 282–292.

——— (1983) The origin and evolution and evolution of life at the molecular level. In: C. Helene (ed.) Structure, dynamics, interactions, and evolution of biological macromolecules. Dordrecht: Reidel. pp. 353–370.

Ekland, E. H., and D. P. Bartel (1996) RNA-catalyzed polymerization using nucleoside triphosphates. Nature **382:** 373–376.

Elitzur, A. C. (1993) The origin of life. Contemporary Physics **34:** 275–278.

——— (1994a) Let there be life: Thermodynamic reflections on biogenesis and evolution. J. Theor. Biol. **168:** 429–459.

——— (1994b) Time and consciousness: New reflections on ancient riddles (in Hebrew). Tel Aviv: The Ministry of Defence.

——— (1995) Life and mind, past and future: Schrödinger's vision fifty years later. Perspectives in Biology and Medicine **8:** 433–457.

——— (1996) Life's emergence is not an axiom: A reply to Yockey. J. Theor. Biol. **180:** 175–180.

Ellington, A. D. (1994) Empirical explorations of sequence space: Host-guest chemistry in the RNA world. Ber. Bunsenges. Phys. Chem. **98:** 1115–1121.

Emmeche, C. (1991) The problem of medium independence in artificial life. In: C. Mosekilde and L. Mosekilde (eds.) Complexity, chaos, and biological evolution. New York: Plenum Press. pp. 247–257.

——— (1994) The garden in the machine: The emerging science of artificial life. Princeton: Princeton University Press.

Engel, M. H., and S. A. Macko (1997) Isotopic evidence for extraterrestrial non-racemic amino acids in the Murchison meteorite. Nature **389:** 265–268.

Engels, F. (ca. 1880; the manuscript was lost for about sixty years) Dialectic of nature. Translated and edited by C. D. Dutt, 1940. New York: International Publishers.

Epstein, I. R. (1995) The consequences of imperfect mixing in autocatalytic chemical and biological systems. Nature **374:** 321–327.

Eriani, G., J. Cavarelli, F. Martin, L. Ador, B. Rees, J.-C. Thierry, J. Gangloff, and D. Moras (1995) The class II aminoacyl-tRNA synthetases and their active site: Evolutionary conservation of an ATP binding site. J. Mol. Evol. **40:** 499–508.

Ertem, G., and J. P. Ferris (1996) Synthesis of RNA oligomers on heterogeneous templates. Nature **379:** 238–240.

Eschenmoser, A. (1994) Chemistry of potentially prebiological natural products. Origins of Life Evol. Biosphere **24:** 389–423.

Farley, J. (1977) The spontaneous generation controversy. Baltimore: John Hopkins University Press.

Ferreira, R., and A. R. O. Cavalcanti (1997) Vestiges of early molecular processes leading to the genetic code. Origins of Life Evol. Biosphere **27:** 397–403.

Ferreira, R., and K. R. Coutinho (1993) Simulation studies of self-replicating oligoribotides, with a proposal for the transition to a peptide-assisted stage. J. Theor. Biol. **164:** 291–305.

Ferris, J. P. (1989) Issues of organic origins. Nature **337:** 609–610.

——— (1993a) Catalysis and prebiotic RNA synthesis. Origins of Life Evol. Biosphere **23:** 307–315.
——— (1993b) Prebiotic synthesis on minerals: RNA oligomer formation. In: J. M. Greenberg, C. X. Mendoza-Gómez, and V. Pirronello (eds.) The chemistry of life's origins. Dordrecht: Kluwer Academic Press. pp. 301–322.
——— (1994a) Chemical replication. Nature **369:** 184–185.
——— (1994b) The prebiotic synthesis and replication of RNA oligomers: The transition from prebiotic molecules to the RNA world. In: G. R. Fleischaker, S. Colonna, and P. L. Luisi (eds.) Self-production of supramolecular structures. Dordrecht: Kluwer Academic Press. pp. 89–98.
——— (1995) Life at the margins. Nature **373:** 659.
Ferris, J. P., A. R. Hill Jr., R. Liu, and L. E. Orgel (1996) Synthesis of long prebiotic oligomers on mineral surfaces. Nature **381:** 59–61.
Fifty years ago (1996) Nature **380,** no. 6576: p. ix.
Fitch, W. M., and K. Upper (1987) The phylogeny of tRNA sequences provides evidence of ambiguity reduction in the origin of the genetic code. Cold Spring Harbor Symp. Quant. Biol. **52:** 759–767.
Fleischaker, G. R. (1988) Autopoiesis: The status of its system logics. BioSystems **22:** 37–49.
——— (1990a) Origins of life: An operational definition. Origins of Life Evol. Biosphere **20:** 127–137.
——— (1990b) Three models of a minimal cell. In: C. Ponnamperuma, and F. R. Eirich (eds.) Biological Self-organization. Hampton, Virginia: Deepak Publishing.
——— (1991) The myth of the putative "organism." Uroboros **1:** 23–43.
——— (1994) A few precautionary words concerning terminology. In: S. Colonna Fleischaker and P. L. Luisi (eds.) Self-production of supramolecular structures. Dordrecht: Kluwer Academic Press. pp. 33–41.
Florkin, M. (1972) A history of biochemistry. In: M. Florkin and E. Stotz (eds.) Comprehensive biochemistry. Vol. **30.** Amsterdam: Elsevier Publishing Company.
Folk, R. L. (1997) Letter to the editor. Science 276: 1777.
Folsome, C. E. (1979) The origin of life. San Francisco: Freeman and Company.
Fong, P. (1973) Thermodynamic statistical theory of life: An outline. In: A. Locker (ed.) Biogenesis, evolution, homeostasis. A symposium by correspondence. Berlin: Springer-Verlag. pp. 93–101.
Fontana, W., and L. W. Buss (1994) What would be conserved if "the tape were played twice"? Proc. Natl. Acad. Sci. U.S.A. **91:** 757–761.
Forterre, P. (1992) New hypotheses about the origin of viruses, prokaryotes and eukaryotes. In: J. and K. Tran Thanh Van, J. C. Mounolou, J. Schneider, and C. McKay (eds.) Frontiers of life. Singapore: Fong & Sons Printers. pp. 221–233.
——— (1996) A hot topic: The origin of hyperthermophiles. Cell **85:** 789–792.
Forterre, P., F. Confalonieri, F. Charbonnier, and M. Duguet (1995) Speculations on the origin of life and thermophily: Review of available information on reverse gyrase suggests that hyperthermophilic procaryotes are not so primitive. Origins of Life Evol. Biosphere **25:** 235–249.
Fox, R. F. (1988) Energy and the evolution of life. New York: W. H. Freeman and Company.
Fox, S. (1988) The emergence of life: Darwinian evolution from the inside. New York: Basic Books.
——— (1995) Experiments congruent with evolution: Forward or backward? Newsletter of ISSOL **22**(3): pp. 10–11.
Fox, S. W., and K. Dose (1977) Molecular evolution and the origin of life. New York: Marcel Dekker.
Francklyn, C., and P. Schimmel (1989) Aminoacylation of RNA minihelices with alanine. Nature **337:** 478–481.
Frank, J., J. Zhu, P. Panczek, Y. Li, S. Srivastava, A. Verschoor, M. Radermacher, R.

Grassucci, R. K. Lata, and R. K. Agrawal (1995) A model of protein synthesis based on cryo-electron microscopy of the E. coli ribosome. Nature **376:** 441–444.

Fry, I. (1995) Are the different hypotheses on the emergence of life as different as they seem? Biology and Philosophy **10:** 389–417.

Gaffey, M. J. (1997) The early solar system. Origins of Life Evol. Biosphere 27: 185–203.

Gamlin, L. (1992) Wallace in brief. New Scientist, 8 February: p. 51.

Gamow, G. (1954) Possible relation between deoxyribonucleic acid and protein structure. Nature **173:** 318–320.

Ganti, T. (1987) The principle of life. Budapest: Omikk.

García-Meza, V., A. González-Rodríguez, and A. Lazcano (1994) Ancient paralogous duplications and the search for archean cells. In: G. R. Fleischaker, S. Colonna, and P. L. Luisi (eds) Self-reproduction of supramolecular structures: From synthetic structures to models of minimal living systems. Dordrecht: Kluwer Academic Publishers. pp. 231–246.

Garcia-Ruiz, J. M. (1994) Inorganic self-organization in precambrian cherts. Origins of Life Evol. Biosphere **24:** 451–467.

Garrison, W. M., D. C. Morrison, J. G. Hamilton, A. A. Benson, A. A., and M. Calvin (1951) Reduction of carbon dioxide in aqueous solutions by ionising radiation. Science **114:** 416–418.

Gaskell, Augusta (1919) Life. The origin of life. Death. The cause of neoplass. Ames Research Center Library, NASA.

Gatlin, L. (1972) Information theory and the living system. New York: Columbia University Press.

Gedulin, B., and G. Arrhenius (1994) Sources and geochemical evolution of RNA precursor molecules: The role of phosphate. In: S. Bengtson (ed.) Early life on Earth. Nobel Symposium 84. New York: Columbia University Press. pp. 91–106.

Gerard, D. R. (1958) Concepts in biology. Behavioral Sciences **3:** 92–215.

Gibbs, D., R. Lohrmann, and L. E. Orgel (1980) Template-directed synthesis and selective adsorption of oligoadenylates on hydroxyapatite. J. Mol. Evol. **15:** 347–354.

Gilbert, W. (1986) The RNA world. Nature **319:** 618.

——— (1987a) The exon theory of genes. Cold Spring Harbor Symp. Quant. Biol. **3:** 901–905.

——— (1987b) Why genes in pieces? Nature **271:** 501.

Girvetz, H., G. Geiger, H. Hantz, and B. Morris (1966) Science, folklore, and philosophy. New York: Harper & Row.

Giver, L., S. Lato, and A. Ellington (1994) Models for the autocatalytic replication of RNA. In: G. R. Fleischaker, S. Colonna, and P. L. Luisi (eds.) Self-production of supramolecular structures. Dordrecht: Kluwer Academic Press. pp. 137–146.

Goddard, D. R. (1958) In: Gerard, R. W. (ed.) Concepts in biology. The Biology Council, Division of Biology and Agriculture, National Academy of Sciences–National Research Council. p. 133.

Gogarten, J. P., H. Kibak, P. Ditrich, L. Taiz, E. J. Bauman, B. J. Bauman, M. F. Manolson, R. J. Poole, T. Date, T. Oshima, J. Konishi, K. Denda, and M. Yoshida (1989) Evolution of the vacuolar H^+-ATPase: Implications for the origin of eukaryotes. Proc. Nat. Acad. Sci. U.S.A. **86:** 6661–6665.

Gogarten, J. P., and L. Taiz (1992) Evolution of proton pumping ATPases: Rooting of the tree of life. Minireview. Photosynthesis Research **33:** 137–146.

Gogarten-Boekels, M., E. Hilario, and P. Gogarten (1995) The effects of heavy meteorite bombardments on the early evolution: The emergence of the three domains of life. Origins of Life Evol. Biosphere **25:** 251–264.

Goldanskii, V. (1992) Chirality, origin of life, and evolution. In: J. and K. Tran Thanh Van, J. C. Mounolou, J. Schneider, and C. McKay (eds.) Frontiers of Life. Singapore Fong & Sons Printers. pp. 67–84.

Goldanskii, V. I., and V. V. Kuzmin (1991) Chirality and cold origin of life. Nature **352:** 114.

Golding, G. B., N. Tsao, and R. E. Pearlman (1994) Evidence for intron capture: An unusual path for the evolution of proteins. Proc. Natl. Acad. Sci. U.S.A. **91:** 7506–7509.

Goodwin, J. T., P. Luo, J. C. Leitzel, and D. G. Lynn (1994) Template-directed synthesis of oligomers: Kinetic versus thermodynamic control. In: G. R. Fleischaker, S. Colonna, and P. L. Luisi (eds.) Self-production of supramolecular structures. Dordrecht: Kluwer Academis Press. pp. 99–104.

Gordon, K. H. J. (1995) Were RNA replication and translation directly coupled in the RNA (+protein?) world? J. Theor. Biol. **173:** 179–193.

Gould, S. J. (1989) This wonderful life. New York: Norton.

Gould, S. J. (1994) The evolution of life on the Earth. Scientific American, October: 85–91.

Granick, S. (1950) The structural relationships between heme and chlorophyll. Harvey Lectures **44:** 220–245.

——— (1957) Speculations on the origins and evolution of photosynthesis. Ann. N.Y. Acad. of Sci. **69:** 292–308.

Green, P., D. Lipman, L. Hillier, R. Waterson, D. States, and J.-M. Claverie (1993) Ancient conserved regions in new gene sequences and the protein databases. Science **259:** 1711–1716.

Green, R., and J. W. Szostak (1992) Selection of a ribozyme that functions as a superior template in a self-copying reaction. Science **258:** 1910–1915.

Greenberg, J. M. (1993) Preface. In: J. M. Greenberg, C. X. Mendoza-Gómez, and V. Pirronello (eds.) The chemistry of life's origins. Dordrecht: Kluwer Academic Press.

——— (1995) Chirality in interstellar dust and in comets: Life from dead stars. In: D. B. Cline (ed.) Proceedings from the Symposium in Santa Monica, "Physical Origin of Homochirality in Life." Santo Monica: AIP Press. pp. 185–210.

Greenberg, J. M., and C. X. Mendoza-Gómez (1993) Interstellar dust evolution: A reservoir of prebiotic molecules. In: J. M. Greenberg, C. X. Mendoza-Gómez, and V. Pirronello (eds.) The Chemistry of Life's Origins. Dordrecht: Kluwer Academic Press. pp. 1–32.

Grotzinger, J. P., and D. H. Rothman (1996) An abiotic model for stromatolite morphogenesis. Nature **383:** 423–425.

Gupta, R. S., and G. B. Golding (1996) The origin of the eukaryotic cell. TiBS **21:** 166–171.

Haeckel, E. (1866) Quoted in Oparin 1957, pp. 77–78.

——— (1919) Kristallseelen: Studien über das anorganische leben. Leipzig: Alfred Kroner Verlag.

Haezrahi, P. (1970) On the perfect being. Akademon, The Hebrew University of Jerusalem. (Hebrew).

Hafenbradl, D., M. Keller, G. Wächtershäuser, and K. O. Setter (1995) Primordial amino acids by reductive amination of α-oxo acids in conjunction with the oxidative formation of pyrite. Tetrahedron Letters **36:** 5179–5181.

Haldane, J. B. S. (1929) The origin of life. Appendix. In: J. D. Bernal (1967), The origin of life. London: Weidenfeld and Nicolson. pp. 242–249.

Happe, R. P., W. Roseboom, A. J. Plerik, S. P. J. Albracht, and K. A. Bagley (1997) Biological activation of hydrogen. Nature **385:** 126.

Harris, J. (1994) When Bernal got plastered with Picasso . . . New Scientist, 23 July: 42.

Härtlein, M., and S. Cusack (1995) Structure, function, and evolution of seryl-tRNA synthetases: Implications for the evolution of aminoacyl-tRNA synthetases and the genetic code. J. Mol. Evol. **40:** 519–530.

Hartman, H. (1975) Speculations on the origin and evolution of metabolism. J. Mol. Evol. **4:** 359–370.

——— (1992a) The eukaryotic cell and the RNA-protein world. In: J. and K. Tran Thanh Van, J. C. Mounolou, J. Schneider, and C. McKay (eds.) Frontiers of life. Singapore: Fong & Sons Printers. pp. 163–174.

——— (1992b) Minireview. Conjectures and reveries. Photosynthesis Research **33:** 171–176.

——— (1995a) Speculations on the evolution of the genetic code IV: The evolution of the aminoacyl-tRNA aynthetases. Origins of Life Evol. Biosphere **25:** 265–269.

——— (1995b) Speculations on the origin of the genetic code. J. Mol. Evol. **40:** 541–544.

Hartman, H., J. G. Lawless, and P. Morrison (eds.) (1985) Search for the universal ancestors. Ames Research Center, NASA SP-477.

Haukioja, E. (1982) Are individuals really subordinate to genes? A theory of living entities. J. Theor. Biol. **99:** 357–375.

Haynes, R. H. (1987) The "purpose" of chance in light of the physical basis of evolution. In: J. M. Robson (ed.) Origin and evolution of the universe: Evidence for design? Kingston, Ontario: McGill-Queen's University Press. pp. 1–31.

Henderson, I. M., M. D. Hendy, and D. Penny (1989) Influenza viruses, comets, and the science of evolutionary trees. J. Theor. Biol. **140:** 289–303.

Henley, R. W. (1996) Chemical and physical context for life in terrestrial hydrothermal systems: Chemical reactors for the early development of life and hydrothermal ecosystems. In: Evolution of hydrothermal ecosystems on Earth (and Mars?). Chichester: John Wiley & Sons.

Hermes-Lima, M., and A. Vieyra (1992) Pyrophosphate synthesis from phospho-(enol)pyruvate catalyzed by precipitated magnesium phosphate with "enzyme-like" activity. J. Mol. Evol. **35:** 277–285.

Hilario, E., and J. P. Gogarten (1993) Horizontal transfer of ATPase genes: the tree of life becomes a net of life. Biosystems **31:** 111–119.

Hill, D. K. (1995) Gathering air schemes for averting asteroid doom. Science **268:** 1562–1563.

Hipps, D., K. Shiba, B. Henderson, and P. Schimmel (1995) Operational RNA code for amino acids: Species-specific aminoacylation of minihelices switched by a single nucleotide. Proc. Natl. Acad. Sci. U.S.A. **92:** 5550–5552.

Hogben, L. (1931) The nature of living matter. New York: Alfred A. Knopf.

Holland, H. D. (1997) Evidence for life on Earth more than 3850 million years ago. Science **275:** 38–39.

Holm, N. G. (1992) Why are hydrothermal systems proposed as plausible environments for the origin of life? Origins of Life Evol. Biosphere **22:** 5–14.

Holm, N. G., and R. J.-C. Hennet (1992) Hydrothermal systems: Their varieties, dynamics, and suitability for prebiotic chemistry. Origins of Life Evol. Biosphere **22:** 15–31.

Hopfield, J. J. (1978) Origin of the genetic code: A testable hypothesis based on tRNA structure, sequence, and kinetic proofreading. Proc. Natl. Acad. Sci. U.S.A. **75:** 4334–4338.

Horgan, J. (1995) From complexity to perplexity. Scientific American, June: 104–109.

Horowitz, N. H. (1945) On the evolution of biochemical biosynthesis. Proc. Natl. Acad. Sci. U.S.A. **31:** 153–157.

——— (1959) On defining life. In: F. Clark and R. L. M. Synge (eds.) The origin of life on Earth. London: Pergamon. pp. 106–107.

——— (1986) To utopia and back: The search for life in the solar system. New York: H. W. Freeman and Co.

Hotchkiss, R. D. (1956) Quoted in Gerard 1958, pp. 95–215.

Hou, Y.-M., C. Francklyn, and P. Schimmel (1989) Molecular selection of a transfer RNA and the basis for its identity. TiBS, 14 June: 233–237.

Hou, Y.-M., and P. Schimmel (1988) A simple structural feature is a major determinant of the identity of a transfer RNA. Nature 333: 140–145.

Hoyle, F., and N. C. Wickramasinghe (1981) Space travelers. Cardiff: University College Cardiff Press.

——— (1986) The case for life as a cosmic phenomenon. Nature **322:** 509–511.

Huber, C., and G. Wächtershúser (1997) Activated acetic acid by carbon fixation on (Fe,Ni)S under primordial conditions. Science **276:** 245–247.
Hucho, F., and K. Buchner (1997) Signal transduction and protein kinases: The long way from plasma membrane into the nucleus. Naturwiss. **84:** 281–290.
Hurst, L. D. (1994) The uncertain origin of introns. Nature **371:** 381–382.
Huxley, J. (1953) Evolution in action. A Signet Scientific Library Book. New York: New American Library.
Huxley, T. H. (1868) Quoted in J. Farley 1977, p. 73.
Inue, T., and L. E. Orgel (1983) A nonenzymatic RNA polymerase model. Science **219:** 859–862.
Irvine, W. M. (1992) Chemistry in the cosmos. In: J. and K. Tran Than Van, J. C. Mounolou, J. Schneider, and C. McKay (eds.) Frontiers of Life. Third "Recontres de Blois." Editions Frontires. Singapore: Fong & Sons Printers.
Iwabe, N., K. Kuma, M. Hasegawa, M. Osawa, and T. Miyta (1989) Evolutionary relationship of archaebacteria, eubacteria, and eukaryotes inferred from phylogenetic trees of duplicated genes. Proc. Natl. Acad. Sci. U.S.A. **86:** 9355–9359.
Jablonka, E., and M. J. Lamb (1995) Epigenetic inheritance and evolution. Oxford: Oxford University Press.
Jacobson, K., E. D. Sheets, and R. Simon (1995) Revisiting the fluid mosaic model of membranes. Science **268:** 1441–1442.
Jaeger, L. (1997) The new world of ribozymes. Current Opinion in Structural Biology **7:** 324–335.
James, K. D., and A. D. Ellington (1995) The search for missing links between self-replicating nucleic acids and the RNA world. Origins of Life Evol. Biosphere **25:** 515–530.
Jantsch, E. (1980) The self-organizing universe. Oxford: Pergamon Press.
Jiménez-Sánchez, A. (1995) On the origin and evolution of the genetic code. J. Mol. Evol. **41:** 712–716.
Joyce, G. F. (1983) The instability of the autogen. J. Mol. Evol. **19:** 192–194.
——— (1987) Non-enzymatic template-directed synthesis of informational molecules. Cold Spring Harbor Symp. Quant. Biol. **52:** 41–51.
——— (1989) RNA evolution and the origin of life. Nature **338:** 217–224.
——— (1992) Directed molecular evolution. Scientific American, December: 90–97.
——— (1994) Foreword. In: D. W. Deamer and G. R. Fleischaker (eds.) Origins of life: The central concepts. Boston: Jones and Bartlet Publishers. pp. xi–xii.
Joyce, G. F., and L. E. Orgel (1993) Prospects for understanding the origin of the RNA world. In: R. F. Gesteland and J. F. Atkins (eds.) The RNA world. New York: Cold Spring Harbor Laboratory Press. pp. 1–25.
Joyce, G. F., G. M. Visser, C. A. A. van Boeckel, J. H. van Boom, L. E. Orgel, and J. van Westrenen (1984) Chiral selection in poly(C)-directed synthesis of oligo(G). Nature **310:** 602–604.
Judson, H. F. (1979) The eighth day of creation. A Touchstone Book. New York: Simon & Schuster.
Jukes, T. H. (1994) Divergent proteins and views. Nature **371:** 734.
Jukes, T. H., and S. Osawa (1993) Evolutionary changes in the genetic code. Comp. Biochem. Physiol. **106B:** 489–494.
Kalapos, M. P. (1997) Possible evolutionary role of methylglyoxolase pathway: Anaplerotic route for reductive citric acid cycle of surface metabolists. J. Theor. Biol. **188:** 201–206.
Kamaluddin, M. Nath, and A. Shamra (1994) Role of metal ferrocyanides in chemical evolution. Origins of Life Evol. Biosphere **24:** 469–477.
Kamminga, H. (1982) Life from space: a history of Panspermia. Vistas in Astronomy **26:** 67–86.
——— (1988a) Historical background of the concept of the origin of life. Report of spe-

cial research project on Evolution of Matter for 1987, University of Tsuhuba, pp. 1–13.
——— (1988b) Historical perspective: The problem of the origin of life in the context of developments in biology. Origin of Life Evol. Biosphere **18:** 1–11.
Kanavarioti, A. (1992) Self-replication of chemical systems based on recognition within a double or a triple helix: A realistic hypothesis. J. Theor. Biol. **158:** 207–219.
——— (1994) Template-directed chemistry and the origin of the RNA world. Origins of Life Evol. Biosphere **24:** 479–494.
——— (1997) Dimerization in highly concentrated solutions of phosphoimidazolide activated mononucleotides. Origins of Life Evol. Biosphere **27:** 257–376.
Kanavarioti, A., and E. E. Baird (1995) Faster rates with less catalyst in template-directed reactions. J. Mol. Evol. **41:** 169–173.
Kanavarioti, A., C. F. Bernasconi, D. J. Alberas, and E. E. Baird (1993) Kinetic dissection of individual steps in the poly(C)-directed oligoguanylate synthesis from guanosine 5′-monophosphate 2-methylimidazolide. J. Am. Chem. Soc. **115:** 8537–8546.
Kanavarioti, A., and D. H. White (1987) Kinetic analysis of the template effect in ribooligoguanylate elongation. Origins of Life Evol. Biosphere **17:** 333–349.
Kaschke, M., M. J. Russell, and W. J. Cole (1994) [FeS/FeS_2]. A redox system for the origin of life. Origins of Life Evol. Biosphere **24:** 43–56.
Kasting, J. F. (1993a) Early evolution of the atmosphere and ocean. In: J. M. Greenberg, C. X. Mendoza-Gómez, and V. Pirronello (eds.) The chemistry of life's origins. Dordrecht: Kluwer Academic Publishers pp. 149–176.
——— (1993b) Earth's early atmosphere. Science **259:** 920–926.
Kasting, J. F., D. C. B. Whittet, and W. R. Sheldon (1997) Ultraviolet radiation from F and K stars and implications for planetary habitability. Origins of Life Evol. Biosphere **27:** 413–420.
Katz, M. J. (1986) Templets and explanation of complex patterns. Cambridge: Cambridge University Press.
Kauffman, S. A. (1993) The origin of order: Self-organization and selection in evolution. New York: Oxford University Press.
Kauffman, S. (1996) Even peptides do it. Nature **382:** 496–497.
Kazakov, S., and S. Altman (1992) A trinucleotide can promote metal ion-dependent specific cleavage of RNA. Proc. Natl. Acad. Sci. U.S.A. **89:** 7939–7943.
Keefe, A. D., A. Lazcano, and S. L. Miller (1995) Evolution of the biosynthesis of the branched-chain amino acids. Origins of Life Evol. Biosphere **25:** 99–110.
Keefe, A. D., and S. L. Miller (1995) Are polyphosphates or phosphate esters prebiotic reagents? J. Mol. Evol. **41:** 693–702.
——— (1996a) Potentially prebiotic syntheses of condensed phosphates. Origins of Life Evol. Biosphere **26:** 15–25.
——— (1996b) Was ferrocyanide a prebiotic reagent? Origins of Life Evol. Biosphere **26:** 111–129.
Keefe, A. D., S. L. Miller, G. McDonald, and J. Bada (1995) Investigation of the prebiotic synthesis of amino acids and RNA bases from CO_2 using FeS/H_2S as a reducing agent. Proc. Natl. Acad. Sci. U.S.A. **92:** 11904–11906.
Keefe, A. D., G. L. Newton, and S. L. Miller (1995) A possible prebiotic synthesis of pantetheine, a precursor to coenzyme A. Nature **373:** 683–685.
Keller, M., E. Blöchl, G. Wächtershäuser, and K. O. Setter (1994) Formation of amide bonds without a condensation agent and implications for origin of life. Nature **368:** 836–838.
Kerr, R. A. (1996) Ancient life on Mars? Science **273:** 864–866.
Keynes, R. D. (1995) Erasmus Darwin's "Temple of Nature." J. Mol. Evol. **40:** 3–6.
King, G. A. M. (1982) Recycling, reproduction, and life's origin. BioSystems **15:** 89–97.
Kissane, J. M. (1994) Mammoth task. Nature **354:** 477.
Kissel, J., F. R. Krueger, and R. Roessler (1997) Organic chemistry in comets from remote

and in situ observations. In: P. J. Thomas, C. F. Chyba, and C. P. McKay (eds.) Comets and the origin and evolution of life. New York: Springer-Verlag, pp. 69–108.

Knudsen, C., R. Feldberg, and S. Rasmussen (1991) Information dynamics of self-programmable matter. In: C. Mosekilde and L. Mosekilde (eds.) Complexity, chaos, and biological evolution. New York: Plenum Press. pp. 223–245.

Kobayashi, K., and C. Ponnamperuma (1985a) Trace elements in chemical evolution, **I.** Origins of Life Evol. Biosphere **16:** 41–55.

——— (1985b) Trace elements in chemical evolution, **II.** Synthesis of amino acids under simulated primitive earth conditions in the presence of trace elements. Origins of Life Evol. Biosphere **16:** 57–67.

Kolb, V. M., J. P. Dworkin, and S. L. Miller (1994) Alternative bases in the RNA world: The prebiotic synthesis of Urazole and its ribosides. J. Mol. Evol. **38:** 549–557.

Kolb, V., S. Zhang, Y. Xu, and G. Arrhenius (1997) Mineral induced phosphorilation of glycolate ion—a metaphor in chemical evolution. Origins of Life Evol. Biosphere **27:** 485–503.

Konecny, J., M. Eckert, M. Schöniger, and G. L. Hofacker (1993) Neutral adaptation of the genetic code to double-strand coding. J. Mol. Evol. **36:** 407–416.

Konecny, J., M. Schöniger, and G. L. Hofacker (1995) Complementary coding conforms to the primeval comma-less code. J. Theor. Biol. **173:** 263–270.

Koza, J. R. (1994) Artificial life: Spontaneous emergence of self-replicating and evolutionary self-improving computer programs. In: C. G. Langton (ed.) SFI Studies in the Science of Complexity. Proc. Vol. 17, pp. 225–262. Reading: Addison-Wesley.

Krishnamurthy, R., S. Pitsch, and G. Arrhenius (1996) Mineral-induced synthesis of ribose phosphates. ISSOL '96. 11th ISSOL International Conference on the Origin of Life. Orleans, France. July. Abstract C2.3.

Kuhn, H. (1976) Model consideration for the origin of life: Environmental structure as stimulus for the evolution of chemical systems. Naturwissenschaften **63:** 68–80.

Kuhn, H., and J. Waser (1994a) Hypothesis on the origin of the genetic code. FEBS Letters **352:** 259–264.

——— (1994b) A model of the origin of life and perspectives in supramolecular engineering. In: J.-P. Behr (ed.) The lock-and-key principle: The state of the art—100 years on. Chichester: John Wiley & Sons. pp. 247–306.

Küppers, B.-O. (1990) Information and the origin of life. Cambridge: MIT Press.

Kutter, G. S. (1987) The universe and life. Boston: Jones and Publishers, Inc.

Lacey, J. C., Jr., L. Hall, and D. W. Mullins Jr. (1985) Rationalization of some genetic anticodonic assignments. Origins of Life Evol. Biosphere **16:** 69–79.

Lacey, J. C., Jr., and D. W. Mullins Jr. (1983) Experimental studies related to the origin of the genetic code and the process of protein synthesis: A review. Origins of Life Evol. Biosphere **13:** 3–42.

Lacey, J. C., Jr., and M. P. Staves (1990) Was there a universal tRNA before specialized tRNAs came into existence? Origins of Life Evol. Biosphere **20:** 303–308.

Lacey, J. C., Jr., N. S. M. D. Wickramasinghe, and G. W. Cook (1992) Experimental studies on the origin of the genetic code and the process of protein synthesis: A review update. Origins of Life Evol. Biosphere **22:** 243–275.

Lahav, N. (1985) The synthesis of primitive "living" forms: Definitions, goals, strategies, and evolution synthesizers. Origins of Life Evol. Biosphere **16:** 129–149.

——— (1989) Exon-intron-like pattern of the first propagating molecule? J. Mol. Evol. **29:** 475–479.

——— (1991) Prebiotic co-evolution of self-replication and translation or RNA world? J. Theor. Biol. **151:** 531–539.

——— (1993) The RNA-world and co-evolution hypotheses and the origin of life: Implications, research strategies, and perspectives. Origins of Life Evol. Biosphere **23:** 329–344.

——— (1994) Minerals and the origin of life: Hypotheses and experiments in heterogeneous chemistry. Heterogeneous Chemistry Reviews **1:** 159–179.

Lahav, N., and S. Chang (1976) The possible role of solid surface area in condensation reactions during chemical evolution: Reevaluation. J. Mol. Evol. **8:** 357–380.

——— (1982) The possible role of soluble salts in chemical evolution. J. Mol. Evol. **19:** 36–46.

Lahav, N., L. M. Coyne, and J. G. Lawless (1982) Prolonged triboluminescence in clays and other minerals. Clays and Clay Minerals **30:** 73–75.

——— (1985) Characterization of dehydration-induced luminescence of kaolinite. Clays and Clay Minerals **33:** 207–213.

Lahav, N., and S. Nir (1997) Emergence of template-and-sequence-directed (TSD) syntheses, I: A bio-geochemical model. Origins of Life Evol. Biosphere **27:** 377–395.

Lahav, N., and H. D. White (1980) A possible role of fluctuating clay-water systems in the production of ordered prebiotic oligomers. J. Mol. Evol. **16:** 11–21.

Lahav, N., H. D. White, and S. Chang (1978) Peptide formation in the prebiotic era: Thermal condensation of glycine in fluctuating clay environments. Science **201:** 67–69.

Lake, J. A. (1991) Tracing origins with molecular sequences: Metazoan and eukaryotic beginnings. TiBS **16:** 46–50.

Lancet, D., O. Kedem, and Y. Pilpel (1994) Emergence of order in small autocatalytic sets maintained far from equilibrium: Application of a probabilistic receptor affinity distribution (RAD) model. Ber. Bunsenges. Phys. Chem. **98:**1–4.

Langton, C. G. (1989) Artificial life. In: C G. Langton (ed.) Artificial life. Redwood City, California: Addison-Wesley. pp. 1–47.

Langton, C. G., C. Taylor, J. D. Farmer, and S. Rasmussen (eds.) (1992) Artificial life II. New York: Addison-Wesley.

Lapides, I. L., and E. E. Lustenberg (1990) Short-range order of cations in natural double-chain silicates: Markov-chain models. Translated from: Doclady Academii Nauk SSSR **303:** 190–193. (Transactions of the USSR Acad. Sci. Scripta Technica.) New York: Wiley.

Larralde, R., M. P. Robertson, and S. L. Miller (1995) Rates of decomposition of ribose and other sugars: Implications for chemical evolution. Proc. Natl. Acad. Sci. U.S.A. **92:** 8158–8160.

Laszlo, P. (1986) Molecular correlations of biological concepts. In: M. Florkin and E. H. Stolz (eds.) Comprehensive biochemistry. Vol. **34A.** Amsterdam: Elsevier Science Publishers.

Lawless, J. G., and N. Levi (1979) The role of metal ions in chemical evolution: Polymerization of alanine and glycine in a cation-exchanged clay environment. J. Mol. Evol. **13:** 281–286.

Lazard, D., N. Lahav, and J. Orenberg (1987) The biogeochemical cycle of the adsorbed template I: Formation of the template. Origins of Life Evol. Biosphere **17:** 135–148.

——— (1988) The biogeochemical cycle of the adsorbed template II: Selective adsorption of mononucleotides on adsorbed polynucleotides templates. Origins of Life Evol. Biosphere **18:** 347–357.

Lazcano, A. (1994a) A. I. Oparin: The man and his theory. In: B. F. Poglazow, B. I. Kurganov, M. S. Kritsky, and K. L. Gladilin (eds.) Evolutionary biochemistry and related areas of physicochemical biology. Moscow: Bach Institute of Biochemistry and ANKO.

——— (1994b) The RNA world, its predecessors, and its descendants. In: S. Bengtson (ed.) Early life on Earth. Nobel Symposium 84. New York: Columbia University Press. pp. 70–80.

——— (1994c) The transition from non-living to living. In: S. Bengtson (ed.) Early life on Earth. Nobel Symposium 84. New York: Columbia University Press. pp. 60–69.

Lazcano, A., G. E. Fox, and J. F. Oró (1992) Life before DNA: The origin and evolution of early archaean cells. In: R. P. Mortlock (ed.) The evolution of metabolic function. Boca Raton, Florida: CRC Press.

Lazcano, A., and S. L. Miller (1994) How long did it take for life to begin and evolve to cyanobacteria? J. Mol. Evol. **39:** 546–554.

——— (1996) The origin and early evolution of life: Prebiotic chemistry, the pre-RNA-world, and time. Cell **85:** 793–798.

Lederman, L. (1993) The God particle. Boston: Houghton Mifflin.

Lee, D. H., J. R. Granja, J. A. Martinez, K. Severin, and M. R. Ghadiri (1996) A self-replicating peptide. Nature **382:** 525–528.

Lee, T., M. L. Arrhenius, S. S.-Y. Hui, K. M. Ring, B. I. Gedulin, L. E. Orgel, and G. Arrhenius (1993) Double layer hydroxide minerals as host structures for bioorganic molecules. 10th International Conference on the Origin of Life, Barcelona, July 4–9.

Lee, Y.-H., L. Dsouza, and G. E. Fox (1993) Experimental investigation of an RNA sequence space. Origins of Life Evol. Biosphere **23:** 365–372.

Lehman, N., and G. F. Joyce (1993) Evolution in vitro of an RNA enzyme. Nature **361:** 182–185.

Lehninger, A. L., D. L. Nelson, and M. M. Cox (1993) Principles of biochemistry. New York: Worth Publishers.

Lenski, R. E., and J. E. Mittler (1993) The directed mutation controversy and neo-Darwinism. Science **259:** 188–194.

Lenski, R. E., and P. D. Sniegowski (1995) "Adaptive mutation": The debate goes on. Science **269:** 285–286.

Lerman, L. (1986) Exploration of the liquid-gas-interface as a reaction zone for condensation processes: The potential role of bubbles and droplets in primordial and planetary chemistry. Department of Geophysics, Stanford University.

——— (1993) The bubble-aerosol-droplet cycle as a natural reactor for prebiotic organic chemistry I, II. 7th ISSOL meeting, Barcelona, July 4–9.

Levine, L. (1993) Gaia: Goddess and idea. BioSystems **31:** 85–92.

L'Haridon, S., A. L.-R. Reysenbach, P. Glénat, D. Prieur, and C. Jeanthon (1995) Hot subterranean biosphere in a continental oil reservoir. Nature **377:** 223–224.

Li, T., and K. C. Nicolaou (1994) Chemical self-replication of palindromic duplex DNA. Nature **369:** 218–221.

Liebig, J. von (1868) Quoted in Engels ca. 1880, p. 190.

Lifson, S. (1987) Chemical selection, diversity, teleonomy, and the second law of thermodynamics. Biophys. Biochem. **26:** 303–311.

——— (1997) On the crucial stages in the origin of animated matter. J. Mol. Evol. **44:** 1–8.

Lindh, A. G. (1992) Natural selection. Nature **358:** 272.

Lipmann, F. (1965) Projecting backward from the present stage of evolution of biosynthesis. In: S. W. Fox (ed.) The origin of prebiological systems and of their molecular matrices. New York: Academic Press. pp. 212–226.

Liu, R., and L. E. Orgel (1997) Oxidative acylation using thioacids. Nature **389:** 52–54.

Lloyd, D., and E. I. Volkov (1991) The ultradian clock: Timekeeping for intracellular dynamics. In: C. Mosekilde and L. Mosekilde (eds.) Complexity, chaos, and biological evolution. New York: Plenum Press. pp. 51–59.

Lodish, H., D. Baltimore, A. Berk, S. L. Zipurzky, P. Matsudaira, and J. Darnell (1995) Molecular cell biology. 3rd ed. New York: Scientific American Books.

Lohse, P. A., and J. W. Szostak (1996) Ribozyme-catalyzed amino-acid transfer reactions. Nature **381:** 442–444.

Long, M., C. Rosenberg, and W. Gilbert (1995) Intron phase correlations and the evolution of the intron/exon structure of genes. Proc. Natl. Acad. Sci. U.S.A. **92:** 12495–12499.

Lorch, J. (1975) The charisma of crystals in biology. In: Y. Elkana (ed.) Interaction between science and philosophy. New York: Humanities Press. pp. 445–461.

Lorsch, J. R., and J. W. Szostak (1994) In vitro evolution of new ribozymes with polynucleotide kinase activity. Nature **371:** 31–36.

Lovelock, J. (1988) The ages of gaia: A biography of our living earth. New York: Norton.
Lovtrup, S. (1987) Darwinism: The refutation of a myth. London: Croom Helm.
Lowe, D. R. (1994) Early environments: Constraints and opportunities for early evolution. In: S. Bengtson (ed.) Early life on Earth. Nobel Symposium **84.** New York: Columbia University Press. pp. 24–35.
Luisi, P. L. (1994) The chemical implementation of autopoiesis. In: G. R. Fleischaker, S. Colonna, and P. L. Luisi (eds.) Self-production of supramolecular structures. Dordrecht: Kluwer Academic Press. pp. 179–197.
Luisi, P. L., and F. J. Varela (1989) Self-replicating micelles: A chemical version of a minimal autopoietic system. Origins of Life Evol. Biosphere **19:** 633–643.
Luisi, P., P. Walde, and T. Oberholzer (1994) Enzymatic RNA synthesis in self-reproducing vesicles: An approach to the construction of a minimal synthetic cell. Ber. Bunsenges. Phys. Chem. **98:** 1160–1165.
Lyne, J. E., and M. Tauber (1995) Origin of the Tunguska event. Nature **375:** 638–639.
Macallum, A. (1908) Quoted in J. Farley, 1977, p. 159.
MacDermott, A. J. (1993) The weak force and the origin of life. In: C. Ponnamperuma and J. Chela-Flores (eds.) Chemical evolution: Origin of life. Hampton, Virginia: A. Deepak Publishing. pp. 85–99.
Maciá, E., M. V. Hernández, and J. Oró (1997) Primary sources of phosphorus and phosphates in chemical evolution. Origins of Life Evol. Biosphere **27:** 459–480.
MacLeod, G., C. McKeown, A. J. Hall, and M. J. Russell (1994) Hydrothermal and oceanic pH conditions of possible relevance to the origin of life. Origins of Life Evol. Biosphere **24:** 19–41.
Maddox, J. (1994) Origin of the first cell membrane? Nature **371:**101.
Maden, B. E. H. (1995) No soup for starters?: Autotrophy and the origins of metabolism. TiBS **20:** 337–341.
Magner, L. N. (1994) A history of the life sciences. 2nd ed. New York: Marcel Dekker, Inc.
Maher, K. A., and J. D. Stevenson (1988) Impact frustration of the origin of life. Nature **331:** 612–614.
Maizels, N., and A. M. Weiner (1993) The genome tag hypothesis: Viruses as molecular fossils of ancient strategies for genomic replication. In: R. F. Gesteland and J. F. Atkins (eds.) The RNA world. New York: Cold Spring Harbor Laboratory Press. pp. 577–602.
Mancinelli, R. L., and C. P. McKay (1988) The evolution of nitrogen cycling. Origins of Life Evol. Biosphere **18:** 311–325.
Maniloff, J. (1996) The minimal cell genome: "On being the right size." Proc. Natl. Acad. Sci. U.S.A. **93:** 10004–10006.
——— (1997) A letter to the editor. Science **276:** 1776.
Marcus, J. N., and M. A. Olsen (1991) Biological implications of organic compounds in comets. In: R. L. Newburn, M. Neugelauer, and J. Rahe (eds.) Comets in the post-Halley era. Dordrecht: Kluwer Academic Publishers. **1:** 439–462.
Margulis, L. (1981) Symbiosis in cell evolution. San Francisco: W. H. Freeman and Company.
——— (1993a) Origins of species: Acquired genomes and individuality. BioSystems **31:** 121–125.
——— (1993b) Symbiosis in cell evolution. New York: W. H. Freeman and Company.
——— (1996) Archaeal-eubacterial mergers in the origin of eukarya: Phylogenetic classification of life. Proc. Natl. Acad. Sci. U.S.A. **93:** 1071–1076.
Maron, V. I., M. D. Nussinov, and S. Santoli (1993) Nanotechnology and nanobiology: Breakthroughs to new concepts in physics, chemistry, and biology. Nanobiology **2:** 189–199.
Marshall, C. R., E. C. Raff, and R. A. Raff (1994) Dollo's law and the death and resurrection of genes. Proc. Natl. Acad. Sci. U.S.A. **91:** 12283–12287.

Martell, E. A. (1992) Radionucleotide-induced evolution of DNA and the origin of life. J. Mol. Evol. **35:** 346–355.
Mason, S. F. (1991) Chemical evolution. Oxford: Clarendon Press.
——— (1997) Extraterrestrial handness. Nature **389:** 804.
Matsuno, K. (1984) Protobiology: A theoretical synthesis. In: K. Matsuno, K. Dose, K. Harada, and D. A. Rholfing (eds.) Molecular evolution and protobiology. New York: Plenum Press. pp. 433–464.
Matthews, B. W., C. S. Craik, and H. Neurath (1994) Can small cyclic peptides have the activity and specificity of proteolytic enzymes? Proc. Natl. Acad. Sci. U.S.A. **91:** 4103–4105.
Matthews, C. N. (1995) Hydrogen cyanide polymers: Prebiotic agents for the simultaneous origin of proteins and nucleic acids. In: J. Chela-Flores, A. chandra, A. Negrón-Mendoza, and T. Oshima (eds.) Chemical evolution: Self-organization of the macromolecules of life. Hampton, Virginia: Deepak Publishing. pp. 41–48.
Maurel, M.-C. (1992a) RNA in evolution: A review. J. Evol. Biol. **5:** 173–188.
——— (1992b) Studies of nucleic acid-like polymers as catalysts. J. Mol. Evol. **35:** 190–195.
Mauzerall, D. (1992) Minireview. Light, iron, Sam Granick, and the origin of life. Photosynthesis Research **33:** 163–170.
Maynard Smith, J. (1975) The theory of evolution. Harmondsworth: Penguin.
Maynard Smith, J., and E. Szathmáry (1995) The major transitions in evolution. Oxford: W. H. Freeman.
Mayr, E. (1982) The growth of biological thought. Cambridge: Harvard University Press.
——— (1988) Toward a new philosophy of biology: Observations of an evolutionist. Cambridge: The Belknap Press of Harvard University Press.
——— (1994) The resistance to Darwinism and the misconception on which it was based. In J. H. Campbell and J. W. Schopf (eds.) Creative evolution?! Boston: Jones and Bartlett Publishers. pp. 35–45.
McKay, C. P. (1997a) The search for life on Mars. Origins of Life Evol. Biosphere **27:** 263–289.
——— (1997b) Life in comets. In: P. J. Thomas, C. F. Chyba, and C. P. McKay (eds.) Comets and the origin and evolution of life. New York: Springer-Verlag. pp. 273–282.
McKay, D. S., E. G. Gibson Jr., K. L. Thomas-Keprta, H. Vali, C. S. Romanek, S. J. Clemett, X. D. F. Chillier, C. R. Maechling, and R. N. Zare (1996) Search for past life on Mars: Possible relic biogenic activity in Martian meteorite ALH84001. Science **273:** 924–930.
McLachlan, A. D. (1987) Gene duplication and the origin of repetitive protein structures. Cold Spring Harbor Symp. Quant. Biol. **52:** 411–420.
Mendelsohn, E. I. (1976) Philosophical biology versus experimental biology: Spontaneous generation in the seventeenth century. In: M. Grene and E. I. Mendelsohn (eds.) Topics in the philosophy of biology. Dordrecht-Holland/Boston-U.S.A. D. Reidel Publishing Company. pp. 36–65.
Mercer, E. H. (1981) The foundation of biological theory. New York: Wiley-Intersciences.
Miller, S. L. (1953) A production of amino acids under possible primitive Earth conditions. Science **117**: 528–529.
——— (1955) Production of some organic compounds under possible primitive earth conditions. Jour. Amer. Chem. Soc. **77:** 2351–2361.
——— (1987) Which organic compounds could have occurred on the prebiotic earth? Cold Spring Harbor Symp. Quant. Biol. **52:** 17–27.
——— (1992) The prebiotic synthesis of organic compounds as a step toward the origin of life. In: J. W. Schopf (ed.): Major events in the history of life. Boston: Jones and Bartlett Publishers. pp. 1–28.
Miller, S. L., and J. L. Bada (1988) Submarine hot springs and the origin of life. Nature **334:** 606–611.

Miller, S. L., and A. Lazcano (1995) The origin of life: Did it occur at high temperatures? J. Mol. Evol. **41:** 689–692.
Miller, S. L., and L. E. Orgel (1974) The origins of life on the Earth. Englewood Cliffs, New Jersey: Prentice-Hall.
Miller, S. L., and M. Paris (1964) Synthesis of pyrophosphate under primitive earth conditions. Nature **204:** 1248–1250.
Miller, S. L., and G. Schlesinger (1993a) Prebiotic syntheses of vitamine coenzymes: I. Cysteamine and 2-mercaptoethanesulfonic acid (coenzyme M). J. Mol. Evol. **36:** 302–307.
——— (1993b) Prebiotic syntheses of vitamine coenzymes: II. Pantoic acid, panthotenic acid, and the composition of coenzyme A. J. Mol. Evol. **36:** 308–314.
Milstein, M. (1995) A glimpse of early life. Science **270:** 226.
Mingers, J. (1995) Self-reproducing systems. New York: Plenum Press.
Misra, K. K. (1992) Origin of life beyond replication. Proc. Zool. Soc. Calcutta **45:** 101–112.
Mojzsis, S. J., G. Arrhenius, K. D. McKeegan, T. M. Harrison, A. P. Nutman, and G. R. L. Friend (1996) Evidence for life on Earth before 3,800 million years ago. Nature **384:** 55–59.
Möller, W., and G. M. C. Janssen (1990) Transfer RNAs for primordial amino acids contain remnants of a primitive code at position 3 to 5. Biochimie **72:** 361–368.
——— (1992) Statistical evidence for remnants of the primordial code in the acceptor stem of prokaryotic transfer RNA. J. Mol. Evol. **34:** 471–477.
Monod, J. (1971) Chance and necessity. New York: Knopf.
Mooers, A. Ø., and R. J. Redfield (1996) Digging up the roots of life. Nature **379:** 587–588.
Moore, M. J. (1995) Exploration by lamp light. Nature **374:** 766–767.
Moore, P. B. (1993) Ribosomes and the RNA World. In: R. F. Gesteland and J. F. Atkins (eds.) The RNA world. New York: Cold Spring Harbor Laboratory Press. pp. 119–135.
Morán, F., A. Moreno, J. J. Merelo, and P. Chacón (1995) In: F. Morán, A. Moreno, J. J. Merelo, and P. Chacon (eds.) Advances in artificial life. Proceedings of the Third European Conference on Artificial Life. Granada, Spain. June. pp. v–iv. Berlin: Springer-Verlag.
Moras, D. (1992) Structural and functional relationships between aminoacyl-tRNA synthetases. TiBS, 17 April: 159–164.
Morell, V. (1996) Proteins "clock" the origins of all creatures—great and small. Science **271:** 448.
Morowitz, H. J. (1992) Beginnings of cellular life: Metabolism recapitulates biogenesis. New Haven: Yale University Press.
——— (1994) Artificial biochemistry: Life before enzymes. In: C. G. Langton (ed.) SFI studies in the sciences of complexity, Proc. Vol. 17. Reading: Addison-Wesley. pp. 381–388.
——— (1996) A letter to the editor. Science **276:** 1639–1640.
Morowitz, H. J., B. Heinz, and D. W. Deamer (1988) The chemical logic of a minimum protocell. Origins of Life Evol. Biosphere **18:** 281–287.
Morowitz, H. J., E. Peterson, and S. Chang (1995) The synthesis of glutamic acid in the absence of enzymes: Implications for biogenesis. Origins of Life Evol. Biosphere **25:** 395–399.
Morrison, D. (1997) The contemporary hazards of cometary impacts. In: P. J. Thomas, C. F. Chyba, and C. P. McKay (eds.) Comets and the origin and evolution of life. New York: Springer-Verlag. pp. 243–258.
Mosekilde, C., and L. Mosekilde (1991) Structure, complexity, and chaos in living systems. In: C. Mosekilde and L. Mosekilde (eds.) Complexity, chaos, and biological evolution. New York: Plenum Press.
Mosqueira, F. G., G. Albarrán, and A. Negrón-Mendoza (1996) A review of conditions

affecting the radiolysis due to ^{40}K on nucleic acid bases and their derivatives adsorbed on clay minerals: Implications in prebiotic chemistry. Origins of Life Evol. Biosphere **26:** 75–94.

Moya, A., E. Domingo, and J. J. Holland (1995) RNA viruses: A bridge between life and artificial life. In: F. Moran, A. Moreno, J. J. Merelo, and P. Chacon (eds.) Advances in artificial life. Proceedings of the Third European Conference on Artificial Life. Granada, Spain. June 1995. Berlin: Springer-Verlag. pp. 170–178.

Mushegian, A. R., and E. V. Koonin (1996) A minimal gene set for cellular life derived by comparison of complete bacterial genomes. Proc. Natl. Acad. Sci. U.S.A. **93:** 10268–10273.

Nagel, G. M., and R. F. Doolittle (1995) Phylogenetic analysis of the aminoacyl-tRNA synthetases. J. Mol. Evol. **40:** 487–498.

Navarro-Gonzalez, R., R. K. Khanna, and C. Ponnamperuma (1993) Chirality and the origins of life. In: C. Ponnamperuma and J. Chela-Flores (eds.) Chemical evolution: Origin of life. Hampton, Virginia: A. Deepak Publishing. pp. 135–155.

Nealson, K. H. (1997) A letter to the editor. Science **276:** 1639.

Negrón-Mendoza, A., and G. Albarrán (1993) Chemical effects of ionizing radiation and sonic energy in the context of chemical evolution. In: C. Ponnamperuma and J. Chela-Flores (eds.) Chemical evolution: Origin of life. Hampton, Virginia: A. Deepak Publishing. pp. 235–247.

Negrón-Mendoza, A., G. Albarrán, and S. Ramos-Bernal (1996) Clays as natural catalyst in prebiotic processes. In: J. Chela-Flores and F. Roulin (eds.) Chemical evolution: Physics of the origin and evolution of life. Dordrecht: Kluwer Academic Publishers, pp. 97–106.

Nemoto, N., and Y. Husimi (1995) A model of the virus-type strategy in the early stages of encoded molecular evolution. J. Theor. Biol. **176:** 67–77.

Nicholas, H. B., Jr., and W. H. McClain (1995) Searching tRNA sequences for relatedness to aminoacyl-tRNA synthetase families. J. Mol. Evol. **40:** 482–486.

Nielsen, P. G. (1993) Peptide nucleic acid (PNA): A model structure for the primordial genetic material? Origins of Life Evol. Biosphere **23:** 323–327.

Ninio, J. (1983) Molecular approaches to evolution. Princeton: Princeton University Press. (Originally published in 1979 in French.)

Nir, S., J. Bentz, J. Wilschut, and N. Duzgunes (1983) Aggregation and fusion of vesicles. Prog. in Surface Science **13:** 1–124.

Nir, S., K. Klappe, and D. Hoekstra (1986) Kinetics of fusion between sendai virus and erythrocyte ghosts: Application of mass action kinetic model. Biochemistry **25:** 2155–2161.

Nir, S., and N. Lahav (1997) Emergence of template- and sequence-directed (TSD) syntheses II: A computer simulation model. Origins of Life Evol. Biosphere **27:** 567–584.

Nir, S., T. Stegmann, and J. Wilschut (1986) Fusion of influenza virus and cardiolipin liposomes at low pH: Mass action analysis of kinetics and extent. Biochemistry **25:** 257–266.

Nisbet, E. G., J. R. Cann, and C. L. Van Dover (1995) Origins of photosynthesis. Nature **373:** 480–481.

Nisbet, E. G., and C. M. R. Fowler (1996) Some like it hot. Nature **382:** 404–405.

Noller, H. F. (1993) On the origin of the ribosome: Coevolution of subdomains of tRNA and rRNA. In: R. F. Gesteland and J. F. Atkins (eds.) The RNA world. New York: Cold Spring Harbor Laboratory Press. pp. 137–156.

Noller, H. F., V. Hoffarth, and L. Zimniak (1992) Unusual resistance of peptidyl transferase to protein extraction procedures. Science **256:** 1416–1419.

Nussinov, M. D., and V. I. Maron (1990) The universe and the origin of life (origin of organics on clays). J. British Interplanetary Soc. **43:** 3–10.

——— (1993) Impulse paradigm of self-organization of matter in the universe, and nanobiological principles. Nanobiology **2:** 215–228.

Nussinov, M. D., V. A. Otroschenko, and S. Santoll (1997) The emergence of the non-cellular phase of life on the fine-grade clayish particles of the early Earth's regolith. BioSystems **42:** 111–118.

Oberbeck, V. R., and G. Fogleman (1989) Estimates of the maximum time required to originate life. Origins of Life Evol. Biosphere **19:** 549–560.

Oberbeck, V. R., J. Marshall, and T. Shen (1991) Prebiotic chemistry in clouds. J. Mol. Evol. **32:** 296–303.

Ohmoto, H., T. Kakegawa, and D. R. Lowe (1993) 3.4-bilion-year-old biogenic pyrite from Barberton, South Africa: Sulfur isotope evidence. Science **262:** 555–557.

Olomucki, M. (1993) The chemistry of life. New York: McGraw-Hill.

Olsen, G. J. (1994) Archaea, Archaea, everywhere. Nature **371:** 657–658.

Oparin, A. I. (1957) The origin of life on the earth. 3rd ed. New York: Academic Press.

——— (1961) Life: Its nature, origin, and development. Edinburgh: Oliver and Boyd.

Orgel, L. E. (1968) Evolution of the genetic apparatus. J. Mol. Biol. **38:** 381–393.

——— (1986) RNA catalysis and the origin of life. J. Theor. Biol. **123:** 127–149.

——— (1987) Evolution of the genetic apparatus: A review. Cold Spring Harbor Symp. Quant. Biol. **52:** 9–16.

——— (1989) The origin of polynucleotide-directed protein synthesis. J. Mol. Evol. **29:** 465–474.

——— (1992) Molecular replication. Nature **358:** 203–209.

——— (1994) The origin of life on the earth. Scientific American, October: 77–83.

——— (1995) Unnatural selection in chemical systems. Acc. Chem. Res. **28:** 109–118.

Oró, J. (1960) Synthesis of adenine from ammonium cyanide. Biochem. Biophys. Res. Commun. **2:** 407–412.

——— (1961) Comets and the formation of biochemical compounds on the primitive earth. Nature **190:** 389–390.

——— (1994) Early chemical stages in the origin of life. In: S. Bengtson (ed.) Early life on Earth. Nobel Symposium 84. New York: Columbia University Press. pp. 48–59.

Oró, J., and A. Lazcano (1984) A minimal living system and the origin of a protocell. Adv. Space Res. **4:** 167–176.

——— (1997) Comets and the origin and evolution of life. In: P. J. Thomas, C. F. Chyba, and C. P. McKay (eds.) Comets and the origin and evolution of life. New York: Springer-Verlag. pp. 3–27.

Oró, J., T. Mills, and A. Lazcano (1992) Comets and the formation of biochemical compounds on the primitive earth—A review. Origins of Life Evol. Biosphere **21:** 267–277.

Österberg, R. (1974) Origins of metal ions in biology. Nature **249:** 382–383.

——— (1997) On the prebiotic role of iron and sulfur. Origins of Life Evol. Biosphere **27:** 481–484.

Ovenden, M. W. (1987) Of stars, planets, and life. In: J. M. Robson (ed.) Origin and evolution of the universe: Evidence for design? Kingston, Ontario: McGill-Queen's University Press. pp. 87–107.

Pace, N. R. (1991) Origin of life: Facing up to the physical setting. Cell **65:** 531–533.

Pattee, H. H. (1995) Artificial life needs a real epistemology. In: F. Morán, A. Moreno, J. J. Merelo, and P. Chacón (eds.) Advances in artificial life. Proceedings of the Third European Conference on Artificial Life. Granada, Spain. June 1995. Berlin: Springer-Verlag. pp. 23–38.

Peaff, G. (1996) Peptide catalyzes its own replication. C&EN, August 12: 8–9.

Perrett, J. (1952) Biochemistry and bacteria. New Biology **12:** 68–69.

Petty, H. R. (1993) Molecular biology of membranes. New York: Plenum Press.

Pfeffer, W. (1897) Quoted in L. von Bertalanffy, 1993, p. 48.

Piccirilli, J. A. (1995) RNA seeks its maker. Nature **376:** 548–549.

Pitsch, S., A. Eschenmoser, B. Gedulin, S. Hui, and G. Arrhenius (1995) Mineral-induced formation of sugar phosphates. Origins of Life Evol. Biosphere **25:** 297–334.

Pledge, H. T. (1959) Science since 1500. New York: Harper Torchbooks/Science Library.

Pohorille, A., and M. W. Wilson (1995) Molecular dynamics studies of simple membrane-water interfaces: Structure and function in the beginning of cellular life. Origins of Life Evol. Biosphere **25:** 21–46.

Polzin, K. L., J. M. Toole, J. R. Ledwell, and R. W. Schmitt (1997) Spatial variability of turbulent mixing in the abyssal ocean. Science **276:** 93–96.

Ponnamperuma, C., and J. Chela-Flores, eds. (1993) Chemical evolution: Origin of life. Hampton, Virginia: A. Deepak Publishing.

Popa, R. (1997) A sequential scenario for the origin of biological chirality. J. Mol. Evol. **44:** 121–127.

Preston, C. M., K. Y. Wu, T. F. Molinski, and D. F. DeLong (1996) A psychrophilic crenarchaeon inhabits a marine sponge: Cenarchaeum symbiosum gen. nov., sp. nov. Proc. Natl. Acad. Sci. U.S.A. **93:** 6241–6246.

Prigogine, I. (1984) Order out of chaos: Man's new dialogue with Nature. Toronto: Bantam Books.

——— (1988) Origin of complexity. In: A. C. Fabian (ed.): Origins. The Darwin College Lectures. Cambridge: Cambridge University Press. pp. 69–88.

Prince, R. C., and M. J. Grossman (1993) Novel iron-sulfur clusters. TiBS **18:** 153–154.

Psenner, R., and M. Loferer (1997) Letters to the editor. Science **276:** 1777.

Putter, A. (1923) Quoted in L. van Bertalanffy, 1933, p. 51.

Rasmussen, S., R. Feldberg, and C. Knudsen (1992) Self-programming of matter and the evolution of proto-biological organization. In: J. and K. Tran Thanh Van, J. C. Mounolou, J. Schneider, and C. McKay (eds.) Frontiers of life. Singapore: Fong & Sons Printers. pp. 133–143.

Raulin, F. (1992) Titan: A prebiotic planet? In: J. Tran Than Van, K. Mounolou, J. Schneider, and C. McKay (eds.) Frontiers of life. Singapore: Fong & Sons Printers. pp. 326–346.

Reanney, D. C. (1977) Aminoacyl thioesters and the origins of genetic specificity. J. Theor. Biol. **65:** 555–569.

Rebek, J., Jr. (1994) Extrabiotic replication and self-assembly. In: G. R. Fleischaker, S. Colonna, and P. L. Luisi (eds.) Self-production of supramolecular structures. Dordrecht: Kluwer Academis Press. pp. 75–87.

Ritterbush, P. C. (1964) Overture to biology. New Haven: Yale University Press.

Robertson, M. P., M. Levy, and S. L. Miller (1996) Prebiotic synthesis of diaminopyrimidine and thiocytosine. J. Mol. Evol. **43:** 543–550.

Robertson, M. P., and S. L. Miller (1995a) An efficient prebiotic synthesis of cytosine and uracil. Nature **375:** 772–774.

——— (1995b) Prebiotic synthesis of 5-substituted uracils: A bridge between the RNA world and the DNA-protein world. Science **268:** 702–705.

Robinson, J. M. (1968) An introduction to early Greek philosophy. Boston: Houghton Mifflin Company.

Rodin, S. N., and S. Ohno (1995) Two types of aminoacyl-tRNA synthetases could be originally encoded by complementary strands of the same nucleic acid. Origins of Life Evol. Biosphere **25:** 565–589.

Rodin, S., A. Rodin, and S. Ohno (1996) The presence of codon-anticodon pairs in the acceptor stem of tRNAs. Proc. Natl. Acad. Sci. U.S.A. **93:** 4537–4542.

Rogers, K. C., and D. Söll (1995) Divergence of glutamate and glutamine aminoacylation pathways: Providing the evolutionary rationale for mischarging. J. Mol. Evol. **40:** 476–481.

Root-Bernstein, R. S. (1982) Amino acid pairing. J. Theor. Biol. **94:** 885–894.

——— (1983) Protein replication by amino acid pairing. J. Theor. Biol. **100:** 99–106.

Root-Bernstein, R. S., and P. F. Dillon (1997) Molecular complementarity: I: The complementarity theory of the origin and evolution of life. J. Theor. Biol. **188:** 447–479.

Rosen, R. (1989) The roles of necessity in biology. In: J. Casti and A. Karlkvist (eds.) Newton to Aristo: Toward a theory of models for living systems. Boston: Birkhauser. pp. 11–39.

Rouault, T.A., and R. D. Klausner (1996) Iron-sulfur clusters as biosensors of oxidants and iron. TiBS **21:** 174–177.
Rowlands, T., P. Baumann, and S. P. Jackson (1994) The TATA-binding protein: A general transcription factor in eukaryotes and archaebacteria. Science **264:** 1326–1329.
Ruse, M. (1979) The Darwinian revolution. Chicago: University of Chicago Press.
——— (1997) The origin of life: Philosophical perspectives. J. Theor. Biol. **187:** 473–482.
Russell, M. J., R. M. Daniel, and A. J. Hall (1993) On the emergence of life via catalytic iron-sulfide membranes. Terra Nova **5:** 343–347.
Russell, M. J., R. M. Daniel, A. J. Hall, and J. A. Sherringham (1994) A hydrothermally precipitated catalytic iron sulfide membrane as a first step toward life. J. Mol. Evol. **39:** 231–243.
Saccone, C., G. Gissi, C. Lanave, and G. Pesole (1995) Molecular classification of living organisms. J. Mol. Evol. **40:** 273–279.
Sagan, C., and S. J. Ostro (1994) Dangers of asteroid deflection. Nature 368: 501.
Saks, M. E., and J. R. Sampson (1995) Evolution of tRNA recognition system and tRNA gene sequences. J. Mol. Evol. **40:** 509–518.
Saks, M. E., J. R. Sampson, and J. N. Abelson (1994) The transfer RNA identity problem: A search for rules. Science **263:** 191–197.
Samaha, R. R., R. Green, and H. F. Noller (1995) A base pair between tRNA and 23S rRNA in the peptidyl transferase centre of the ribosome. Nature **377:** 309–314.
Santoli, S. (1997) Introduction. BioSystems **42:** 77–84.
Sassanfar, M., and J. W. Szostak (1993) An RNA motif that binds ATP. Nature **364:** 550–553.
Sassone-Corsi, P. (1994) Rhythmic transcription and autoregulatory loops: Winding up the biological clock. Cell **78:** 361–364.
Sattler, R. (1986) Bio-philosophy. Berlin: Springer-Verlag.
Schidlowski, M. (1988) A 3,800-million-year isotopic record of life from carbon in sedimentary rocks. Nature **333:** 313–318.
——— (1993) The beginnings of life on earth: Evidence from the geological record. In: J. M. Greenberg, C. X. Mendosa-Gómez, and V. Pirronello (eds.) The chemistry of life's origins. Dordrecht: Kluwer Academic Publishers pp. 389–414.
Schimmel, P. (1995) An operational RNA code for amino acids and variations in critical nucleotide sequences in evolution. J. Mol. Evol. **40:** 531–536.
——— (1996) Origin of genetic code: A needle in the haystack of tRNA sequences. Proc. Natl. Acad. Sci. U.S.A. 93: 4521–4522.
Schimmel, P., R. Giege, D. Moras, and S. Yokoyama (1993) An operational RNA code for amino acids and possible relationship to genetic code. Proc. Natl. Acad. Sci. U.S.A. **90:** 8763–8768.
Schimmel, P., and B. Henderson (1994) Possible role of aminoacyl-RNA complexes in noncoded peptide synthesis and origin of coded synthesis. Proc. Natl. Acad. Sci. U.S.A. **91:** 11283–11286.
Schimmel, P., and L. R. de Pouplana (1995) Transfer RNA: From minihelix to genetic code. Cell **81:** 983–986.
Schimmel, P., and E. Schmidt (1995) Making connections: RNA-dependent amino acid recognition. TiBS **20:** 1–2.
Schleper, C., G. Pühler, B. Kühlmorgen, and W. Zillig (1995) Life at extremely low pH. Nature **375:** 741–742.
Schmidt, J. G., P. E. Nielsen, and L. E. Orgel (1997) Enantiomeric cross-inhibition in the synthesis of oligonucleotides on a non-chiral template. J. Am. Chem. Soc. **119:** 1494–1495.
Schopf, J. W. (1992a) The oldest evidence of life. In: J. and K. Tran Than Van, J. C. Monolou, J. Schneider, and C. McKay (eds.) Frontiers of life. Singapore: Fong & Sons Printers. pp. 235–262.
——— (1992b) The oldest fossils and what they mean. In: J. W. Schopf (ed.) Major events in the history of life. Boston: Jones and Bartlett Publishers. pp. 29–63.

——— (1993) Microfossils of the early fossil archean apex chert: New evidence of the antiquity of life. Science **260:** 640–646.

Schrödinger, E. (1944) What is life?: Mind and matter. Cambridge: Cambridge University Press.

Schuster, P. (1984) Evolution between chemistry and biology. Origins of Life **14:** 3–14.

——— (1993) RNA based evolutionary optimization. Origins of Life Evol. Biosphere **23:** 373–391.

——— (1995) Artificial life and molecular evolutionary biology. In: F. Moran, A. Moreno, J. J. Merelo, and P. Chacon (eds.) Advances in artificial life. Proceedings of the Third European Conference on Artificial Life. Granada, Spain. June 1995. Berlin: Springer-Verlag. pp. 3–19.

Schwabe, C. (1985) On the basis of the studies of the origins of life. Origins of Life Evol. Biosphere **15:** 213–216.

Schwartz, A. W. (1993) Biology and theory: RNA and the origin of life. In: J. M. Greenberg, C. X. Mendoza-Gómez, and V. Pirronello (eds.) The Chemistry of Life's Origins. Dordrecht: Kluwer Academic Press. pp. 323–344.

——— (1997a) Speculation on the RNA precursor problem. J. Theor. Biol. **187:** 523–527.

——— (1997b) Prebiotic phosphorus chemistry reconsidered. Origins of Life Evol. Biosphere **27:** 505–512.

Schwartz, R. M., and M. O. Dayhoff (1978) Origins of prokaryotes, eukaryotes, mitochondria. and chloroplasts. Science **199:** 395–403.

Schwartz, A. W., and L. E. Orgel (1985) Template-directed synthesis of novel, nucleic acid-like structures. Science **228:** 585–587.

Schwartz, A. W., and M. J. van Vliet (1994) Chirality and the first self-replicating molecules. In: G. R. Fleischaker, S. Colonna, and P. L. Luisi (eds.) Self-production of supramolecular structures. Dordrecht: Kluwer Academic Press. pp. 107–114.

Schwartzman, D. W., S. N. Shore, T. Volk, and M. McMenamin (1994) Self-organization of the earth's biosphere —Geochemical or geophysical? Origins of Life Evol. Biosphere **24:** 435–450.

Schwendinger, M. G., and B. M. Rode (1989) Possible role of copper and sodium chloride in prebiotic evolution of peptides. Analytical Sciences **5:** 411–414.

Scott, J. (1981) Natural selection in the primordial soup. New Scientist, 15 January: 153–154.

Segré, D., Y. Pilpel, and D. Lancet (1998) Mutual catalysis in sets of prebiotic organic molecules: Evolution through computer simulated chemical kinetics. Physica A **249:** 558–564.

Senapathy, P. (1995) Introns and the origin of protein-coding genes. Science **268:** 1366–1367.

Shanahan, M. (1995) Evolutionary automata. In: R. A. Brooks and P. Maes (eds.) Artificial life IV. Proceedings of the Fourth International Workshop on the Synthesis and Simulation of Living Systems. A Bradford Book. Cambridge: MIT Press. pp. 388–393.

Shapiro, R. (1988) Prebiotic ribose synthesis: A critical analysis. Origins of Life Evol. Biosphere **18:** 71–85.

——— (1995) The prebiotic role of adenine: A critical analysis. Origins of Life Evol. Biosphere **25:** 83–89.

Sheldrake, R. (1988) The presence of the past. London: Collins.

Shih, M.-C., P. Heinrich, and H. M. Goodman (1988) Intron existence predated the divergence of eukaryotes and prekaryotes. Science **242:** 1165–1166.

Shimizu, M. (1982) Molecular basis for the genetic code. J. Mol. Evol. **18:** 297–303.

Shock, E. L. (1990) Do amino acids equilibrate in hydrothermal fluids? Geochim. Cosmochim. Acta **54:** 1185–1189.

——— (1992) Chemical environments of submarine hydrothermal systems. Origins of Life Evol. Biosphere **22:** 67–108.

Shock, E. L., T. McCollom, and M. D. Schulte (1995) Geochemical constrains on

chemolithoautotrophic reactions in hydrothermal systems. Origins of Life Evol. Biosphere **25:** 141–159.
Siegel, V. (1997) Recognition of a transmembrane domain: Another role for the ribosome? Cell **90:** 5–8.
Sievers, D., T. Achilles, J. Burmeister, S. Jordan, A. Terfort, and G. von Kiedrowski (1994) Molecular replication: From minimal to complex systems. In: G. R. Fleischaker, S. Colonna, and P. L. Luisi (eds.) Self-production of supramolecular structures. Dordrecht: Kluwer Academic Press. pp. 45–64.
Sievers, D., and G. von Kiedrowski (1994) Self-replication of complementary nucleotide-based oligomers. Nature **369:** 221–224.
Simoneit, B. R. T. (1995) Evidence for organic synthesis in high temperature aqueous media—Facts and prognosis. Origins of Life Evol. Biosphere **25:**119–140.
Sleep, N. H., K. J. Zahnle, J. F. Kasting, and H. J. Morowitz (1989) Annihilation of ecosystems by large asteroid impacts on the early earth. Nature **342:** 139–142.
Smith, M. W., D.-F. Feng, and R. F. Doolittle (1992) Evolution by acquisition: The case for horizontal gene transfer. TiBS **17:** 489–493.
Smith, T. F., and H. J. Morowitz (1982) Between history and physics. J. Mol. Evol. **18:** 265–282.
Sober, E. R. (1993) Philosophy of biology. Boulder, Colorado: Westview Press.
Soll, D. (1993) Transfer RNA: An RNA for all seasons. In: R. F. Gesteland and J. F. Atkins (eds.) The RNA world. New York: Cold Spring Harbor Laboratory Press. pp. 157–184.
Spencer, H. (1884) The principles of biology. New York: D. Appleton and Company.
Staves, M. P., D. P. Bloch, and J. C. Lacey Jr. (1988) Evolution of E. coli $tRNA^{Trp}$. Origins of Life Evol. Biosphere **18:** 97–105.
Steel, D. (1997) Cometary impacts on the biosphere. In: P. J. Thomas, C. F. Chyba, and C. P. McKay (eds.) Comets and the origin and evolution of life. New York: Springer-Verlag, pp. 209–242.
Stein, J. L., and M. I. Simon (1996) Archaeal ubiquity. Proc. Natl. Acad. Sci. U.S.A. **93:** 6228–6230.
Stetter, K. O. (1992) Life at the upper temperature border. In: J. and K. Tran Thanh Van, J. C. Monolou, J. Schneider, and C. McKay (eds.) Frontiers of life. Singapore: Fong & Sons Printers. pp. 195–219.
——— (1994) The lesson of archaebacteria. In: S. Bengtson (ed.) Early life on Earth. Nobel Symposium 84. New York: Columbia University Press. pp. 143–151.
Stevens, O. T., and J. P. McKinley (1995) Lithoautotrophic microbial ecosystems in deep basalt aquifers. Science **270:** 450–454.
Stilwell, W. (1980) Facilitated diffusion as a method for selective accumulation of materials from the primordial ocean by a lipid-vesicle protocell. Origins of Life Evol. Biosphere **10:** 277–292.
Stoltzfus, A., D. F. Spencer, M. Zuker, G. M. Logsdon Jr., and W. F. Doolittle (1994) Testing the exon theory of genes: The evidence from protein structure. Science **265:** 202–207.
——— (1995) Introns and the origin of protein-coding genes. Science **269:** 1367–1369.
Stribling, R., and S. L. Miller (1991) Template-directed synthesis of oligonucleotides under eutectic conditions. J. Mol. Evol. **32:** 289–295.
Stryer, L. (1995) Biochemistry. New York: W. H. Freeman and Company.
Swanson, C. P., and P. L. Webster (1985) The cell. Englewood Cliffs, New Jersey: Prentice-Hall.
Szathmáry, E. (1993) Coding coenzyme handles: A hypothesis for the origin of the genetic code. Proc. Natl. Acad. Sci. U.S.A. **90:** 9916–9920.
——— (1994) Self-replication and reproduction: From molecules to protocells. In: G. P. Fleischaker, S. Colonna, and P. L. Luisi (eds.) Self-production of supramolecular structures. Dordrecht: Kluwer Academic Press. pp. 65–73.

Szathmáry, E., and J. Maynard Smith (1997) From replicators to reproducers: The first major transitions leading to life. J. Theor. Biol. **187:** 555–571.
Taylor, G. J. (1994) The scientific legacy of Apollo. Scientific American, July: 40–47.
Taylor, F. J. R., and D. Coates (1989) The code within the codons. BioSystems **22:** 177–187.
Thom, R. (1989) Causality and finality in theoretical biology: A possible picture. In: J. Casti and A. Karlkvist (eds.) Newton to Aristo: Toward a theory of models for living systems. Boston: Birkhauser. pp. 39–45.
Thomas, J. M. (1994) Turning points in catalysis. Angew. Chem. Int. Ed. Engl. **33:** 917–937.
Tickell, C. (1993) Gaia: Goddess or thermostat. Biosystems **31:** 93–98.
Tiedemann, H. (1997) "Killer" impacts and life's origin. Science **277:** 1687–1688.
Travis, J. (1994) Hints of first amino acid outside solar system. Science **264:** 1668.
Turney, J. (1988) Jacques Loeb: Pioneer of test-tube life. New Scientist, 14 January: 63.
Turner, D. H., and P. C. Bevilacqua (1993) Thermodynamic considerations for evolution by RNA. In: R. F. Gesteland and J. F. Atkins (eds.) The RNA world. New York: Cold Spring Harbor Laboratory Press. pp. 447–464.
Ueda, T., and K. Watanabe (1993) The evolutionary change of the genetic code as restricted by the anticodon and identity of transfer RNA. Origins of Life Evol. Biosphere **23:** 345–364.
Urey, H. C. (1952) On the early chemical history of the earth and the origin of life. Proc. Natl. Acad. Sci. U.S.A. **38:** 351–363.
van Bladeren, A., R. Ruel, and P. Wiltzius (1997) Template-directed colloidal crystalization. Nature **385:** 321–324.
van Vliet, M. J., J. Visser, and A. W. Schwartz (1994) An achiral (oligo) nucleotide analog. J. Mol. Evol. **38:** 438–442.
——— (1995) Hydrogen bonding in the template-directed olgomerization of a pyrimidine nucleotide analogue. J. Mol. Evol. **41:** 257–261.
Varela, F. (1994) On defining life. In G. R. Fleischaker, S. Colonna, and P. L. Luisi (eds.) Self-production of supramolecular structures. Dordrecht: Kluwer Academic Press, pp. 23–31.
Varela, F., H. R. Maturana, and R. Uribe (1974) Autopoiesis: The organization of living systems, its characterization, and a model. BioSystems **5:** 187–196.
Vaughan, V. C. (1927) A chemical concept of the origin and development of life. Chem. Rev. **4:** 167–188.
Velikovsky, I. (1950) Worlds in collision. Garden City, New York: Doubleday.
Virchov, R. (1855) Quoted in J. Farley, 1977, p. 54.
Von Damm, K. L., S. E. Oosting, R. Kozlowski, L. G. Buttermore, D. C. Colodner, H. N. Edmonds, J. M. Edmond, and J. M. Grebmeler (1995) Evolution of east Pacific rise hydrothermal vent fluids following a volcanic action. Nature **375:** 47–50.
von Kiedrowski, G. (1986) A self-replicating hexadeoxynucleotide. Angew. Chem. Int. Ed. Engl. **25:** 932–935.
——— (1993) Minimal replicator theory I: Parabolic versus exponential growth. Bioorganic Chemistry Frontier. **3:** 113–146.
——— (1996) Primordial soup or crêpes? Nature **381:** 20–21.
Wächtershäuser, G. (1988) Before enzymes and templates: Theory of surface metabolism. Microbiological Rev. **52:** 452–484.
——— (1990) Evolution of the first metabolic cycles. Proc. Natl. Acad. Sci. U.S.A. **87:** 200–204.
——— (1992a) Groundworks for an evolutionary biochemistry: The iron-sulfur world. Prog. Biophys. Molec. Biol. **58:** 85–201.
——— (1992b) Order out of order: Heritage of the iron-sulfur world. In: J. and K. Tran Than Van, J. C. Monolou, J. Schneider, and C. McKay (eds.) Frontiers of Life. Singapore: Fong & Sons Printers. pp. 21–39.

——— (1993) The cradle chemistry of life: On the origin of natural products in the pyrite-pulled chemoautotrophic origin of life. Pure & Appl. Chem. **65:** 1343–1348.

——— (1994a) Life in a ligand sphere. Proc. Natl. Acad. U.S.A. **91:** 4283–4287.

——— (1994b) Vitalysts and virulists: A theory of self-expanding reproduction. In: S. Bengtson (ed.) Early life on Earth. Nobel Symposium 84. New York: Columbia University Press. pp. 124–132.

——— (1996) The uses of Karl Popper. In: A. O'Hear (ed) Karl Popper: Philosophy and problems. Cambridge: Cambridge University Press. pp. 177–189.

——— (1997) The origin of life and its methodological challenge. J. Theor. Biol. **187:** 483–494.

Waddell, T. G., B. S. Henderson, R. T. Morris, C. M. Lewis, and A. C. Zimmermann (1987) Chemical evolution of the citric acid cycle: Sunlight photolysis of α-ketoglutaric acid. Origins of Life **17:** 149–153.

Waldrop, M. M. (1992) Finding RNA makes proteins gives "RNA World" a big boost. Science **256:** 1396–1397.

Wallis, M. K., N. C. Wickramasinghe, and F. Hoyle (1992) Cometary habitats for primitive life. Adv. Space Res. **12:** 281–285.

Walse, B., M. Ullner, C. Lendbladh, L. Bülow, T. Drakenberg, and O. Teleman (1996) Structure of a cyclic peptide with a catalytic triad, determined by computer simulation and NMR spectroscopy. J. Computer-Aided Mol. Design **10:** 11–22.

Walter, M. (1996) Old fossils could be fractal frauds. Nature **383:** 385–386.

Weber, A. L. (1987) The triose model: Glyceraldehyde as a source of energy and monomers for prebiotic condensation reactions. Origins of Life Evol. Biosphere **17:** 107–119.

——— (1995) Prebiotic polymerization: Oxidative polymerization of 2,3-dimercapto-1-propanol on the surface of iron(III) hydroxide oxide. Origins of Life Evol. Biosphere **25:** 53–60.

——— (1997) Energy from redox disproportionation of sugar carbon drives biotic and abiotic synthesis. Origins of Life Evol. Biosphere **44:** 354–360.

Weber, A. L., and S. L. Miller (1981) Reasons for the occurrence of the twenty coded amino acids. J. Mol. Evol. **17:** 273–284.

Weber, A. L., and L. E. Orgel (1979) The formation of peptides from glycine thioesters. J. Mol. Evol. **13:** 193–202.

Weiner, A. M., and N. Maizels (1987) tRNA-like structures tag the 3′ ends of genomic RNA molecules for replication: Implications for the origin of protein synthesis. Proc. Natl. Acad. Sci. U.S.A. **84:** 7383–7387.

——— (1991) The genomic tag model for the origin of protein synthesis. Further evidence from the molecular fossil record. In: S. Osawa and T. Honjo (eds.) Evolution of life: Fossils, molecules, and culture. Tokyo: Springer Verlag.

Weiss, R., and J. Cherry (1993) Speculations on the origin of the ribosomal translocation. In: R. F. Gesteland and J. F. Atkins (eds.) The RNA world. New York: Cold Spring Harbor Laboratory Press. pp. 71–89.

Weissbuch, I., R. Popovitz-Biro, L. Leizerowitz, and M. Lahav (1994) Lock-and-key processes at crystalline interfaces: Relevance to the spontaneous generation of chirality. In: J.-P. Behr (ed.) The lock-and-key principle. Chichester: John Wiley & Sons Ltd. pp. 173–246.

Welch, G. R. (1995) Schrödinger's What Is Life ?: A 50-year reflection. TiBS, 20 January 1995: pp. 45–48.

Welch, G. R., and J. S. Easterby (1994) Metabolic channeling versus free diffusion: Transition time analysis. TiBS **19:** 193–197.

Westheimer, F. H. (1987) Why nature chose phosphates? Science **235:** 1173–1178.

Wetzel, R. (1995) Evolution of the aminoacyl-tRNA synthetases and the origin of the genetic code. J. Mol. Evol. **40:** 545–550.

Wheelis, M. L., O. Kandler, and C. R. Woese (1992) On the nature of global classification. Proc. Natl. Acad. Sci. U.S.A. **89:** 2930–2934.

White, H. B. (1976) Coenzymes as fossils of an earlier metabolic state. J. Mol. Evol. **7:** 101–104.

White, H. D. (1980) A theory for the origin of a self-replicating chemical system. **I:** Natural selection of the autogen from short, random oligomers. J. Mol. Evol. **16:** 121–147.

White, H. D., and J. C. Erickson (1980) Catalysis of peptide bond formation by histidyl-histidine in a fluctuating clay environment. J. Mol. Evol. **16:** 279–290.

——— (1981) Enhancement of peptide bond formation by polyribonucleotides on clay surfaces in fluctuating environments. J. Mol. Evol. **17:** 19–26.

White, H. D., and M. S. Raab (1982) A theory for the origin of a self-replicating chemical system. II: Computer simulation of the autogen. J. Mol. Evol. **18:** 207–216.

Whittet, D. C. B. (1997) Is extraterrestrial organic matter relevant to the origin of life on Earth? Origins of Life Evol. Biosphere **27:** 247–262.

Wicken, J. S. (1987) Evolution, thermodynamics, and information: Extending the Darwin program. New York: Oxford University Press.

Williams, R. J. P. (1993) Are enzymes mechanical devices? TiBS, 18 April: 115–117.

Wills, P. R. (1994) Does information acquire meaning naturally? Ber. Bunsenges. Phys. Chem. **98:** 1129–1134.

Wilson, C., and J. W. Szostak (1995) In vitro evolution of a self-alkylating ribozyme. Nature **374:** 777–782.

Winter, D., and G. Zubay (1995) Binding of adenine and adenine-related compounds to the clay montmorillonite and the mineral hydroxylapatite. Origins of Life Evol. Biosphere **25:** 61–81.

Wittung, P., P. E. Nielsen, O. Buchardt, M. Egholm, and B. Norden (1994) DNA-like double helix formed by peptide nucleic acid. Nature **368:** 561–563.

Woese, C. R. (1979) A proposal concerning the origin of life on the planet earth. J. Mol. Evol. **13:** 95–101.

——— (1980a) An alternative to the Oparin view of the primeval sequence. In: H. O. Halvorson and K. E. Van Holde (eds.) The origins of life and evolution. New York: Alan R. Liss, Inc. pp. 65–76.

——— (1980b) Just so stories and Rube Goldberg machines: Speculations on the origin of the protein synthesis machine. In: G. Chambliss, G. R. Grave, J. Davies, K. Davis, L. Kahan, and M. Nomura (eds.) Ribosomes: Structure, function, and genesis. Baltimore: University Park Press. pp. 357–373.

——— (1987) Bacterial evolution. Microbiol. Reviews **51:** 221–271.

Woese, C. R. (1991) The use of ribosomal RNA in reconstructing evolutionary relationships among bacteria. In: R. K. Selander, A. G. Clark, and T. S. Whittam (eds.) Evolution at the molecular level. Sunderland, Massachusetts: Sinauer Associates.

Woese, C. R., and G. E. Fox (1977) Phylogenetic structure of the prokaryotic domain: The primary kingdoms. Proc. Natl. Acad. Sci. U.S.A. **74:** 5088–5090.

Woese, C. R., O. Kandler, and M. L. Wheelis (1990) Towards a natural system of organisms: Proposal for the domains Archaea, Bacteria, and Eucarya. Proc. Natl. Acad. Sci. U.S.A. **87:** 4576–4579.

Woese, C. R., and N. R. Pace (1993) Probing RNA structure, function, and history by comparative analysis. In: R. F. Gesteland and J. F. Atkins (eds.) The RNA world. New York: Cold Spring Harbor Laboratory Press. pp. 91–117.

Wong, J. Tze-Fei (1975) A co-evolution theory of the genetic code. Proc. Natl. Acad. Sci. U.S.A. **72:** 1909–1912.

——— (1976) The evolution of a universal genetic code. Proc. Natl. Acad. Sci. U.S.A. **73:** 2336–2340.

——— (1980) Role of minimization of chemical distances between amino acids in the evolution of the genetic code. Proc. Natl. Acad. Sci. U.S.A. **77:** 1083–1086.

Wood, P. N. (1991) The use of bonding between tRNAs to implement early peptide synthesis. J. Mol. Evol. **33:** 464–469.

Woodger, J. H. (1929) Biological principles: A critical study. London: Kegan Paul, French, Trubner & Co.

Yamagata, Y., H. Watanabe, M. Saitoh, and T. Namba (1991) Volcanic production of polyphosphates and its relevance to chemical evolution. Nature **352:** 516–519.

Yariv, S. (1975) Infrared study of grinding kaolinite with alkali metal chlorides. Powder Technology **12:** 132–138.

Yarus, M. (1988) A specific amino acid binding site composed of RNA. Science **240:** 1751–1758.

——— (1993) An RNA–amino acid affinity. In: R. F. Gesteland and J. F. Atkins (eds.) The RNA world. New York: Cold Spring Harbor Laboratory Press. pp. 205–217.

Yarus, M., and E. L. Christian (1989) Genetic code origins. Nature **342:** 349–350.

Yčas, M. (1955) A note on the origin of life. Proc. Natl. Acad. Sci. U.S.A. **41:** 714–716.

Yockey, H. P. (1973) Information theory with application to biogenesis and evolution. In: A. Locker (ed.) Biogenesis, evolution, homeostasis: A symposium by correspondence. Berlin: Springer-Verlag.

——— (1992) Information theory and molecular biology. Cambridge: Cambridge University Press.

——— (1995) Comments on "Let there be life: Thermodynamic reflections on biogenesis and evolution" by Avshalom C. Elitzur. J. Theor. Biol. **176:** 349–355.

Zahnle, K. J., and N. H. Sleep (1997) Impacts and the early evolution of life. In: P. J. Thomas, C. F. Chyba, and C. P. McKay (eds.) Comets and the origin and evoluton of life. New York: Springer-Verlag. pp. 175–208.

Zamaraev, K. I., V. N. Romannikov, R. I. Salganik, W. A. Wlassoff, and V. V. Kharmatsov (1997) Modeling of the prebiotic synthesis of oligopeptides: Silicate catalysts help to overcome the critical stage. Origins of Life Evol. Biosphere **27:** 325–337.

Zhang, B., and T. R. Cech (1997) Peptide bond formation in vivo selected ribozymes. Nature **390:** 96–100.

Zhang, S., and M. Egli (1994) A hypothesis: Reciprocal information transfer between oligonucleotides and oligopeptides in prebiotic molecular evolution. Origins of Life Evol. Biosphere **24:** 495–505.

Zhang, S., T. Holmes, C. Lockshin, and A. Rich (1993) Spontaneous assembly of self-complementary oligopeptide to form a stable macroscopic membrane. Proc. Natl. Acad. Sci. U.S.A. **90:** 3334–3338.

Zieboll, G., and L. E. Orgel (1994) The use of gel electrophoresis to study the reactions of activated amino acids with oligonucleotides. J. Mol. Evol. **38:** 561–565.

Zillig, W., P. Palm, and H.-P. Klenk (1992) The nature of the common ancestor of the three domains of life and the origin of the eucarya. In: J. and K. Tran Thanh Van, J. C. Mounolou, J. Schneider, and C. McKay (eds.) Frontiers of life. Singapore: Fong & Sons Printers. pp. 181–193.

Zimmermann, R. A. (1995) Ins and outs of the ribosome. Nature **376:** 391–392.

Zubay, G. (1996) Origins of life on the earth and in the cosmos. Dubuque, Iowa: Wm. C. Brown Publishers.

Zuckerkandl, E., and L. Pauling (1965) Molecules as documents of evolutionary history. J. Theor. Biol. 8: 357–366.

Zull, J. E., and S. K. Smith (1990) Is genetic code redundancy related to retention of structural information in both DNA strands? TiBS **15:** 257–261.

Index